Fred Caswell, B.A.
Coventry University

John Murray

First published 1982
by John Murray (Publishers) Ltd
50 Albemarle Street, London W1S 4BD
Reprinted (revised) 1985
Reprinted 1986, 1988
Second edition 1989
Reprinted 1990, 1991, 1993
Third edition 1995
Reprinted 1996, 1999, 2001

Typeset by Colset Private Ltd, Singapore
Printed and bound in Great Britain by
Athenaeum Press Ltd, Gateshead, Tyne & Wear

A CIP catalogue record for this book is available from the British Library

ISBN 0 7195 7202 9

Contents

Success Studybooks

Advertising and Promotion
Book-keeping and Accounts
Business Calculations
Chemistry
Commerce
Commerce: West African Edition
Communication
Economics
Electronics
European History 1815–1941
Information Processing
Insurance
Law
Managing People
Marketing
Politics
Principles of Accounting
Principles of Accounting: Answer Book
Psychology
Sociology
Statistics
Twentieth Century World Affairs
World History since 1945

Foreword

In our modern society it has become increasingly necessary to be numerate. Numeracy at one time implied merely agility at formal arithmetic but the arrival of electronic calculators and computers has lifted this burden and enables us to concentrate on the interpretation of numerical data rather than on numerical procedures.

The rate of inflation, index linked savings, forecasts of trends in the economy, public opinion polls, suggestions of links between environmental factors and illness—all require some knowledge of statistical techniques in order to be fully understood and interpreted. It is hoped that this book will provide a useful introduction to statistics for anyone who wishes to appreciate published data in greater depth.

There are many people, however, for whom a study of statistics is an essential element of their working lives and *Success in Statistics* caters for them: accountants, market executives, managers, economists, scientists and others. The aim is not to produce specialist statisticians but to give an insight into basic statistical methods which can be applied in many diverse fields. Unit 18, Computers and Statistics, shows that the collection, calculation and presentation of statistics is brought well within the scope of such non-specialists by the latest technology.

This Studybook is appropriate for all BTEC National and Higher National modules with a statistical content. The Suggestions for Projects, which can be adapted for use as BTEC assignments, and Unit 16, Interest and Investment Appraisal, are particularly intended for such courses. The examinations of many professional bodies now have a statistical element at an early stage and *Success in Statistics* provides the necessary material to master such syllabuses. It also covers the requirements for 16+ statistics examinations and provides a useful introduction to statistics for those intending to study the subject at a higher level.

F. C. and H. W.

Acknowledgments

Thanks are due to the many people who have helped in the preparation of this book, and especially to Helen Wright, who acted as consultant editor for the first edition. For help with the first edition I am also grateful to George O'Sullivan of the University of Central England, Robin Gardner of Lambeth College, Anne Knight and other statistics lecturers of West Herts College, Anne Webster of John Murray, Jean Macqueen, the West Herts College librarians, and Ann Caswell and Ronald Reed who typed the manuscript. For help with revisions for the second and third editions I would like to thank Geoff Stratford, John Wilkin, and Harry Lennon of Coventry University.

Questions from past examination papers are reproduced by permission of the following: Associated Examining Board; Association of Certified Accountants; Institute of Chartered Accountants; Institute of Chartered Secretaries and Administrators; Institute of Cost and Management Accountants; Institute of Personnel Management; Institute of Statisticians; Institution of Industrial Managers. Answers, where given, are the responsibility of the author, not the examining body.

Data from the *Annual Abstract of Statistics, Census Monitors,* the *Family Expenditure Survey,* the *Monthly Digest of Statistics* and *Social Trends* are reproduced by permission of the Controller of Her Majesty's Stationery Office.

Figure 18.1 is reproduced by permission of Caspe Research and with thanks to Kendata Peripherals Ltd. Figure 18.4 is based on a diagram in the Lotus 123 User's Manual (Release 1A) by permission of Lotus Development (UK) Ltd. Figures 18.4, 18.5 and 18.8 are reproduced by permission of Microsoft Corporation.

Mathematical tables are reproduced by permission of the following: Addison-Wesley (Table D, pages 354–7, from A. Haber and R. P. Runyon, *General Statistics*); McGraw-Hill (Table E, page 358, from M. R. Spiegel, *Statistics*); Pitman (Tables F and G, pages 359 and 360, from J. E. Freund and F. J. Williams, *Modern Business Statistics*); Department of Statistics, University College, London (Table H, page 362, from *Tracts for Computers, 24*); Institute of Cost and Management Accountants (Table I, pages 364–5, from *The Profitable Use of Capital in Industry*); Biometrika Trustees (Table J, page 366, from *Tables for Statisticians*).

F. C.

UNIT ONE

Introduction to statistics

1.1 Why statistics matter

Are you aware of the extent to which statistics enter your life? You will probably have read in the newspaper or heard on news programmes, of such statistics as 'average earnings' and the 'Index of Retail Prices'. These kinds of statistics are used in negotiations that lead to decisions that can have major effects on living standards.

Have you ever seen people standing around, usually in busy shopping centres, clutching clip-boards and stopping the occasional passer-by to ask questions? Perhaps you may have had someone calling at your home asking for your help with a survey of some kind, or have been asked to complete a questionnaire. The results of such surveys may be used to influence the types and qualities of goods and services that you are offered in the future.

Every ten years in Great Britain all households must complete a census return which is used to compile statistics on population patterns and developments. These statistics have an effect on government planning for the provision of housing, services such as schools and hospitals, and the development of industry.

In an environment where statistics play such an important role it is in the interests of us all to know more about them: how they are collected, analysed and used.

1.2 The meaning of statistics

The word statistics has two meanings:

(*a*) In the plural sense it means collections of numerical facts and is widely used when reference is made to facts and figures on such things as population, crime and education. Statisticians call the figures which have been collected *data*.

(*b*) In the singular sense it means the science (or art) of dealing with statistical data. The collection, analysis and interpretation of data is called *statistical method*, and it is with this sense of the word that we are mainly concerned in this book.

There are two subdivisions of statistical method:

(i) Descriptive statistics This deals with the compilation and presentation of data in various forms such as tables, graphs and diagrams. The purpose of descriptive statistics is to display and pass on information from which conclusions can be drawn and decisions made. Businesses, for example, use descriptive statistics when presenting their annual accounts and reports, and the Government is a particularly prolific provider of descriptive statistics.

(ii) Mathematical or inductive statistics This deals with the tools of statistics, the techniques that are used to analyse the data and to make estimates or draw conclusions from the data.

1.3 The growth and importance of statistics

Statistics have been collected since the earliest times. Rulers needed to have information about the population and their possessions so that taxes could be levied to maintain the state and the court, and it was also essential for them to be aware of the military strength of the nation. In the sixteenth century the word 'statist' was used to describe a politician – a dealer in facts about the state, its Government and its people.

With the growth of the population and the advent, in the eighteenth and nineteenth centuries, of the industrial revolution and the accompanying agricultural revolution there was a need for a greater volume of statistics on an increasing variety of subjects. In this country the Government also began to intervene more and more in the affairs of the people and of business and to attempt to control the workings of the economy. It therefore required information on:

(i) production;
(ii) earnings;
(iii) expenditure;
(iv) imports and exports.

As time went on, the Government took over many of the activities that had been part of the private sector of the economy, such as education and health services. It was vital therefore for information to be available on:

(v) population growth or decline;
(vi) disease and its incidence;
(vii) housing conditions.

All this led to an enormous expansion in the volume of statistics that have been and are being collected by Governments over the last few decades. For Governments to make sensible decisions, however, these data need to be correctly collected, processed and analysed.

1.4 Statistics and the business world

It is not only Governments that have required more and more statistics: companies have grown to such an extent that some have an annual turnover as great as the annual budget of a national economy. Large firms therefore have to make decisions on the basis of data.

The statistics that are collected by the Government are vital to businesses, and in the United Kingdom the Government's Central Statistical Office provides a specialist statistical service for businesses.

Businesses also collect their own statistics or pay specialist companies to collect the data for them. They require information on:

(i) the reaction of other companies to their products;
(ii) the reaction of customers to their products;
(iii) the effect of other companies' activities on their own;
(iv) the need for new products.

A large company will also need to compile internal statistics on sales, production, purchases, costings, and personnel matters. The accounting records of a company can be considered as a set of statistics displayed in a particular manner.

The tools of statistics are essential to modern business in areas such as planning, forecasting, and quality control. The techniques of statistical method are analogous to those of any other business tool: they are designed to do particular jobs. The methods of tackling any job can be summarized in the following steps:

(i) assessing the problem;
(ii) selecting the correct tool;
(iii) collecting the materials to enable the tools to be used.

Problems that involve statistical method should be approached in the same way. Business people have to use their knowledge of statistics to select the correct tools to deal with the jobs they are faced with.

1.5 Misuse of statistics

No doubt you have heard an argument supported by the words 'statistics show that . . .' Many people seem to believe that a case is proven if statistics can be produced to support it. Thus an endless stream of data is thrown at us in an attempt to impress, persuade or even coerce us into believing that a particular political party is wise and good, or that we should buy such and such a product or that we should hold certain opinions. What we are seldom told is how the statistics which are meant to impress us have been collected, where they were collected or from whom they were collected. The old quip about 'lies, damned lies and statistics' has some truth in it. The figures themselves cannot mislead, but the statisticians who present the figures certainly can.

It is hoped that this book will alert you to the pitfalls associated with the use of statistics and that you will be ready to question claims that are supported merely by statistics. Data can be misused in the following ways:

(i) They can be used for the wrong purpose, that is, one that is different from the purpose for which they were collected.
(ii) They can be collected incorrectly so that they are biased.
(iii) They can be analysed carelessly so that the results obtained from them are misleading.

Be ready to find out more about these matters before believing any claims. A good statistician will be eager to show you how data were obtained and dealt with.

1.6 Using this book

To understand the subject-matter fully, you must be prepared to *work* through this book rather than merely to read it from cover to cover. You will find it useful to have paper and pencil to hand. Squared paper or graph paper, plus a ruler, are required for some units. You should check the working in the text and attempt as many of the exercises as possible. The answers to numerical questions are at the end of the book. Any necessary mathematical and statistical tables are also to be found there.

You will find an electronic calculator an invaluable aid. Before buying one, study carefully the regulations regarding calculators for your particular examination. For statistical work a calculator should have a square root facility and, if possible, a memory, an exponential key (e^x) and an exponent key (x^y). 'Statistical' calculators giving standard deviations, for example, are available and can be used in some examinations. It is not essential, however, to have such a model since examiners at this level want to see full details of your workings—they are not solely interested in the final answers.

The material in the first eleven units is essential reading for most first examinations in statistics, whether for a BTEC award, a professional qualification or 16 + statistics. Later units are not required by all examining bodies so look carefully at your syllabus and past examination papers to see which sections you need. Units 12–15, in particular, are essential if you are to continue your statistical studies at a higher level. The sections on official statistics refer mainly to UK sources. If you are not resident in the UK, knowledge of these publications may not be appropriate for your examinations. In this case you should try to obtain information from your nearest library about relevant publications on such topics as population, prices, employment, wages and production.

In 'real life', statistical work often involves extensive calculations but the prime objective here is to enable you to learn principles and methods rather than to do lots of arithmetic. Most of the examples and exercises contain far fewer numbers than would be encountered in practice. After Unit 17, Suggestions for Projects are given. You are recommended to attempt some of them so that you acquire an appreciation of the practical difficulties that are met with in actual statistical investigations.

Some knowledge of mathematical techniques is needed in statistical work. It may be some time since you did any formal mathematical study and you may find it useful to revise specific topics as and when they are needed in the book.

A Further Reading list is to be found at the end of the book to help you to pursue statistical topics of particular interest to you in greater depth.

UNIT TWO

The collection of data

Statistics is concerned with the analysis of numerical data, so the first stage in statistical method must be the collection of the data to be analysed. It is an important stage, since unless data are collected carefully and correctly they will be of little value.

2.1 Populations

The data that the statistician works with are a series of observations from a *population*. For example, a statistical quality control exercise may be conducted in a factory producing electric light bulbs. At regular intervals a light bulb is picked at random from the production line and tested to see how long it will burn. The population here is the lifetimes of all the electric light bulbs on that production line. The lifetimes of those bulbs selected for testing are a *sample* from that population.

Another example might be an auctioneer at a livestock market who keeps records of all the cattle that pass under his hammer, including details of weights and prices, so that he can produce statistics in the future. Here there are two populations: the weights of all the cattle and the prices of all the cattle.

Sometimes populations concern people. When the results of a general election are being predicted, the population under scrutiny will be the opinions of all the people eligible to vote in that election. Another example would be when statisticians are trying to find out how families at a certain level in the economy spend their weekly income so that the increase in the cost of living can be calculated; the population here is the expenditure of every family at that economy level. We shall often refer loosely to the people and the families as populations, although, strictly speaking, it is their opinions and their expenditures respectively which form the populations.

If an attempt is made to observe a complete population, human or otherwise, then we say a *census* is being carried out. The decennial Census of Population in Great Britain is an increasingly important source of data for commercial organizations as well as supplying vital information on the size, distribution and living conditions of the population for central government and local authorities. The annual Census of Production collects data about UK productive activity. The Census of Distribution (now replaced by *Retailing* in the Business Monitor series) covered retail trade. A general election attempts to record the votes of all those on the electoral list. Whereas in the Census of Population each head of household is legally required to complete a census return, there is no compulsion to vote in a general election in the United Kingdom. Thus, 100% coverage of the population is not achieved. The overall percentage of those on the electoral roll who actually vote in a general election is around 80%. Some of the discrepancy is accounted for by deaths and migration in addition to those people who do not wish to record their votes.

We can readily see why samples have to be taken from populations: it reduces the data to manageable proportions. When data are taken from every family in an economy, it costs so much in time (and money) to collect and analyse them that the statistics are out of date by the time they are published. It might *just* be possible for the livestock auctioneer to record details of all the cattle he deals with and to compile statistics from the whole population, but it would clearly be impracticable to test every light bulb on the production line to extinction: there would be none left to sell. So, no matter what the size of the population, samples are taken and the observations which are made from the samples provide the data to be used to calculate the statistics of interest. It should not be thought that a sample survey is necessarily inferior to a census. Complete coverage of a population is seldom achieved unless the population is very small. Sample surveys take less time and can be more closely controlled. Consequently, trained investigators can make a more thorough examination of a sample than would be possible with a census.

2.2 Problems of bias

When we collect data for analysis, we try to take a *random* sample, that is, one taken in such a way that every member of the population stands an equal chance of being selected as part of the sample. There must be no *bias* towards any individual or any groups of individuals in the population. We wish the sample truly to represent the population from which it is taken so that any conclusions drawn from it can be extended to the population as a whole. The sample must also be large enough: the larger the sample the more accurately findings from it should reflect the findings that would be obtained from the whole population.

Of course, if we want to prove an argument by fair means or foul we could deliberately set out to obtain a biased sample, collecting the data in such a way that they will support our argument. Any conclusions, however, could soon be faulted by someone who looked carefully at the way the data were collected. When you are presented with statistics in support of an argument, remember to inquire how the data from which the statistics are drawn were collected.

Unfortunately, obtaining a random and unbiased sample is probably the biggest problem we face as statisticians. We have to rely on the skill of the people who collect the data, we have to hope that the sample fills in any questionnaires accurately and honestly, we have to hope that we get a good response from the sample, and, finally, we have to pick a suitable sampling method.

2.3 Sampling frames

The first thing which has to be done before choosing the sample is to try to identify the population from which the sample is to be taken. To help in making this identification we can use a *sampling frame*, if an appropriate one is available. This is a list of the entire population from which the sample is to be selected. Some sampling frames come easily to mind: one of the most commonly used is the telephone directory. If a random sample from the population of Norwich is required, names can be selected at random from the Norwich telephone directory. However, the telephone directory does not contain the names of everyone in Norwich: it omits people who do not have a telephone, a large and important section of the population. If we were surveying the lifestyle of the population of Norwich we should obviously need to include people without a telephone if we did not want our findings to be biased. The directory would probably only be an ideal sampling frame if we wanted to sample telephone owners in Norwich.

The electoral roll is another commonly used sampling frame. It contains the names of all members of the population who are eligible to vote—ideal for a survey of how people will vote in an election, or for assessing the impact of a new brand of beer perhaps, but not suitable for assessing the impact of a new kind of bubble-gum. The frame must suit the survey being conducted.

Another sampling frame for the population of a town is the local street map. This can be used to select streets at random, and then some system of selecting houses within those streets is devised—every tenth house—in order to collect a sample. Here, residences are being selected for the sample, rather than people, and the data are collected from the occupants of those residences.

There are many more possible sampling frames: trade directories, professional registers, club membership lists and the Council Tax valuation list of dwellings at the local council offices. It is essential to try to find a suitable sampling frame before beginning to select a sample, although it will not always be possible to find such a frame. Suppose you are conducting a survey which involves interviewing housewives with two children: there are no directories listing housewives with two children. The only way of finding that population is to go out and look for it, either knocking on doors or stopping people in the street and asking them if they possess those characteristics.

The final, and obvious, point to be made about sampling frames is that they should be as up-to-date as possible.

2.4 Sampling methods

There are various ways of actually selecting a sample. For probability, i.e. random sampling, a sampling frame is required. Quota sampling, which is an example of non-random sampling, does not need a sampling frame.

(i) Simple random sampling, sometimes known as the *lottery method*. This is the best method from a theoretical viewpoint of selecting a truly random sample. All the items in the population are given a number and pieces of paper each with one number on are placed in a drum or hat. The numbers are selected one at a time, with the drum being revolved or the hat shaken between each selection, until the required sample size is reached. If the population is large, then an alternative method using a computer-generated table of random numbers (see Table H) is preferable. For example, suppose we have an electoral register containing 56 000 names. We could label each name with a five-digit number 00 001, 00 002, 00 003, . . ., 00 100, 00 101, 00 102, . . ., 01 000, 01 001, 01 002, . . ., 55 998, 55 999, 56 000. We have to use five digits to ensure that each item on the list is labelled in an exactly similar manner so that no bias is shown to any part of the population. We then consult the random number table which is so constructed that each digit has the same chance of appearing at any position on the page so the fact that there is a 4 in one particular spot has no effect on the next digit to appear: it is equally likely to be any one of the digits 0, 1, 2, . . . 9. The grouping of the random numbers on the page is of no consequence but merely an aid to easy reference. Having decided upon our starting point in the table, we read off digits systematically in groups of five (since our items are labelled with five digits), preferably along the rows. Suppose we have decided to start at the third major column and the fourth block down. We read off the numbers 34 343, 22 260, 53 911, 73 326, 44 709, 31 499, 70 490, 57 448, 10 553 etc. Those items in the population corresponding to these numbers are included in the sample. We continue reading numbers from the table until the required sample size has been reached. We discard any numbers such as 73 326 which are too large and which do not refer to any item in the sampling frame. This method is very tedious, particularly for very large populations. In addition, although the method of selection is free from personal bias, there is no guarantee that the resulting sample is unbiased. For example, our random sample from the electoral roll could, by chance, contain people from only one area of town, but this is not very likely.

(ii) Systematic sampling, also known as the *constant skip method*. As the name implies, some kind of system is used for selecting a sample. You could, for example, let the telephone directory fall open at random and then select every twentieth name from there onwards until the required sample has been selected. Every two-hundredth light bulb from the production line or every tenth house in a street are further examples. Thus only the first item in the sample is chosen randomly and this reduces considerably the work involved. Problems can arise, however, if there is a periodicity in the sampling frame which corresponds to the interval between items in the sample. For instance, suppose we wish to find the opinions on noise levels of families in a multistorey block of flats. If there are ten flats on each floor and we decide to take a sample systematically of every tenth flat, then we would obtain results from flats all directly above one another. These flats would face the same direction and their occupants' views on noise would be likely to be more in agreement than if we had taken a simple random sample from the block. As a second example, let us consider items on a production line being sampled at regular intervals of time. The starting time for measurements each day should be varied since fluctuations of attention by operators and in the electrical supply to machines could change in a regular pattern through the day.

(iii) Stratified sampling This method entails the use of the natural divisions of a sampling frame, such as sex, age, income group, social class or occupation. These are the *strata* and they can be used to ensure all sections of the population are adequately represented in any sample.

When using stratified methods of sampling it is essential to know in advance the proportion of the population in each natural stratum and to take account of this when selecting the sample. Suppose, for example, that a survey is being conducted in a college to see if its population is in favour of proposed changes in refreshment and meal services in the canteen. It could be that there is a difference of opinion which is dependent on a student's department or on whether the student is on a full-time or a part-time course. Each student has an enrolment or registration card which has an enrolment number and identifies the department and type of course. A simple random sample would use a random selection from all enrolment numbers and contact the corresponding students. To ensure, however, that the sample is representative of all departments and types of course, stratification before selection of the sample might be advisable. Table 2.1 shows the distribution of students among departments and types of course.

The authorities have decided that they have the resources available to interview 160 students i.e. 10% of the population. To obtain a stratified sample they contact 10% of the students in each category. For example, they will interview 10% of the 300 full-time business students i.e. 30 students. These 30 students will be chosen randomly from the 300 full-time business students using the simple random sampling methods of subsection (i). The procedure is repeated for every category. To select a small sample from each stratum in turn is not so tedious and time-consuming as selecting one large sample from the complete population. The numbers of students to be interviewed in each stratum are shown in Table 2.2.

The final sample of 160 will be fully representative of the population in respect of department and type of course. It might still be unrepresentative in other respects. For example, the sample might not reflect the sex distribution

Table 2.1 Distribution of college students

Department	Full-time	Part-time	Totals
Business studies	300	300	600
Management	50	150	200
Engineering	300	0	300
Science, maths and computing	50	50	100
Printing	300	100	400
TOTALS	1000	600	1600

Table 2.2 Stratified sample sizes for a sample of 160 college students

Department	Full-time	Part-time	Totals
Business studies	30	30	60
Management	5	15	20
Engineering	30	0	30
Science, maths and computing	5	5	10
Printing	30	10	40
TOTALS	100	60	160

in the college or the type of student accommodation, whether hostel, self-catering, home-based, etc., both of which factors might influence a student's attitude towards changes in meal services. The more strata included, the more organizational work there is involved and it could be, for example, that information on the accommodation of part-time students is not kept and such a stratification would then be impossible.

The strata chosen using this sampling method must be readily determinable. They should be exhaustive and mutually exclusive, i.e. each item in the population belongs to one and only one stratum. Stratification by area of residence or electoral ward could be appropriate in a survey of a large town, whereas stratification by age or income would not be possible since information on age or income is not usually available to a researcher before a survey is conducted.

(iv) Cluster sampling In cluster sampling the sampling frame consists of a list of groups of individuals rather than the individuals themselves. A random sample of these groups or *clusters* is taken and then observations are made on every individual within these selected groups. For example, to take a cluster sample in a large town, we would begin by listing all the streets containing residential property. These streets are the clusters. We then take a random sample of the streets. This is relatively easy to do since there are many fewer streets than there are individuals. We then obtain the views of every resident in our chosen streets. The advantage of this method is that no complete list of *individuals* is needed. This is particularly attractive in developing countries where official statistics are sparse and where suitable lists of populations may not be available. Cluster samples are also popular with geographers and biologists. The method in such cases is to cover the survey area with a grid of numbered squares. A random sample of the squares is taken and then a complete investigation is made of the selected squares whether it be counting the number and type of industrial establishments, the number of bacteria or the number and species of plants. The cluster sampling method has the disadvantage that the sample might not be representative of the population as a whole since nothing is known of those clusters not sampled. Thus it is preferable to divide the population into a large number of small clusters rather than a small number of large clusters.

(v) Multi-stage sampling This method is designed to reduce the time and cost of surveying samples from very large populations. Suppose we wanted to sample 2000 members of the population of England at random. We could select four counties at random, then select five towns at random from each of these counties and finally we could select 100 people at random within each of the 20 towns to make up our sample of 2000. We can then collect the data in 20 relatively small areas instead of having to visit 2000 scattered locations. At the last level of the multi-stage selection, the whole of the last level, one street in each town for example, could be used as the sample, so that the last level could in fact be a cluster sample. Stratification can be used in conjunction with multi-stage methods. Particularly with opinion polls it would be desirable to divide the country into its parliamentary constituencies according to voting patterns in the last election. A stratified random sample of constituencies would ensure that constituencies supporting all political parties were considered. The second stage would be a random sample of electoral wards and the final stage a random sample of streets or of individuals.

(vi) Quota sampling This method allows the interviewer a certain amount of discretion when collecting the data. Quotas for different sections of the population are set and the interviewer is allowed to select the sample according to these quotas. The sections can be as described under (iii) above—sex, social class and so on. For example, an interviewer could be sent out to take a sample that includes 10 professional men, 30 middle managers and 60 shop-floor workers. Quota sampling is used, therefore, to ensure that the sample contains members of the population in the desired proportions. But just how the sample is selected is left to the interviewer's discretion. This can be a source of bias.

Quota sampling is not a method of random sampling. It is an example of *non-random* or *non-probability* sampling. No sampling frame is needed so this considerably reduces the time and cost of a survey and explains its popularity with many research organizations. It has the disadvantage, however, that not everyone in the population has a chance of being included in the sample. Only the views of those chosen by the interviewer and who wish to cooperate are recorded. For this reason the results can be biased and this method is not to be recommended from a theoretical viewpoint.

In selecting a sample it is unlikely that one method alone will be used, but combinations of methods are more likely. You may use a multi-stage method to select the areas in which to collect the data and then send out interviewers in those areas to select the sample according to quotas. You could also take some of the personal bias out of the quota selection by using professional registers or other lists to select the persons in each part of the quota.

Of the commercial UK survey organizations Marplan, National Opinion Polls, Gallup and MORI use both quota and random sampling methods. Quota methods are used for cheapness and speed, particularly in surveys of political opinions. Random methods are used to minimize selection bias and to provide estimates to a given precision.

(vii) Panel methods This type of survey collects data from the same sample on more than one occasion. The initial sample can be chosen by any of the random sampling methods mentioned earlier. This method has several advantages. Long-term trends in spending patterns, for example, can be monitored. The response of the panel to an advertising campaign can be assessed. Changes can be measured more precisely than with several independent random samples. It is also possible to identify the type of person whose behaviour has altered. The cost of this sample design is relatively cheap since the selection of the sample occurs once only. There are problems, however, in persuading people to fill in questionnaires or to be interviewed on a regular basis. It is found that many of those who are enthusiastic initially drop out at a later stage and replacements from a similar background have to be found. It can also happen that merely being a member of a panel affects the results since the panel is made aware of products and advertisements which formerly they might have ignored. This can be overcome by gradually replacing a panel over a number of months. If the panel studies continue over a period of years at infrequent intervals, it may become difficult to trace members and although originally the panel was representative, changes or additions to the panel may be needed to reflect the current population.

2.5 Survey preliminaries

Before any investigation can be started, whether it be a census or a sample survey, the exact problem and the scope of the enquiry must be clearly defined. In an investigation into incomes of executives, what do we include under the heading of 'income'? Is it just salary or do we include use of a company car, provision of subsidized meals and other fringe benefits? Do we wish to look at executives in our own company or in similar-sized companies or in companies manufacturing a similar product? Do we wish to consider only those companies in our immediate area?

The units of measurement together with the desired degree of precision must be specified. Are incomes to be measured in pounds sterling? If we are making comparisons with overseas companies, what values of exchange rates are we to use to convert incomes to pounds sterling? Are we intending to record incomes to the nearest £1 or to the nearest £100?

It is important at this planning stage to ensure that the task can be completed in a reasonable length of time. For example, comparison of incomes would have little value if the survey work had taken a year to complete since one would expect salaries to rise over a period of twelve months.

If a sampling method is to be used, then a decision about the size of sample must be made at this stage. The time available, the cost of the survey, the number of trained personnel at one's disposal all have a bearing on this. Most surveys result in sample figures being used to estimate population values (*parameters*) and to be able to estimate these parameters to the desired precision with a specified degree of confidence there is a minimum number of readings which has to be taken. We shall look at this problem in more depth later.

2.6 Collecting the data

Once a sample has been selected the data must be collected from that sample. There are basically three principal methods of collecting data and we will consider the advantages and disadvantages of each.

(a) The personal interview

This is a commonly used method of collecting data from the general public, but it involves placing great reliance upon the integrity and skill of the interviewer. The main advantages are:

(i) A high response rate: the skilled interviewer can persuade all but the most reluctant to answer the questions.
(ii) The interviewer can explain any question which the interviewee cannot understand.
(iii) The interviewer can check the answers to some questions – by making a visual check of age, for example. The interviewee will be less inclined to exaggerate when answering questions about house, car, income or life-style if any exaggerations will be visibly obvious.
(iv) More information can be collected than with other methods as generally more time will be devoted by an interviewee to a personal interview.

The main disadvantages are:

(i) The interviewer may, even unknowingly, introduce bias by the way that questions are asked or answers recorded.
(ii) The interviewer may be tempted not to stick to the sample selected – missing out interviewees who could not readily be contacted, or making substitutions, or failing to complete interviews because of time restraints. There is a great danger of bias here. Suppose, for example, the interviewer makes visits during the working day and if an interviewee is not at home calls at the house next door and continues down the street until somebody is found at home. The sample is likely to contain an abundance of housewives and the elderly but to be lacking in families where both adult members are out at work. The interviewer should call back, in the evening if necessary, until the sampling schedule is completed. There may also be a danger of the interviewer avoiding houses that look forbidding, or people that look offputting. Of course, if quota sampling is used, this problem is avoided.
(iii) Some people may be too embarrassed to give confidential information in a personal interview.
(iv) A tactless interviewer may obtain inaccurate responses through upsetting or angering the interviewee.
(v) It is a very expensive method of collecting data. Interviewers must be

recruited and paid and should, as the above disadvantages show, be well trained. Cost may prevent a big enough sample from being taken.

Interviews can also be conducted by telephone which gives some of the advantages of the personal interview and is cheaper, but it is easier for people to refuse a telephone interview by simply replacing the receiver and many are reluctant to disclose personal details as they have no means of checking that it is a genuine call. The interview is limited by time as well, and of course a sample restricted to telephone renters is liable to be biased. In any event, the telephone renter, i.e. the name in the directory, is not necessarily the person who answers the call. The telephone interview has perhaps more relevance to industrial market research where problems of confidentiality can be overcome by preliminary correspondence. If survey work develops as in the United States of America, then use of the telephone interview is likely to increase. Recent research has shown, perhaps surprisingly, that telephone methods give similar results to those obtained from personal interviews in so far as expenditure on food, tobacco and consumer durables is concerned.

(b) The postal questionnaire

This method of data collection is commonly used for collecting official statistics. Its main advantages are:

(i) It is a less costly method than the personal interview, although the number of useful responses received may make it expensive in the long run.
(ii) The sample can usually be collected from a much wider area.
(iii) The bias of an interviewer is removed.
(iv) The respondent is not asked to give instant replies and so answers can be considered carefully and records consulted if necessary.

The main disadvantages are:

(i) There is usually a very poor response rate. It may be worth offering 'carrots' in the way of gifts or rewards for returning the questionnaire completed. A reply-paid envelope should always be included. The Government can, of course, use compulsion to ensure that its questionnaires such as tax returns, electoral registration forms, etc. are returned.
(ii) Bias may be introduced because only a particular type of person may reply, perhaps people such as pensioners with plenty of time on their hands or people who simply enjoy filling in questionnaires. A great danger is that only those who have a particular interest in the subject being investigated will reply, causing a very biased response.
(iii) Nobody is on hand to explain questions, so some may be incorrectly answered or not answered at all. Only very simple questions can be asked.
(iv) Questions may not be answered in any particular sequence because the respondent can see the whole form before he fills it in. For some questionnaires this may be of great importance.
(v) Some people may send in facetious answers.

(vi) Questionnaires may be filled in as team efforts, so that the opinions of several people are embodied in one form.
(vii) There may be considerable delay before enough replies are received from which valuable statistics can be drawn.

(c) Observation

We have looked at two methods of collecting data widely used in market research and social surveys. In the world of commerce and business, however, the most commonly used method of collecting statistical data is by observation. We have already considered a quality control exercise in a factory making light bulbs in which the data are collected by taking bulbs from the production line and observing the number of hours they last. There is no other way of collecting this type of data.

Observation is used in work study and in organization and methods exercises. It may be used by systems analysts who are finding out how a department works and how information flows and the time it takes to flow through departments. Observation is also used by social scientists learning the customs and habits of communities.

Each employee in an organization has a record card and many statistics, e.g. details of pay and absences, can be taken directly from the cards without reference to the employee. If you were asked how many days' leave from work you had had in the last twelve months, you would probably not be able to give an immediate answer and your eventual answer might not be correct since it relies on your memory. Direct observation of your record card would give the answer immediately and accurately.

The main advantage of observation is that it records what actually happened rather than what people say would have happened or did happen. For this reason it is often undesirable for people to know that they are being observed for an experiment.

The main disadvantages are:

(i) The observer needs to be highly skilled and unbiased.
(ii) Observation tells us nothing about people's attitudes.
(iii) Present observation tells us nothing about past or future happenings. Some forms of behaviour can only be fully observed over long periods of time, which means that it is very difficult to hide the fact that observations are being made.

2.7 The design of questionnaires

The principal concern in the design of a questionnaire is clarity. The layout must be arranged so that the questions are easily seen and their associated responses are easily matched to the questions. The form must be well designed and attractive so that the respondent is not put off simply by the look of the form. This applies particularly to postal questionnaires. The questions must be clear, simple, easy to read and to understand, while the questionnaire should be as short as possible. The following is a list of rules to consider when designing questionnaires:

(i) Make sure that the questions are easy to understand and that the space for an answer goes clearly with its question. To this end it helps if the questions are numbered and if each question and response are separated by a space from the next.

(ii) Do not ask for long answers to your questions (you are sure to get them). 'Yes/No' responses are most suitable if possible.

(iii) Do not ask questions that may lead the respondent to an answer. Questions that begin with the words 'Do you agree that . . .' are examples of this type of question.

(iv) Do not ask questions that call for calculations. People may find it easy to tell you the amount they spend on cigarettes in a week, but it is more difficult to say how much they spend in a year.

(v) Similarly, do not ask questions that rely on memory for answers, such as 'Did you spend more on cigarettes last year than this year?'

(vi) Make sure that 'multiple-choice' questions have viable alternatives for their answers. The following is an example of a multiple-choice question:

How much per week do you spend on cigarettes? (Tick the appropriate box):

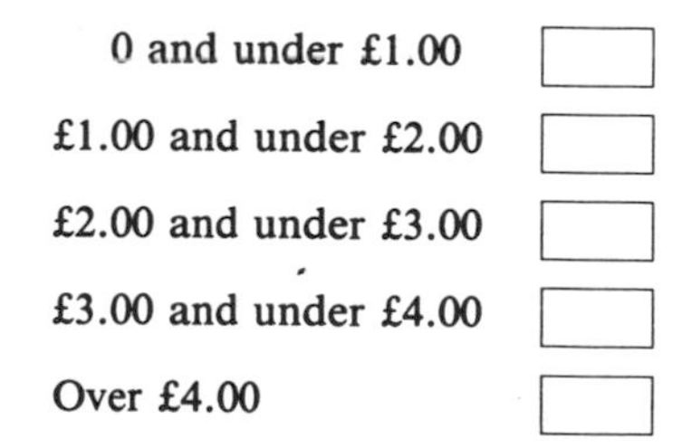

0 and under £1.00

£1.00 and under £2.00

£2.00 and under £3.00

£3.00 and under £4.00

Over £4.00

This question has categories which cover all possibilities. The use of such questions helps in the analysis of the results since the data have been placed in one of five specified categories.

(vii) Make sure that questions are unambiguous. 'Do you smoke?' is better than 'Do you make a habit of smoking?'

2.8 Pilot surveys

Finally, if the survey you are conducting is large and the results are important, it may be a good idea to conduct a pilot survey before the survey proper, if resources are available. A pilot survey is conducted to test the technique of the survey and not necessarily to collect viable data. Its object is to find out if the questionnaire is valid and whether interviewers and respondents can understand and answer the questions. The methods to be used in analysing the data to be collected can also be tested.

2.9 Errors from sampling

There are two distinct ways in which error or bias can arise when sampling methods are used.

Firstly, because a sample is part of a population it can never be completely representative. Our estimates obtained from the sample will not be the same as those we would have gained from a study of the whole population. As we take more sample readings, we expect our sample estimates to approach the population values and for our *sampling errors* to decrease. Provided a random sampling method is used, these errors can be estimated. To improve the precision of our estimate by a factor of two, however, we must quadruple the sample size and to improve it tenfold we must take one hundred times as many readings—so you can see why this process is said to be subject to 'the law of diminishing returns'.

Sampling errors are inevitable but the second way in which errors occur in sampling could, in theory, be prevented if we had perfect control of the population and the sampling method. These errors are sometimes referred to as *non-sampling errors*. Some of these errors were mentioned in Unit 2.6—i.e. interviewer bias and bias arising from substitution. Failure to cover the whole of the sample is yet another problem. Deaths, moves and refusals affect the best-planned investigations. Non-response commonly accounts for about a 10% reduction in the effective sample size. When a postal questionnaire is not returned, a follow-up of non-respondents with a personal interview might be thought desirable. When personal interviews are involved, it is usual to call back three times if necessary to attempt to make a contact. It might be that the time of day chosen for the call was inappropriate.

Non-response, for whatever reason, can introduce bias into the results if it is not representative. If possible, non-respondents should be classified by comparing the sample 'profile' as regards age, income, educational attainment, etc. with a population profile obtained from the results of the Census of Population. In this way any group under-represented in the sample can be identified.

Mistakes in recording answers and in processing the results are another source of non-sampling errors. All non-sampling errors are likely to increase as the size of the sample increases since there are more non-respondents, more chances for an interviewer to influence the results and a greater number of mistakes likely at the processing stage. This emphasizes the point made earlier in this Unit—that sampling methods should not be thought of as necessarily inferior to the census approach.

2.10 Exercises

1. A toothpaste manufacturer wishes to do a survey to assess demand for and tastes in toothpaste before a new product is introduced. Suggest the way it may set about undertaking this survey and the questions it may be interested in.

2. 'Statisticians would be lost without good sampling techniques.' Discuss

this statement and describe two methods of sampling that are commonly used.

3. A forward-looking company wishes to allocate more money to the activities of its sports and social club. The club already runs teams for football, cricket and women's netball. On the social side there are whist drives, regular dances and bingo. The chairman of the club decides to find out if there is interest in new activities and whether the present activities are satisfactory before the funds are allocated. Design a questionnaire that could be distributed to all employees to help the chairman to assess the situation.

4. Discuss the relative advantages of postal questionnaires and personal interviews as methods of obtaining data for a survey into the drinking habits of the adult male population.

5. Examine carefully the steps you would take in a statistical investigation into the likely demand for a new brand of breakfast cereal. In your answer examine any problems you may encounter.

6. What do you consider as suitable sampling frames for the survey in Exercise 5? Give reasons for your answer. Suggest a method of selecting the sample.

7. Design a questionnaire in order to find out how much families spend on a selection of items per week and to see if spending patterns differ according to the class of the family.

8. A local authority wishes to assess public opinion about the proposed building of a costly sports centre. Suggest how it may go about taking a survey to do this. What sections of the population is it important to include in the survey and why?

UNIT THREE

Secondary data

3.1 Introduction

In Unit 2 we looked at techniques that can be used for collecting primary data. Ideally it would always be preferable for us to collect our own data, since we can then be confident that the information has been collected and analysed to suit the specific purpose of the survey in hand.

Unfortunately we cannot always collect our own data, as resources are usually limited. If you are a student they are likely to be very limited, unless you have managed to obtain generous sponsorship for a project. Because of limited resources, any sample you collect cannot be very large. For example, a survey would probably have to be restricted to a very small geographical area, though this may be satisfactory for some purposes.

Many firms and other organizations find the cost of collecting and analysing data so expensive that it is cheaper for them to buy the information they require. Market research companies have created lucrative businesses collecting and analysing data for other companies.

It also takes time to collect and analyse data: very often that time is not available. A firm may need information about a market or economic situation quickly, and there may be no time to recruit and train interviewers, design and print forms and set up the techniques of analysing the data, which may involve the lengthy process of writing computer programs. A market research company already has the resources necessary for the collection and analysis of data.

It is not always necessary to collect primary data, however, because there is a great deal of data already published that can be used. These published data are *secondary data* i.e. they are data which have been collected at an earlier date and for some other purpose. The prime example of secondary data is the official statistics that are published by the Government (see Unit 3.3).

The Government is not the only organization publishing useful sources of data, however. All the large commercial banks publish quarterly reviews which contain articles of current interest, often with up-to-date statistics. If you would like copies sent to you, call in at any branch of the banks and ask to be put on their mailing list.

There are numerous journals of learned bodies in which you can find statistics on many diverse subjects, mostly resulting from academic research. If you can visit the library of a university, do so. You may be surprised at the vast number of such journals that exist; the subject-matter may be esoteric but the amount of data available is extensive.

You can also seek data from books, although up-to-date figures from this source are hard to come by. Magazines such as the *Economist* and the newspapers such as the *Financial Times may* offer more current data and occasionally publish special reports on recent developments in the UK regions, countries of the world and particular industries (but see Unit 3.2).

The electronic storage of data is starting to revolutionize access to secondary data. Television screens linked, for example, to British Telecom's *Prestel* system allow such statistics as the Index of Retail Prices to be displayed immediately on the day of publication. This instant availability of up-to-date statistics both from the Government and from commercial organizations is of great importance. *Prestel* can now be used in some libraries, although not necessarily without charge, and you should make yourself aware of the data at present available and keep abreast of developments in this field. The latest copy of *Connexions* (incorporating the Prestel Directory), usually available in your local reference library, will tell you the data available.

3.2 The dangers of secondary data

If we have to use secondary data, there are dangers to be aware of:

(i) The data available may not be very up-to-date. The collection and compilation of data take a great deal of time; add to that the time required for publication and it could be years before the data become available. This will be a particular problem if the item you are investigating changes rapidly.

(ii) We do not necessarily know how the data were collected and analysed or for what reason. They may be biased because of poor collection techniques or simply because they were collected for a different purpose. They may, of course, be deliberately biased to prove a point in the original argument for which they were used. Definitions of terms may not be clearly given and the units of measurement may not be appropriate for the current investigation.

(iii) We may not be able to find a complete set of data for our purposes in one place. This could mean we would have to collate data from several sources with the chance of making errors while doing so. Obtaining the data from more than one source may also compound the chances of bias discussed in (ii).

(iv) There is the distinct possibility of transcription or printing errors in published data. This is a particular problem with newspapers and magazines, some very prestigious newspapers being dire offenders.

You should be aware of these dangers. If you are using secondary data to support arguments in reports, articles or essays it is advisable to try to find out more about how the data were collected and analysed and why they were collected.

3.3 Official UK statistical sources

The Government has a seemingly unquenchable thirst for statistics. The Government Statistical Service (GSS) exists to service the statistical needs of Government and to provide useful information for business and industry. It consists of the statistics division of all major Government departments along with two large data-collection agencies – the Central Statistical Office and the Office of Population Censuses and Surveys. The Central Statistical Office (CSO) co-ordinates the system. The data compiled by the CSO are published for it by Her Majesty's Stationery Office (HMSO). Other Government departments and ministries also produce statistics. There is really only one way to find out the contents of the published volumes of official statistics and that is to seek them out and examine them for yourself in your nearest reference library. The exercises at the end of this unit will indicate areas for you to investigate.

A useful booklet is *Government Statistics – a Brief Guide to Sources*. It will give you an insight into the publications available and the ground they cover and can be obtained free of charge from Central Statistical Office, Press, Publications and Publicity, Great George Street, London SW1P 3AQ.

An invaluable publication is the *Guide to Official Statistics* which gives an extremely comprehensive review of the information available. It lists many non-official sources of statistics, including articles and reports that contain useful data, as well as all the official sources. You will find a copy of this publication in any good reference library, and there you should also find copies of many of the publications referred to in the *Guide*.

Another Government publication which is a valuable source of up-to-date information is *Economic Progress Report*. This is published bi-monthly and is available free from Economic Progress Report (distribution), COI, Hercules Road, London SE1 7DU. *Statistical News*, published quarterly, is also useful.

It is beyond the scope of this book to describe all the publications available and their contents. The *Guide to Official Statistics* is a large volume by itself. The following is a brief description of some of the more important publications that you should be sure to look at.

(a) *Annual Abstract of Statistics* and *Monthly Digest of Statistics*

These important compilations of statistics should be easily available to you, since most reference libraries have copies. The *Monthly Digest* is a monthly publication covering much the same topics as the *Annual Abstract* but its tables give the most recent and often provisional figures rather than being a record over the long term. A wide range of statistics is covered by the *Annual Abstract*. The chapter headings for the *Annual Abstract* for 1993 will give you some idea of how wide that range is: area and climate; population and vital statistics; social conditions; law enforcement; education; employment; defence; production; agriculture, fisheries and food; transport and communications; distributive trades, research and development; external trade; balance of payments; national income and expenditure; personal income, expenditure and wealth; home finance; banking, insurance etc.; prices. You should examine the *Monthly Digest* and the *Annual Abstract* and note their general contents. They are valuable sources of data, especially the *Annual Abstract* with its more comprehensive coverage and comparative figures for other years.

(b) *Family Expenditure Survey*

Compiled by the Department of Employment, this survey contains a wide range of statistics on household income and expenditure (see Unit 10.9 (*c*)). Students should find the information about spending patterns interesting in itself and certainly it is of interest to businesses planning for future production. The results of this annual survey form the basis of the weighting system for the Retail Prices Index (see Unit 10.9).

(c) *Employment Gazette*

This is a monthly publication of the Department of Employment containing articles on employment and labour, as well as up-to-date news on related legislation, of considerable value to businesses since labour legislation has expanded rapidly in recent years. There is also a summary of the latest statistics on labour, unemployment, job vacancies, hours worked, stoppages and wages, and the latest value of the Retail Prices Index (see Unit 10.9).

(d) *Eurostatistics*

As the UK becomes increasingly integrated within the European Union, it is appropriate to consider comparative statistics with other EU members. This publication is produced by Eurostat, the EU's statistical office in Luxembourg. It is a monthly publication which provides the latest statistical data on the European Union as a whole, each member state, and the United States and Japan. It is complemented by an annual review, which shows in a single volume how the EU has developed over the past 10 years.

(e) *Economic Trends*

This is a monthly publication from the Central Statistical Office that contains statistics on many subjects that affect the economy, including production, labour, prices, wages, investment and finance. It is of particular interest to students of statistics because of the many examples of diagrams and graphs used in presenting the data.

(f) *Regional Trends*

Also produced by the Central Statistical Office, this annual compilation breaks down national statistics to show regional variations covering economic, social and demographic topics.

(g) *Social Trends*

This is an annual compilation from the Central Statistical Office of information that describes UK society and the way in which it is changing. The material covered includes population, education, health, housing and transport, although the coverage differs each year. The publication includes articles and usually a social commentary in addition to many colour charts, tables and graphs.

(h) ***Annual Census of Production***

The first Census of Production was held in 1907 and other surveys of productive activity were made at irregular intervals before the Second World War. This Census is now conducted annually by the Department of Trade and Industry and its main objective is to provide information about the condition of productive industry that is useful to the Government, economic analysts and industrialists. It enables comparisons to be made between different industries, between years, and between UK industries and others within the European Community. From this census *Business Monitors* are published for individual industries. In 1984, a new system of annual censuses was introduced. A series of 'slimline' censuses are conducted, with a larger 'benchmark' census being made every five years or so. This method was introduced to reduce the burden of form-filling for the business community.

(i) ***United Kingdom National Accounts***

Commonly referred to simply as the 'Blue Book', this publication is essential reading for students of economics, since national income and expenditure statistics have become an important indicator to the performance of the economy. The Blue Book gives a comprehensive breakdown of national income, national expenditure and national product, as well as further tables concerned with such topics as the distribution of income, investment, and government and local authority expenditure.

(j) ***Population Monitors***

These publications from the Office of Population Censuses and Surveys (OPCS) give information regularly on vital statistics, that is, the numbers of births, marriages and deaths, as well as on such topics as fertility rates and morbidity (sickness) rates, at less frequent intervals. OPCS also provides details on all aspects of the decennial Census of Population from the content of the questionnaire and method of analysis of results to the final Census tables. Further details of UK population statistics are to be found in Unit 17.

This list of official publications is far from exhaustive: there are others relating to the United Kingdom and many international compilations as well, including a major volume from the United Nations giving demographic and economic statistics for most countries in the world.

3.4 Publications of the European Communities

Since the entry of the United Kingdom into the European Community (EC), familiarity with EC publications has become increasingly necessary. Details of these official publications are available from the Information Offices of the Commission of the European Communities, Jean Monnet House, 8 Storey's Gate, London SW1P 3AT, and at 4 Cathedral Road, Cardiff CF1 9SG. The publications can be ordered from HMSO and are available in English and in some or all of the other Community languages.

The range of statistics available from the Statistical Office of the European Commission (SOEC) is vast. A useful starting point is the Eurostat Index, published by Capital Planning Information Limited, 52 High Street, Stamford, Lincolnshire PE9 2LG. This is a detailed keyword subject index to the entire statistical series published by the SOEC.

Details of publications can also be obtained from the Office for Official Publications of the European Communities, 2 Rue Mercier, Luxembourg L 2985. From this address several catalogues of EC statistics are available free of charge. These include *Eurostat Catalogue*: publications and electronic services 1991, the monthly publication; *Eurostat*: just published; and *Sigma*: the bulletin of European statistics.

3.5 Exercises

1. Give examples of official statistics and their sources that may be of great use to businesses in a competitive industry. Give reasons for their usefulness.

2. It is essential that the statistician should distinguish between primary and secondary data. Explain the meanings of these terms and discuss the dangers of using secondary data.

3. If you were asked by the management of the company for which you work to try to find out how the economic environment might affect the sales and production of the company, how might you use official statistics to help with the exercise? Outline the types of facts that might interest you.

4. Write briefly on the following sources of economic statistics, paying particular attention to their contents:

   (a) *Social Trends*;
   (b) *Annual Abstract of Statistics*;
   (c) *Family Expenditure Survey*;
   (d) *United Kingdom National Accounts* (the Blue Book);
   (e) *Employment Gazette*.

5. You wish to set up in business as an import/export warehousing firm. Explain how you could use published statistics to help you decide on the location and scope of operations of your firm, showing the type of information you could use and the journals that will give you that information.

6. What information does the Government publish on labour statistics and which are the main publications containing the figures?

7. Outline the information of interest to businesses that is available in publications of UK banks.

8. If you were given the task of establishing a statistical service for the Government of a developing country, suggest three subjects on which you think statistical information should be collected. Give reasons for your choice.

UNIT FOUR

Variables and frequency distributions

4.1 Variables defined

Every time we make an observation we are interested in the behaviour of a *variable*, that is, a characteristic that we know will have different values within a population. We shall consider three types of variable.

(i) Attribute or categorical variables These variables are descriptive: for example, colour of hair, colour of eyes or defectiveness of a product. For this reason we often have difficulty identifying an object that possesses a particular attribute. Sometimes it is a matter of opinion or judgement whether an object possesses the attribute. Who, for example, is a 'professional' person?

(ii) Discrete variables These are numerical variables but the numerical values taken can only be particular numbers. One example is the number of rooms in a house: rooms can only be counted in single units. Another example is the number of cars passing a certain spot in a given period of time.

(iii) Continuous variables These again are numerical variables but, unlike discrete variables, they are not restricted to specific values. Instead, the variable is measured on a continuous scale. Examples are the heights, weights or ages of objects or people. When dealing with continuous variables we have problems of precision of measurement and as a result we are usually involved with some degree of approximation.

4.2 Frequency distributions

Suppose the variable we are interested in is the number of visits per year made to their local doctor's surgery by mothers with children under 16 living in a certain district. We study the surgery record cards of 10 mothers and find the following numbers of visits:

8 6 5 5 7 4 5 9 7 4

We have taken a sample of 10 mothers from the population of mothers in the district, observed the variable (number of visits to the surgery) and now have some values of the variable. However, 10 families is a small sample from which to draw firm conclusions. To make inferences about a whole population from a sample drawn from it we would usually try to take a larger sample, if possible.

Suppose instead we sample 100 mothers and find the numbers of visits to be:

```
 9  6  7  7  8  6  8 10  6  5  5  4  9  9  8  8  4 10  6 10
10  8  7  5  4  8  9  9  5  7  7  8  7  4  9  8  8  9  7  8
10  4  7 10  7  9  6  5  9  7  7  6  4  9  7 10  5  8  9  7
 7  6  8  6  4  9  7  8  6  5 10  7  8  4  6  5  7  8  6  7
 8  6  9 10  7  7  5  7  7  5  6  5 10  7  6  7  8  5  7  6
```

We now have more of a problem because we have so many values of the variable that we must count the number of times each value of the variable occurs. A convenient method of doing this is to arrange the variables as shown in Table 4.1, and count the number of mothers by means of tally marks. The stroke through the previous four marks each time a fifth is counted makes the totting-up process at the end of the count much easier.

Table 4.1 Numbers of visits made to doctor's surgery by 100 mothers

Number of visits (variable)	Tally	Number of mothers (frequency)
4	𝍸 III	8
5	𝍸 𝍸 II	12
6	𝍸 𝍸 𝍸	15
7	𝍸 𝍸 𝍸 𝍸 𝍸	25
8	𝍸 𝍸 𝍸 II	17
9	𝍸 𝍸 III	13
10	𝍸 𝍸	10
TOTAL		100

The resulting table is a *frequency distribution*: it shows the frequency with which each value of the variable occurs. We shall be using frequency distributions a great deal in the next Units.

4.3 The classification of data

We have now seen how to construct a simple frequency distribution. It was a simple example because the spread of the data was small: the value of the variable only ranged from 4 to 10. We shall now look at a different example, in which the variable is the examination percentage mark of an A-level statistics candidate. A sample of 100 candidates produced the following marks:

81 85 62 71 70 81 86 67 96 51 63 71 75
69 48 34 87 86 73 75 42 91 58 93 52 82
90 95 82 72 53 38 77 93 85 47 70 68 57
71 96 40 70 92 68 88 58 51 90 74 52 63
96 77 83 76 48 92 81 83 92 73 84 78
78 72 60 84 78 60 43 70 83 64 96 93
55 73 58 40 88 96 72 53 87 92 73 77
63 58 71 80 38 63 56 76 82 61 76 63

To count the numbers of students with each mark would not really be practical, unless, of course, you need to know that specific fact. To condense the marks it is better to allocate them to *classes* and to count the numbers of students in each class. In this instance we can conveniently divide the marks into groups of 10, i.e. 31 to 40, 41 to 50, and so on. Each of these groups is known as a *class*. The frequency distribution is constructed as shown in Table 4.2. This type of distribution is known as a *grouped frequency distribution*.

Table 4.2 Percentage marks of 100 A-level statistics candidates

Percentage mark (variable)	Tally	Number of students (frequency)
31–40	IIII	4
41–50	~~IIII~~	5
51–60	~~IIII~~ ~~IIII~~ ~~IIII~~	15
61–70	~~IIII~~ ~~IIII~~ ~~IIII~~ I	16
71–80	~~IIII~~ ~~IIII~~ ~~IIII~~ ~~IIII~~ IIII	24
81–90	~~IIII~~ ~~IIII~~ ~~IIII~~ ~~IIII~~ I	21
91–100	~~IIII~~ ~~IIII~~ ~~IIII~~	15
TOTAL		100

4.4 Class limits and class boundaries

The frequency distributions we have seen so far have been concerned with discrete variables (numbers of visits and percentage marks). The starting-point and finishing-point of each class are quite clear when you are working with a discrete variable. The smallest and largest values of a variable that can actually occur in a class are called the *class limits*, so the class limits of the 31–40 class in Table 4.2 are 31 and 40. However, when a frequency distribution is drawn up from a set of observations of a continuous variable, there may be problems in identifying the starting- and finishing-points of each class.

Here is an example on the weights in kilograms of 75 pigs:

36.42	24.23	50.36	50.72	27.20	53.36	30.63
40.27	103.62	84.72	35.20	42.63	47.83	28.60
22.27	19.67	105.36	60.37	109.46	63.72	64.83
52.36	50.72	42.78	73.70	37.63	42.74	50.63
64.90	65.60	28.30	70.23	67.20	63.83	57.60
58.73	97.23	63.40	65.23	58.60	51.46	55.60
58.63	51.30	55.50	48.33	70.58	80.93	38.60
29.36	27.32	37.40	33.20	33.40	74-60	87.20
83.23	63.76	77.60	89.40	43.60	47.60	52.30
43.57	74.60	72.20	93.60	95.43	91.30	54.87
	48.98	57.76	45.67	52.89	59.89	

Weight is a continuous variable since there is no restriction on the value a weight can take within a given range. Our measuring equipment, however good, can never record the weight exactly. The recorded weight will always be subject to some error. In this case we shall assume that the weights are correct to the nearest 0.01 kg since they are given correct to two decimal places.

Suppose we formed a frequency distribution from these figures using the following classes:

11–20 kg
21–30 kg
31–40 kg, etc.

The problem here is that there is a gap between each class. There is one whole kilogram between 20 kg and 21 kg but this does not appear to have been allowed for in the classes. Into which class would we put the weight of 50.36 kg for example? Since it must go somewhere, we must make a more careful definition of our classes so that we cover all possible values of the variable. In this example it would be far better to draw-up the frequency distribution as shown in Table 4.3. It is quite clear from this into which class any weight falls.

Table 4.3 Weights of 75 pigs

Weight (kg) (variable)	Tally	Number of pigs (frequency)
Under 20	I	1
20 and under 30	卌 II	7
30 and under 40	卌 III	8
40 and under 50	卌 卌 I	11
50 and under 60	卌 卌 卌 IIII	19
60 and under 70	卌 卌	10
70 and under 80	卌 II	7
80 and under 90	卌	5
90 and under 100	IIII	4
Over 100	III	3
TOTAL		75

As in the discrete case, the class limits are the smallest and largest values that can occur in a class so in the 20 and under 30 kg class, the lower class limit is 20.00 kg and the upper class limit is 29.99 kg. The class limits should be distinguished from the *class boundaries*. The class boundaries are the dividing lines between the classes so the class boundaries of the 20 and under 30 kg group are at 19.995 kg and 29.995 kg. Such values cannot actually be measured, but any weight in the range 19.995 to 29.995 kg will be put in the 20 and under 30 kg class whereas weights outside this range, for example 19.95 kg and 30.02 kg, would go in other classes. You must always look very carefully at any frequency distribution to determine the true class boundaries and limits.

The difference between the class boundaries is called the *class interval* or the *length of the class*. In our example it is 29.995 minus 19.995 kg which is 10 kg. You will sometimes find the midpoint of a class referred to as the *class mark*. For the 20 and under 30 kg group the midpoint is at 19.995 + 10/2, which equals 24.995 precisely, although in practice this might be rounded to 25 kg.

When compiling a frequency distribution, we have to decide on the number of classes to be used. As a general rule we use between five and twenty classes. Five classes would be appropriate for a small number of observations (30) whereas twenty classes would be needed when the data consist of thousands of readings. Too few groups simplify the data too much and information is lost, whereas too many groups can be cumbersome. It is desirable to have equal class interval lengths if possible and to choose the boundaries so that the compiler can easily assign each reading to its appropriate class.

4.5 Open-ended classes

You will notice that the frequency distribution of weights used in the above example has two open-ended classes – under 20 kg and over 100 kg. Often at the ends of distributions we have classes, such as those above, where the class interval is indeterminate. It is usually convenient to treat open-ended classes as having the same class interval length as the following or preceding classes. In the weight example, the open-ended classes should be treated as though they had an interval of 10 kg, the same as the other classes. The 'under 20' class is treated as '10 and under 20', therefore, and the 'over 100' is treated as '100 and under 110'.

4.6 Graphical representation of a frequency distribution

(a) Histogram

A frequency distribution can be depicted in the form of a diagram known as a *histogram*. The drawing of a histogram is straightforward, but it is important to remember the principle behind the construction.

To draw a histogram two axes are required. The horizontal axis shows the values, rounded appropriately, of the class boundaries. The vertical axis records the class frequencies. On each class interval a column is drawn that is

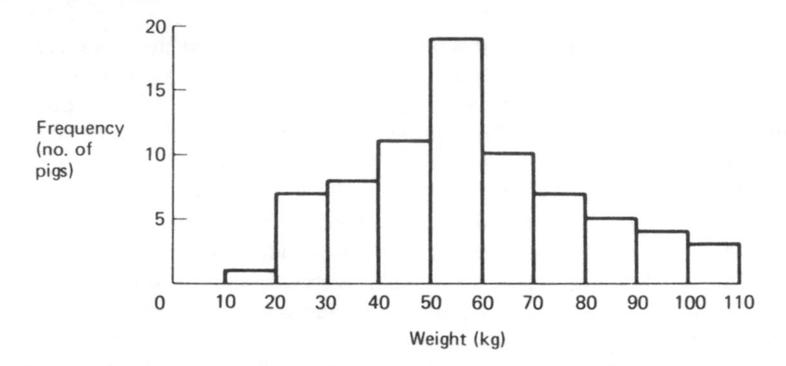

Fig. 4.1 Histogram showing weights of 75 pigs

as high as the frequency recorded for that class. As an example we shall use the grouped frequency distribution (Table 4.3) of the weights of 75 pigs. The resulting histogram is shown in Fig. 4.1.

The important point about the histogram is that the *area* of the rectangle formed by each column represents the frequency with which that value of the variable occurs. In Fig. 4.1 the histogram is drawn for a frequency distribution where the interval for each class is a constant 10 kg. We only had to draw the top of each column at the recorded frequency for the areas of the columns to represent the class frequencies correctly.

In some frequency distributions, however, the class intervals are not constant. Table 4.4 shows the weekly wage distribution in a small firm. You might think that the histogram should look like Fig. 4.2. You would be wrong. Remember that the *area* of each column should represent the frequency, so on this histogram the area of the column for the '0 and under 80' class, which represents a frequency of 8, should be four-fifths of the area of the column

Table 4.4 Frequency distribution of weekly wages in a small firm

Weekly wage (£)	Number of workers (frequency)
0 and under 80	8
80 and under 100	10
100 and under 120	14
120 and under 140	20
140 and under 160	15
160 and under 180	6
180 and under 260	4
TOTAL	77

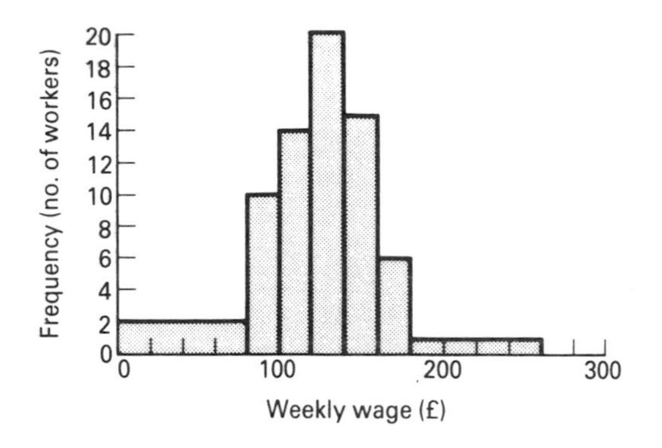

Fig. 4.2 Histogram showing weekly wage distribution in a small firm (correct)

for the '80 and under 100' class which represents a frequency of 10. The interval of the first class, however, is four times the length of the second. By drawing the top of its column level with the frequency of 6 we have made the area of the column larger than that of the second column. The column for the '180 and under 260' class is also wrong. The class interval is £80, and by making the top of the column level with a frequency of 4 we have drawn a column with an area four times greater than it should be.

How do we draw this histogram, then? The answer is simple: we do not change the scale on the horizontal axis, but adjust the frequencies to allow for the uneven class intervals. The standard practice is to use the most common class interval as a 'base' and to adjust the frequencies of the other classes to conform with that base. In this case the most common class interval is £20. We need to adjust the frequency of the first class, which has an interval of £80, by dividing it by 4: the adjusted frequency is then 2. When we have made all the adjustments, the frequency distribution can be shown as in Table 4.5. The adjusted frequency is equivalent to the original frequency divided by the

Table 4.5 Adjusted frequency distribution of weekly wages

Weekly wage (£)	Number of workers (frequency)	Adjusted frequency for £20 interval
0 and under 80	8	2
80 and under 100	10	10
100 and under 120	14	14
120 and under 140	20	20
140 and under 160	15	15
160 and under 180	6	6
180 and under 260	4	1

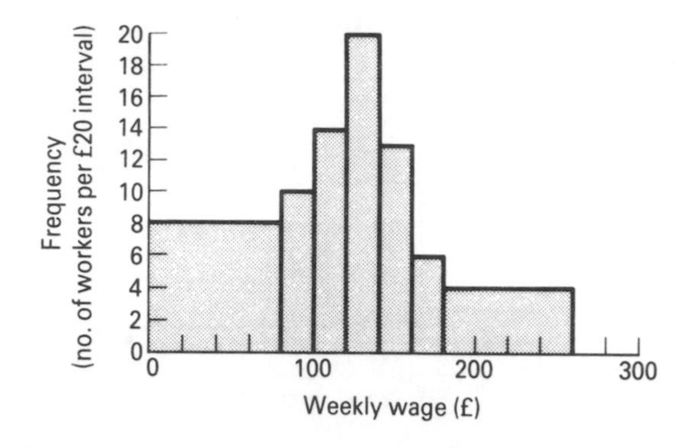

Fig. 4.3 Histogram showing weekly wage distribution in a small firm (incorrect)

number of base class intervals the original interval represents. We can now draw the correct histogram (Fig. 4.3). The areas of the columns represent their frequencies, that is, they are in the correct relationship according to their frequencies.

Histograms can also be drawn for a discrete variable such as the percentage marks in the distribution shown in Table 4.2. The class limits are given in that Table. The class boundaries are at 30.5, 40.5 etc. since any mark between 30.5 and 40.5 goes into the 31–40 class whereas a mark such as 41 would go into another class. The class interval is ten marks (i.e. 40.5 minus 30.5). When drawing the histogram, you must take great care to ensure that the columns start and finish at the class *boundaries*, not at the class limits.

(b) Frequency polygon

Another way of drawing the frequency distribution is with a *frequency polygon*. We can see how this is done using the histogram of pig weights (Fig. 4.1). The frequency polygon is constructed by drawing ruled lines connecting the midpoints of the tops of each column. This line should be continued down to cut the horizontal axis at what would be the midpoint of an extra class having zero frequency, at each end of the distribution (Fig. 4.4).

It is usual to draw either a histogram or a frequency polygon for a set of data. We have put them on the same graph only to illustrate the method of drawing.

The frequency polygon gives us some idea of the shape of the distribution and you will see later (Unit 7.7) that it is sometimes useful to know this shape.

(c) Frequency curve

The shape of the frequency distribution is usually shown by a *frequency curve*. This can be approximated from the histogram or the frequency polygon by superimposing a curve as in Fig. 4.4. To draw a frequency curve properly the length of the class intervals used to construct the frequency distribution must be very small and the number of observations correspondingly large. A

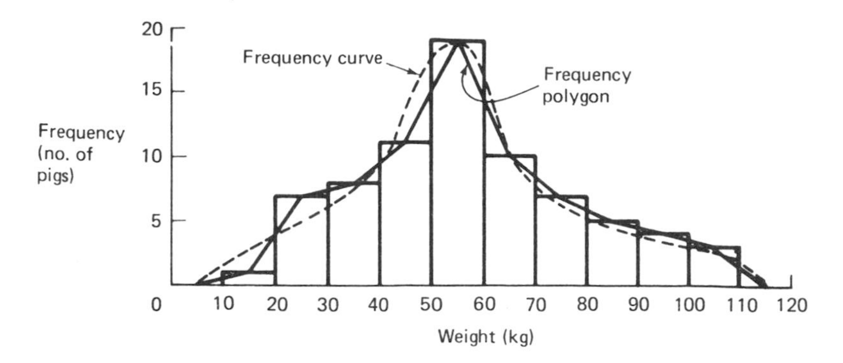

Fig. 4.4 Frequency polygram and approximate frequency curve for weights of 75 pigs

histogram drawn from this distribution would then have long, very narrow columns and, if the midpoints of the tops of these columns were joined by a line as though a frequency polygon were being drawn, the polygon would almost be a smooth curve. But to do this properly we would need an impracticably large sample from the population; for most purposes, an approximation to the frequency curve is enough to indicate the shape of the frequency distribution.

Frequency curves give us an idea of the way in which the variable is distributed in the population. Distributions can be of various shapes, but there are three important types.

(i) Positively skewed distributions In these distributions, the low values of the variable have the highest frequencies. An example of a frequency curve for a positively skewed distribution is shown in Fig. 4.5(*a*).

One of the best examples of a positively skewed distribution is that of the distribution of wages and salaries in the population. The bulk of the population earn wages and salaries at the lower end of the scale.

(ii) Negatively skewed distributions An example of a frequency curve for a negatively skewed distribution is shown in Fig. 4.5(*b*). The high values of the variable have the highest frequencies. Age-related data are sometimes negatively skewed – for example, the distribution of false teeth in the population by age group.

(iii) Normal distributions In these, the frequency curve is symmetrical and has a characteristic 'bell' shape. An example of a frequency curve for a normal distribution is shown in Fig. 4.5(*c*).

The normal distribution is very important as far as statistical theory is concerned. The word 'normal' is being used in a technical sense to denote one specific type of frequency curve. Its use does not in any way imply that this is a 'usual' or an 'expected' distribution. In fact, a perfectly normal distribution is unusual in practice, but some distributions approximate to this shape – for

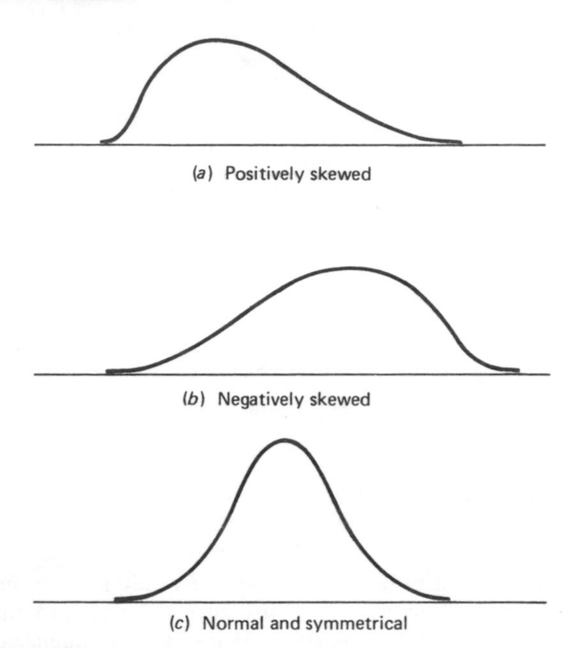

Fig. 4.5 Three frequency curves

example, the heights and weights of an adult population and the weights and dimensions of mass-produced articles.

(d) Ogive

One other important way of representing the frequency distribution is with an *ogive* or *cumulative frequency curve*. This can be drawn in either the 'less than' or 'or more' form. The 'less than' curve shows the number of items in a population that are *less than* any value of the variable. An ogive is drawn as a simple graph. Again using the example of the weights of 75 pigs, we form the cumulative frequency distribution as shown in Table 4.6.

All we have done is to make a running total of the frequencies: we have added the second frequency to the first to show that there were 8 pigs with weights less than 30 kg; we then added the third frequency to the total of the first two, showing that 16 pigs weighed less than 40 kg, and so on. Each 'less than' cumulative frequency therefore shows how many pigs weighed less than the upper boundary of a particular class. To draw the ogive we again need a horizontal axis on which the values, rounded appropriately, of the class boundaries are recorded, but in this graph the vertical axis records the 'less than' cumulative frequency. The 'less than' cumulative frequencies are plotted against the upper boundary of each class (*not* the midpoint) and all the points are then joined with a curve (Fig. 4.6) or with ruled lines, in which case the graph is termed a cumulative frequency polygon rather than a cumulative frequency curve. In drawing the ogive we have made the same assumptions about

Table 4.6 Cumulative frequency distribution of weights of 75 pigs

Weight (kg)	Frequency	'Less than' cumulative frequency
Under 20	1	1
20 and under 30	7	8
30 and under 40	8	16
40 and under 50	11	27
50 and under 60	19	46
60 and under 70	10	56
70 and under 80	7	63
80 and under 90	5	68
90 and under 100	4	72
Over 100	3	75

the boundaries of the open-ended classes as we did when drawing the histogram and the frequency polygon.

Once we have drawn the ogive we can use it for estimating. If we wanted to estimate how many pigs weighed less than say, 75 kg, we could draw a vertical line from that value on the horizontal axis up to the ogive, and a horizontal line from that point on the ogive to the vertical axis and that frequency would be the estimate: Fig. 4.6 shows that approximately 59 pigs weighed less than 75 kg. We shall find the cumulative frequency curve of use in later units (see Units 7.10 and 8.3).

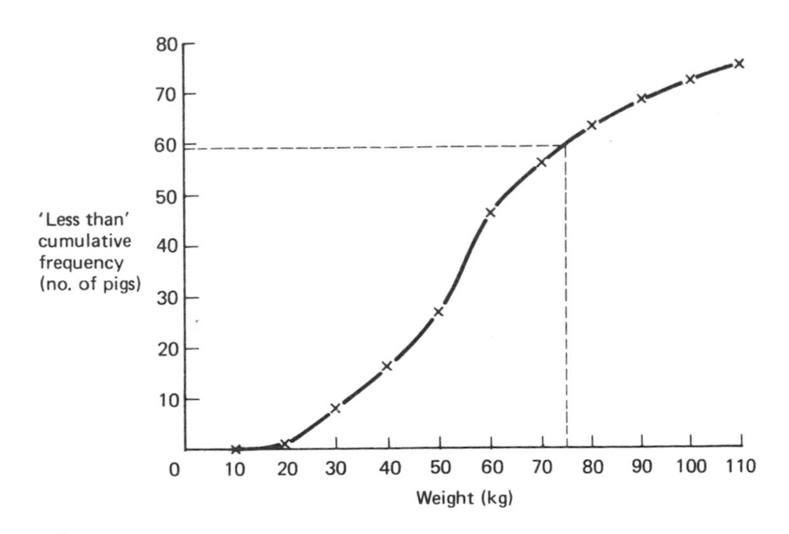

Fig. 4.6 Ogive or cumulative frequency curve of weights of 75 pigs

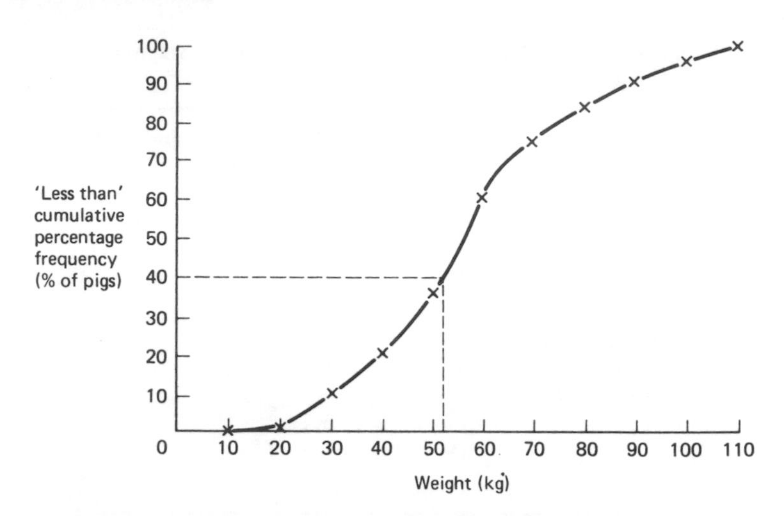

Fig. 4.7 Percentage cumulative frequency curve of weights of 75 pigs

It is often convenient to have the vertical scale of an ogive drawn in percentage frequency form so that it is easy to estimate directly, for example, the weight below which 40% of the pigs were. To plot the curve in this form we replace the actual 'less than' cumulative frequencies by their values as percentages of the total frequency, in this case 75. The resulting percentage frequencies, correct to the nearest whole number, are shown in Table 4.7 and the corresponding ogive in Fig. 4.7.

Our ogives so far have been drawn in 'less than' form. It is, on occasions, more convenient to draw an 'or more' ogive—for instance, if we wished to

Table 4.7 Cumulative percentage frequency distribution of weights of 75 pigs

Weight (kg)	Frequency	'Less than' cumulative frequency	'Less than' cumulative percentage frequency
Under 20	1	1	1
20 and under 30	7	8	11
30 and under 40	8	16	21
40 and under 50	11	27	36
50 and under 60	19	46	61
60 and under 70	10	56	75
70 and under 80	7	63	84
80 and under 90	5	68	91
90 and under 100	4	72	96
Over 100	3	75	100

Table 4.8 'Or more' cumulative frequency distribution of weights of 75 pigs

Weight (kg)	Frequency	'Or more' cumulative frequency
10 and under 20	1	75
20 and under 30	7	74
30 and under 40	8	67
40 and under 50	11	59
50 and under 60	19	48
60 and under 70	10	29
70 and under 80	7	19
80 and under 90	5	12
90 and under 100	4	7
100 and under 110	3	3
110 and over	0	0

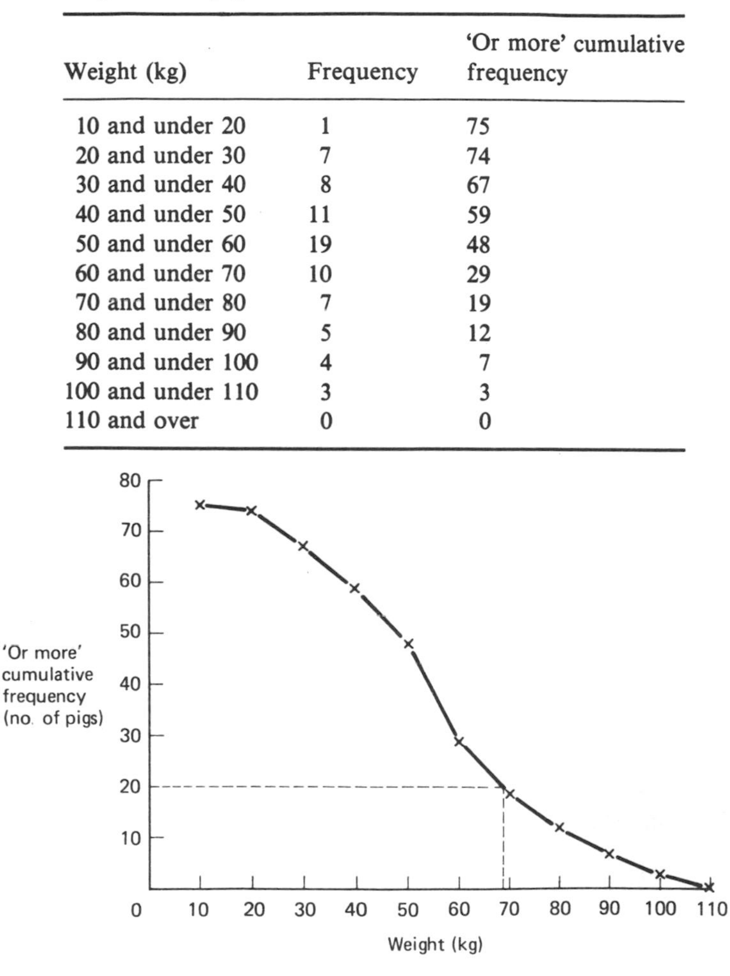

Fig. 4.8 'Or more' ogive of weights of 75 pigs

know the weight exceeded by the 20 heaviest pigs. We again draw up a cumulative frequency table, Table 4.8, but this time we record the numbers of pigs whose weights were the *lower* class boundary or more, so our cumulative frequencies gradually *decrease*. In Table 4.8 we have made the same assumptions as previously about the boundaries of the open-ended classes. Thus 75 pigs weighed 10 kg or more. One pig weighed between 10 and 20 kg so 74 pigs weighed 20 kg or more, 7 pigs weighed between 20 and 30 kg so 67 weighed 30 kg or more and so on, until we reach zero pigs weighing 110 kg or more. The corresponding 'or more' ogive is shown in Fig. 4.8.

To find the weight exceeded by the 20 heaviest pigs we draw a horizontal line at a cumulative frequency of 20. From the point it crosses the ogive we draw a vertical line to the horizontal axis. From Fig. 4.8 we find that the 20 heaviest pigs weighed 69 kg or more. An 'or more' ogive can also be drawn in percentage form if desired.

4.7 Exercises

1. Which ofthe following are discrete and which are continuous variables?
   (*a*) the number of students in a college;
   (*b*) the net wages of a group of workers;
   (*c*) the expansion of a rod of metal when heated;
   (*d*) the number of pints of beer drunk in the public houses of a town;
   (*e*) the number of rooms in the houses in the same town;
   (*f*) the heights of the men in the same town.

2. Outline the steps you would take to draw up a frequency distribution from a series of 500 observations of the weights of a population.

3. What are meant by class boundaries and class limits? Why might they be different in the same frequency distribution?

4. (*a*) What is meant by the term frequency distribution?
   (*b*) Describe the following types of frequency distributions: (i) positively skewed, (ii) negatively skewed, and (iii) normal.
   (*c*) Give examples illustrating where each type may occur.

5. The company for which you work employs 1 500 staff. You are asked to draw up the following frequency distributions:
   (*a*) the number of staff that have taken one day, two days, etc. of sick leave during the past year;
   (*b*) the distribution of salaries earned by the staff.

   Explain why your approach to each of these tasks is different.

6. The following are the heights in centimetres of 100 men:

   169 179 183 192 166 181 177 173 167 193 176 183 162 170 186
   174 188 165 174 170 176 186 177 185 175 179 166 190 182 174
   182 183 180 194 177 184 168 181 180 172 178 192 175 189 180
   175 183 191 172 172 188 180 176 185 178 179 173 165 170 178
   181 181 181 189 187 191 191 179 196 179 182 171 169 171 171
   184 198 182 175 190 187 176 164 187 167 185 177 184 178 176
   169 179 183 186 166 181 177 173 167 193

   Form the data into a frequency distribution and draw a histogram from the distribution.

7. The following data are the time intervals (in seconds) between customers entering a store:

141	43	203	104	82	63	24	84	41	86	47	43	100	53	139
194	124	175	177	162	129	128	219	40	105	48	65	100	53	139
147	137	186	214	106	150	109	170	172	105	154	154	35	149	54
104	109	119	74	140	104	168	53	145	29	112	143	49	199	130
52	109	77	142	75	146	105	125	112	40	126	67	49	90	112
140	132	118	134	133	159	123	161	112	157	104	92	112	151	98
156	117	156	190	122	135	116	96	163	116	186	155	106	153	69
105	136	106	131	118	94	121	127	191	30	109	88	207	152	120

Using the data: (*a*) form a grouped frequency distribution; (*b*) draw a histogram and impose an approximate frequency curve on that histogram; (*c*) draw a cumulative frequency curve.

8. A survey of 125 motorists selected at random were asked the value of their cars. The answers (in pounds) were as follows:

1500	1100	760	2300	2400	3700	10500	2000	2300	4000
1800	1700	2000	2100	1900	3500	4500	2700	2200	2200
800	5600	6000	1800	2000	2300	2300	2200	1900	1950
2100	900	1200	1300	1500	2100	3500	3000	2200	2600
2500	1750	750	1500	1200	1400	1600	2000	1700	1500
4600	3600	1200	2100	1300	2200	3100	2500	1500	5100
800	1200	1300	1500	1500	1800	2000	2100	2500	3000
7400	4500	3500	2100	2200	2000	1800	1750	1100	900
500	1700	1600	2700	2000	2600	4800	2600	2800	2750
2000	2100	4800	2850	2900	3400	4300	3000	1850	1750
2300	2650	1800	4750	7500	2800	6500	2300	1700	5300
3500	2600	3000	5400	300	1900	850	1250	4250	2350
4000	2200	2550	2000	3650					

From these data:
(*a*) construct a frequency distribution; (*b*) draw a histogram from the distribution; (*c*) say what type of distribution it is; and (*d*) draw an ogive for the distribution.

9. The following data are the weekly amounts (£) spent on food by a random sample of families composed of two adults and two children:

48.78	52.56	44.82	53.36	47.37	45.96	45.06	42.76	54.27	44.15
50.23	45.02	41.26	47.06	48.36	41.23	39.06	48.63	52.63	52.36
41.37	46.37	40.87	45.63	50.73	48.07	46.64	43.07	48.63	39.56
42.37	45.56	61.23	50.74	45.36	49.06	43.03	44.73	39.16	51.06
44.02	38.23	41.00	48.67	46.36	50.00	41.63	47.84	45.14	44.14
41.83	45.24	42.00	49.01	45.06	46.37	50.72	53.66	44.99	52.36
48.78	45.63	44.73	45.37	47.00	46.28	44.23	44.16	39.63	41.16
44.23	46.50	37.50	44.98	61.86	42.00				

(*a*) Form a frequency distribution from the data.
(*b*) Draw a histogram and a cumulative frequency curve from the distribution.

UNIT FIVE

Presentation of data

5.1 Introduction

There are four principal methods of presenting data:

(i) a report in words;
(ii) tables;
(iii) diagrams and charts;
(iv) graphs.

Each method has its advantages and disadvantages, which you should be aware of when choosing an appropriate form of presentation of your data. In much published work a combination of some or all of these methods is required.

5.2 Reports

By far the most common method of presentation is a report in words. Press articles, company reports and research papers are all principally in this form. The use of words rather than the more technical-looking tables and charts is more attractive to those who are put off by arrays of numbers and symbols.

A report allows explanation, emphasis and interpretation to be made. However, personal bias can distort the data, and data which do not fit the author's preconceived ideas may be omitted or not dealt with in sufficient depth.

When drawing up a report, you should check the terms of reference so that you know exactly for whom the report is intended and the precise topic. The statistical investigation involving the collection, classification, analysis and interpretation of the data then follows. You must decide how much detail of this you wish to include in the report. Usually details of calculations are not appropriate although a description of the methods used would be given. Many reports fall into the following format:

Terms of reference
Methods of data collection and analysis
Findings
Conclusions
Recommendations

If it is felt that detailed numerical information should be given, then this may suitably be placed in an appendix. You will be able to practise writing reports when tackling the projects suggested at the end of the book.

5.3 Tables

Tabulation is another common method of presenting statistical data. Most publications contain tables of some sort and any source of official statistics will provide examples for you to look at. An accountant presents figures in a layout which is nothing more than a formalized table. A table displays detailed numerical information concisely. Results of an investigation can be recorded accurately so that others may use these figures in their own work. It is immediately obvious if there is an omission of a piece of data. No interpretation of the results can be given in the table. It is also difficult to emphasize any particularly interesting reading unless heavier type, underlining or a footnote is used.

A table is an arrangement of data in horizontal rows and vertical columns. Generally it is better if possible to arrange the table so that there are more rows than columns, but there is no hard and fast rule about this: often a table has to be arranged with more columns than rows if the data are to be made clear. Columns can be subdivided to show a further breakdown of the data if necessary; while it is possible to break down rows into subdivisions, this is not as easy or as convenient as the subdivision of columns and often obscures the message of a table.

Table 5.1 demonstrates the principles stated above. It shows the method of travelling to work used by various age groups of office employees.

You can see that there are more rows than columns. Suppose we had drawn the table the other way round, as in Table 5.2: the information is much less easy to understand.

If we have further information to show, Table 5.3 illustrates how this can be done. The subdivision of the columns is simple, but if we had drawn the table the other way round, as in Table 5.2, the necessary subdivision of the rows would have been difficult and the table would not have been as clear. Try it for yourself and see.

Table 5.1 Method of travel to work: correct layout

	Method of transport		
Age group	Own means	Public transport	Totals
Under 18	12	6	18
18–25	15	10	25
25–40	25	18	43
40–55	13	32	45
Over 55	6	12	18
TOTALS	71	78	149

Table 5.2 Method of travel to work: incorrect layout

Method of transport	Age group Under 18	18–25	25–40	40–55	Over 55	TOTALS
Own means	12	15	25	13	6	71
Public transport	6	10	18	32	12	78
TOTALS	18	25	43	45	18	149

Table 5.3 Method of travel to work: subdivision of columns

	Method of transport				
	Own means		Public transport		
Age group	Motorized	Non-motorized	Bus	Rail	TOTALS
Under 18	9	3	3	3	18
18–25	12	3	8	2	25
25–40	15	10	9	9	43
40–55	8	5	20	12	45
Over 55	3	3	7	5	18
TOTALS	47	24	47	31	149

5.4 Rules for tabulation

When constructing a table the following rules should be applied:

(i) Plan the layout of the table in rough before it is committed to paper. Make sure the columns and rows are large enough for the data and that the table will fit on to the page you want to use.
(ii) Give the table a title that is brief but descriptive.
(iii) Give brief descriptive headings to rows and columns showing clearly the units in which quantities are recorded.
(iv) Use rulings to divide columns and rows, with double or bolder rulings to make main rows and columns, such as headings and totals, stand out. (This rule need not be applied when a table is set in type as in this book, when the use of rulings can be kept to a minimum without loss of clarity.)
(v) Use footnotes to explain entries further if required, using asterisks or other symbols to relate the values in the table to the footnotes. This may be necessary if, for example, the data for the table have been gathered from more than one source or there has been a change in the method of data collection.
(vi) Always state the source of your data.
(vii) Enclose the table within a box frame to make it look neater.
(viii) Do not overcrowd your tables. It is far better to draw another table than

to crowd figures so closely together that they are impossible to read. Worse still, it might discourage people from trying to read the table. Look at Table 5.3 again. Suppose we wanted to show even more information, such as the number of employees that used cars and the number that used motor cycles in the 'Motorized' column and the number who walked and the number who cycled in the 'Non-motorized' column. If this was shown on the same table, there would be overcrowding of information: a further table would be necessary to show the breakdown.

(ix) Above all, be neat: a messy table conveys nothing.

5.5 Diagrams and charts

There are various diagrams and charts that can be used to give a pictorial presentation of data. A chart cannot present information as precisely as a table but it can give a quick overall impression of the findings. A chart or diagram often attracts a reader to a report initially. The reader glances at the chart and then may decide to read the accompanying report which gives a more complete account. Unfortunately many charts seen in the press and on television are over-simplified or sensational and can distort the actual observations. When drawing diagrams and graphs, there are some general rules that should be applied:

(i) Graphs and charts should be given clear and brief titles.
(ii) Axes should be clearly labelled.
(iii) Scales of values and their origins should be clearly marked.
(iv) Reference to the source of the data should be made or an accompanying table provided.
(v) Excessive detail should be avoided.

We shall now consider the most popular types of charts and their construction.

(a) Bar charts

A *bar* is simply another name for a thick line and, in a bar chart, the length of the bar represents either the frequency with which the different values of a variable occur or, more generally, the value taken by a variable. It is an easy chart both to draw and to interpret and can be used both for categorical and numerical variables (see Unit 4.1). The basic format can be modified in various ways. Bar charts are also referred to as column graphs.

(i) Simple bar chart As a first example we shall use the household expenditure data in Table 5.4. The data are displayed on a bar chart in Fig. 5.1 with the different categories of goods and services on the horizontal axis. The thickness or width of the bar is of no consequence. Usually a regular gap is left between successive bars although occasionally you will find the bars placed side by side. It is important that the vertical scale starts at zero since it is the height of the bar which represents the amount.

Table 5.4 Average weekly household expenditure on goods and services 1992

Goods and services	Amount (£)
Housing	47.36
Fuel, light and power	13.02
Food	47.66
Alcoholic drink	11.06
Tobacco	5.38
Clothing and footwear	16.39
Household goods and services	35.3
Personal other goods and services	10.18
Motoring and other transport costs	42.86
Leisure goods and services	40.88
Miscellaneous	1.75
TOTAL	271.84

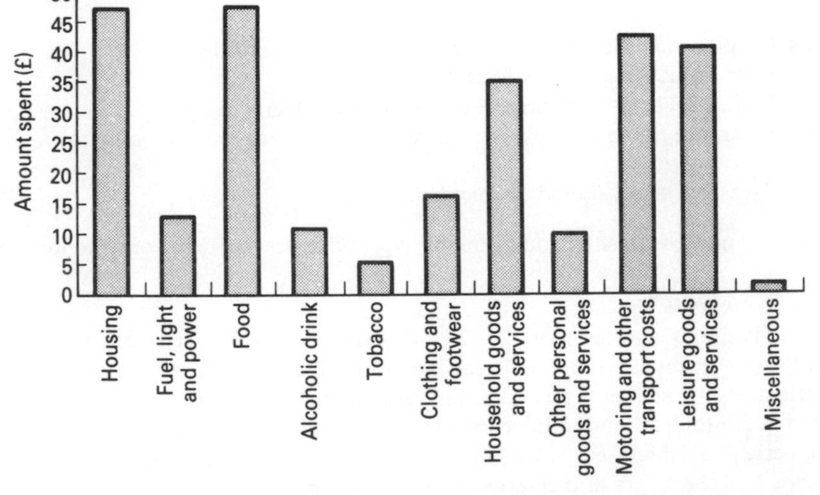

Fig. 5.1 Bar chart showing average weekly household expenditure on goods and services, 1992 (Source: *Family Expenditure Survey*)

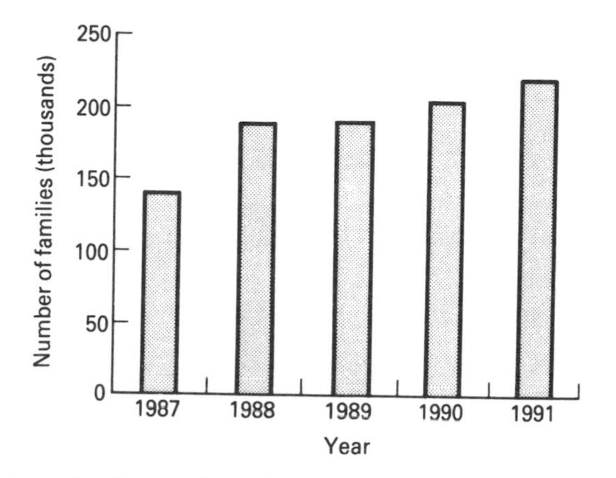

Fig. 5.2 Bar chart showing number of two-parent families receiving Family Income Supplement/Family Credit, 1987–91 (Source: *Annual Abstract of Statistics*)

As a second example consider the numbers (in thousands) of two-parent families receiving Family Income Supplement/Family Credit from 1987 to 1991:

1987	1984	1989	1990	1991
140	189.9	190	201.1	221.5

Source: *Annual Abstract of Statistics*

These values are illustrated in a bar chart in Fig. 5.2. In this example we have a numerical variable on the horizontal axis. A frequency bar chart such as this can be used to represent the frequency distribution of a discrete variable, for example that of the number of visits made by 100 mothers to the local doctor's surgery (Unit 4.2). You should remember that the function of a histogram is to represent the grouped frequency distribution of a *continuous* variable. Try not to confuse a bar chart with a histogram. A bar chart has a much wider application that is not confined to frequency distributions. With a histogram it is the *area* of the columns that represents the frequency and not the height alone.

Bar charts can also be presented in horizontal form, if desired, as in Fig. 5.3. If time is a variable, however, it is preferable to put time on the horizontal axis with the earlier readings to the left as in Fig. 5.2. If the description of a variable is too wordy to write beneath the axis, it is acceptable to place the description beside the bars as is done in Fig. 5.3, or alternatively on the bars themselves.

(ii) Component bar charts You can use a *component* bar chart when, as well as the total amount of a variable, you also want to show the component parts which make up that total. A further analysis of the data on families receiving Family Income Supplement/Family Credit (see (i)) will show the number of children in the families in each of the years (see Table 5.5). Fig. 5.4 shows the data on a component bar chart. Each column is subdivided to show the number of families of each size making up the total number of families receiving

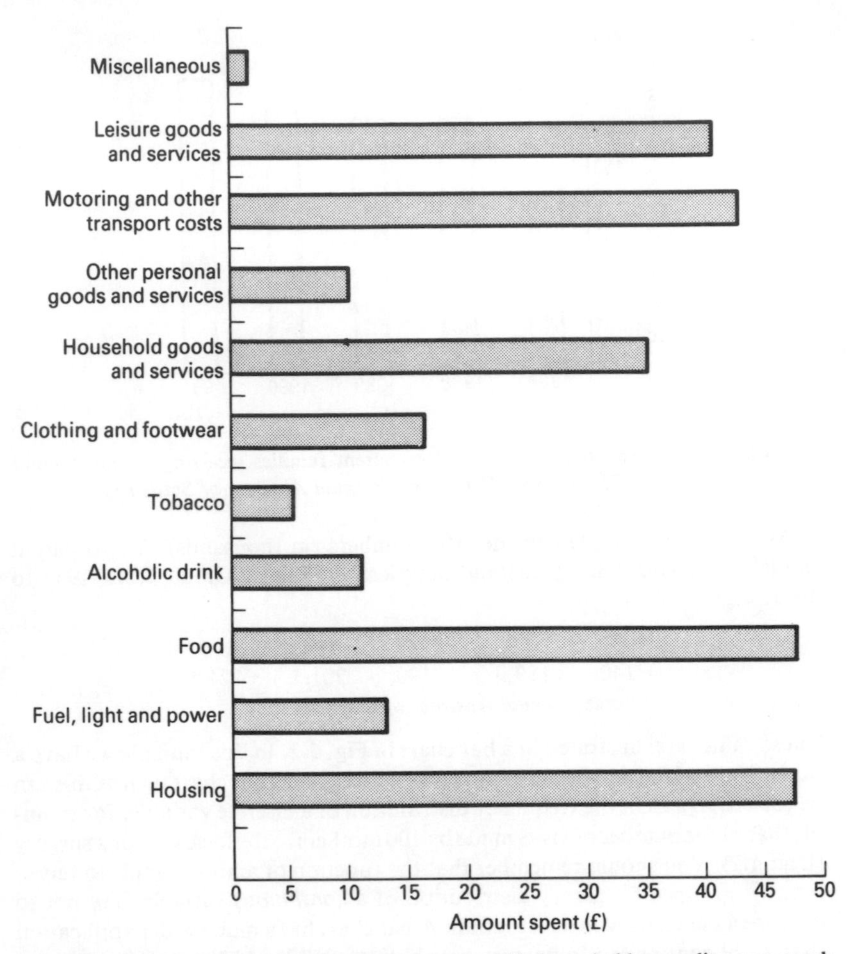

Fig. 5.3 Horizontal bar chart showing average weekly household expenditure on goods and services, 1992 (Source: *Family Expenditure Survey*)

benefit in each of the years. In a component bar chart, although we do portray the breakdown of the total into its component parts, it is quite difficult to compare those components higher up the bars as they do not start at the same level. The actual amounts these components represent cannot be read from the chart directly as we have to work out the difference between the top and base of each component.

(iii) Percentage component bar charts The data on families receiving Family Income Supplement/Family Credit could also be presented in a *percentage* component bar chart. The columns in a percentage bar chart are all of the same height, each representing a total of 100%. The component parts must be

Table 5.5 Two-parent families receiving Family Income Supplement/ Family Credit, 1987–91 (thousands)

Number of children in family	1987	1988	1989	1990	1991
1	30	44.8	43.3	44.5	52.4
2	51	67.1	65.9	71.3	79.3
3	34	44.8	47.1	49.1	52.4
4	16	21.9	21.4	22.7	23.5
5	6	7.8	8.2	8.5	9.1
6 or more	3	3.6	4.1	5.0	4.8
TOTAL	140	190	190	201.1	221.5

Source: *Annual Abstract of Statistics*

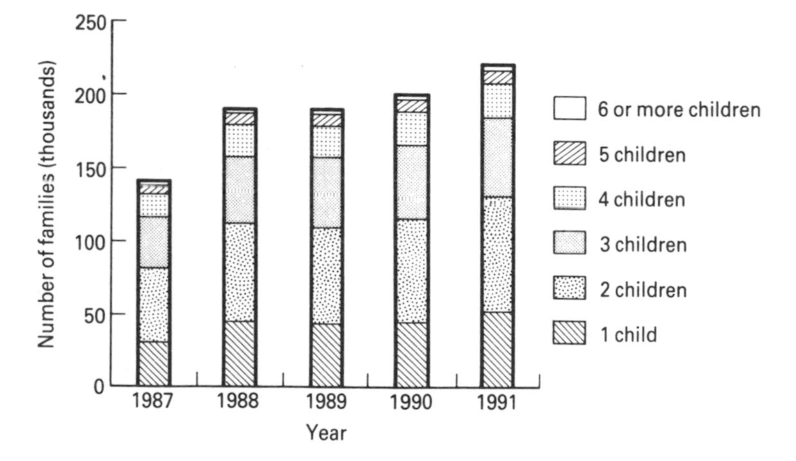

Fig. 5.4 Component bar chart showing family size of two-parent families receiving Family Income Supplement/Family Credit, 1987–91 (Source: *Annual Abstract of Statistics*)

converted to percentages of the total for each column: the percentages for the families receiving supplement are given in Table 5.6. In Table 5.6 the percentages have been calculated to the nearest whole number so, because of rounding errors, some columns do not add up to exactly 100%. The percentage component bar chart for these data is shown in Fig. 5.5. This form of presentation tells us nothing about the total numbers of families each year but shows the breakdown into components as a proportion of the whole.

(iv) Multiple bar charts A bar chart can be adapted to compare changes in more than one variable, in which case it is termed a *multiple* or *compound* bar

Table 5.6 Family size as a percentage of all two-parent families receiving Family Income Supplement/Family Credit, 1987–91

Number of children in family	1987	1988	1989	1990	1991
1	21	24	23	22	24
2	36	35	35	35	36
3	24	24	25	24	24
4	11	12	11	11	11
5	4	4	4	4	4
6 or more	2	2	2	2	2

Source: *Annual Abstract of Statistics*

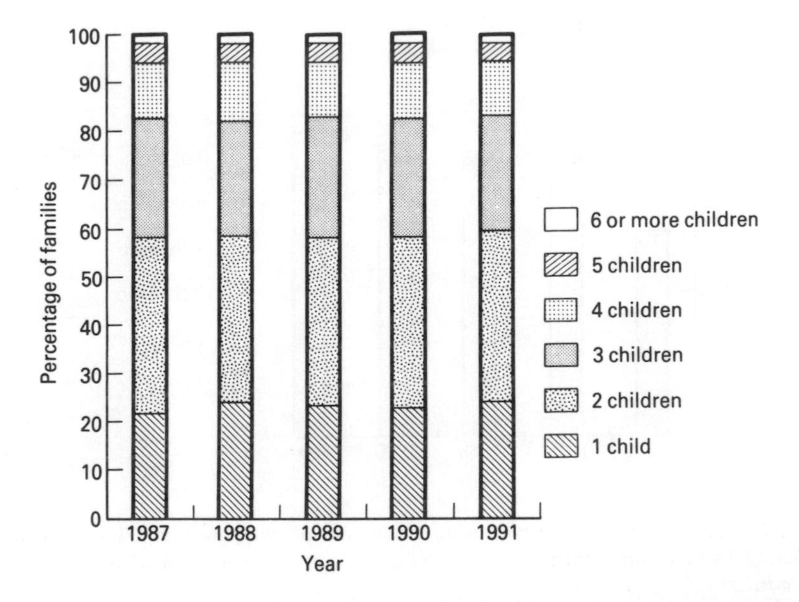

Fig. 5.5 Percentage component bar chart showing families receiving Family Income Supplement/Family Credit, 1987–91 (Source: *Annual Abstract of Statistics*)

chart. For example, with the introduction of legislation and much campaigning for equal treatment of men and women in employment, it may be interesting to display the average weekly earnings of men and women on the same bar chart (see Table 5.7 and Fig. 5.6). The multiple bar chart is drawn with two columns for each year, one column for men's wages and the other for those of the women. The values compared are those for men and women in full-time employment.

In Fig. 5.6 we can compare differences in earnings between men and women

Table 5.7 Average weekly earnings (£) in all industries, full-time adult rates

	1986	1987	1988	1989	1990	1991	1992
Men	207.5	224.0	245.8	269.5	295.6	318.9	340.1
Women	137.2	148.1	164.2	182.3	201.5	222.4	241.1

Source: *Annual Abstract of Statistics*

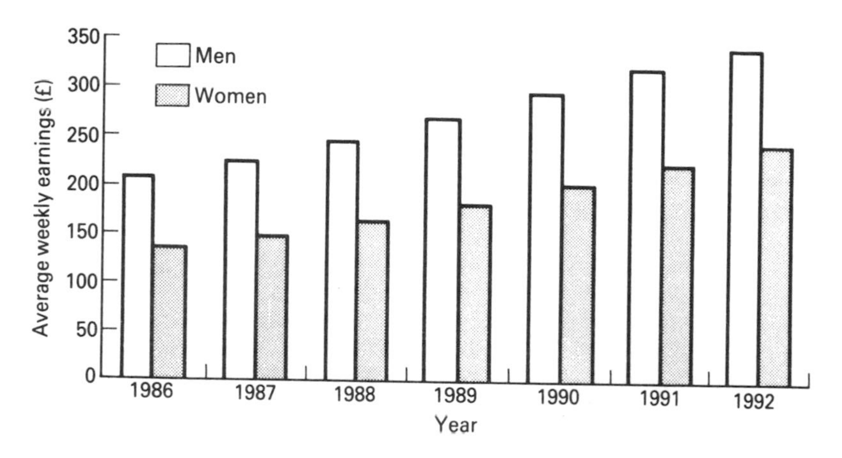

Fig. 5.6 Multiple (dual) bar chart showing men's and women's average weekly earnings (Source: *Annual Abstract of Statistics*)

each year as well as being able to compare how men's earnings have changed with time and how women's earnings have changed with time.

In Fig. 5.6 we have *two* columns for each year and this type of chart is termed a *dual* bar chart. Multiple bar charts can be constructed containing groups of three, four or five bars side-by-side, but comparison becomes difficult if more bars than these are drawn.

(v) Change charts Another useful variation of the bar chart is sometimes known as a *change chart*. It shows changes in the values of variables between two points in time, and is usually drawn as a horizontal bar chart with positive changes being shown to the right of a central vertical axis (which represents the first point in time) and negative changes being shown to the left.

The changes in employment between 1990 and 1991 in certain industries are shown in Table 5.8. A convenient way to illustrate these data is with a change chart (Fig. 5.7).

(vi) Age pyramids A special type of horizontal bar chart that is often used to show population statistics is the *age pyramid*. This is a double bar chart, the data for the male population providing one chart and those for the female the

Table 5.8 Changes in employment in certain industries, 1990–91

Industry	Numbers employed (000s) June 1990	June 1991	Change
Agriculture, forestry and fishing	278	272	−6
Energy and water supply	441	431	−10
Chemical industry	323	303	−21
Motor vehicles and parts	244	220	−24
Food, drink and tobacco	527	544	+17
Retail distribution	2237	2143	−94
Banking, finance, insurance, etc.	2710	2658	−54
Public administration and defence	1545	1565	+20
Other services	855	858	+3

Source: *Annual Abstract of Statistics*

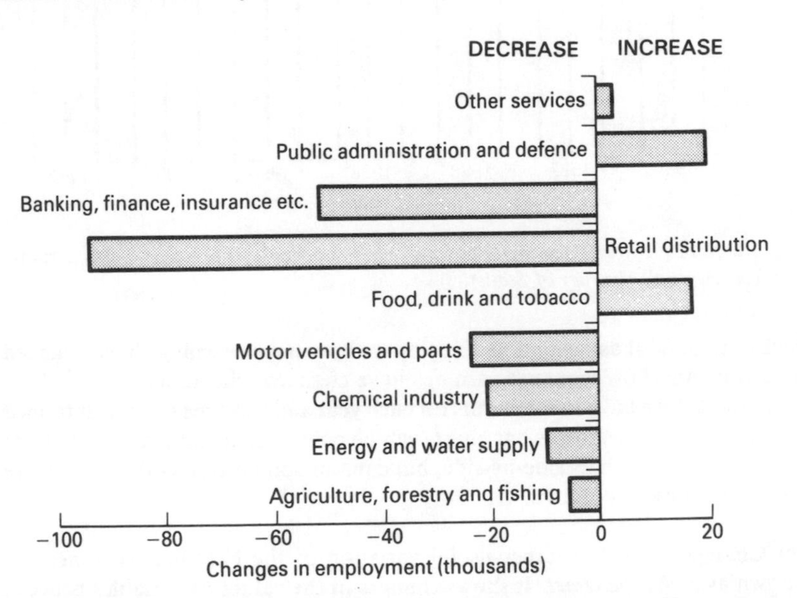

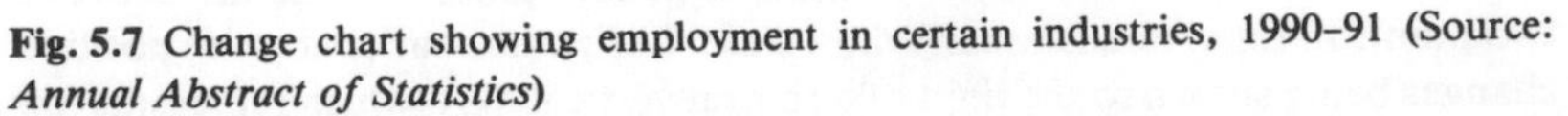

Fig. 5.7 Change chart showing employment in certain industries, 1990–91 (Source: *Annual Abstract of Statistics*)

other; the charts are arranged 'back-to-back'. Each chart shows the numbers in each age group of the male or the female population. The whole diagram takes the shape of a rough pyramid, hence its name. Age pyramids are very informative and it is interesting to compare those for different dates or for different countries. The age pyramid for the United Kingdom in 1991 is shown in Fig. 5.8.

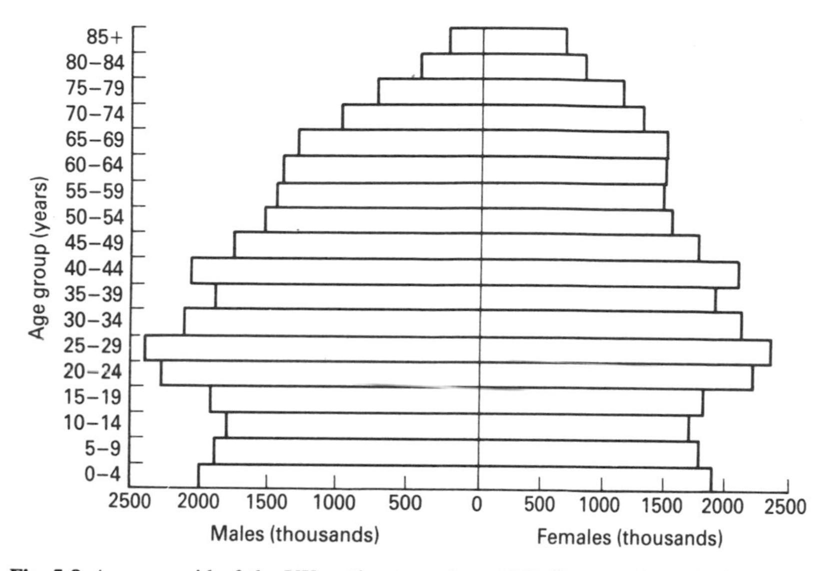

Fig. 5.8 Age pyramid of the UK, estimate at June 1991 (Source: *Annual Abstract of Statistics*)

There is really no end to the variations of the bar chart that can be invented. For instance, the simple bar chart can be used to help in project planning. When a project of any type, e.g. construction work, manufacturing an article, processing orders, is being undertaken, the stages in completing the project can be broken down into smaller units. An analysis of the times taken to complete each stage and the order in which the tasks must be done, allows the scheduling and resource planning for the project to be made systematically. A *Gantt chart* can be drawn up in which the various activities making up a project are represented by bars proportional to the duration of each activity. A *Gantt progress chart* enables planned output or completion times to be compared with the actual results achieved. You will find further reference to the topic of project planning and resource allocation in the Further Reading section at the end of the book.

You would find it interesting to seek out further examples of bar charts for yourself. Always remember, however, that pictorial methods are meant to give a truthful presentation or impression of the facts. It is easy to use a bar chart to give a false picture or to exaggerate an effect. Consider, for example, the effect of shortening the vertical axis and lengthening the horizontal axis on the change chart (Fig. 5.7): the bars would be longer and thinner and the impression would be greatly exaggerated. There are excellent examples of the misuse of pictorial methods in D. Huff's book *How to Lie with Statistics* (for details see Further Reading).

For the presentation of statistics in reports and essays, the bar chart is often the most suitable method. It can be used for both qualitative and quantitative data. If properly drawn, it is clear and easy to understand, the relevant values can be read from the axis and it gives an immediate visual impression of the relationship between the variables. Comparisons can readily be made using bar charts. Above all, a bar chart is easily drawn, nothing more than a pencil and ruler being required.

(b) Pie charts

Another commonly used method of pictorial presentation is the *pie chart*. The purpose of the pie chart is, like that of the component bar chart, to show the component parts of a total. Local authorities, for example, use pie charts to show how each pound of council tax collected is spent.

We can use a pie chart to show the size breakdown of the 221,500 families receiving Family Credit in 1991. Each family size is a 'slice' of the total-number-of-families 'pie'; we have to calculate the angle of each slice (or sector) at the centre of the pie by working out the proportion it bears to the whole (see Table 5.9). The pie chart representing the values in the table is shown in Fig. 5.9.

The pie chart is a useful method of presenting data of this kind. It shows the relationship between the values in such a way that comparisons between the parts and the whole can easily be made—unless, that is, too much information is crowded onto the chart, making some sectors so small that no differences between them are distinguishable. A pie chart performs the same function as a component or a percentage component bar chart. The bar chart would generally be preferred when a comparison of the size of the components is the main object, whereas the pie chart gives a clearer impression of the relationship between each part and the whole, with comparisons between sectors being difficult as the numbers represented cannot be read directly from the chart. If the amounts are to be shown, they must be written on the chart or listed separately.

Table 5.9 Two-parent families receiving Family Credit, 1991: pie chart calculation

Number of children in family	Number of families (thousands)	Angle at centre (°)
1	52.4	$\frac{52.4}{221.5} \times 360 = 85$
2	79.3	$\frac{79.3}{221.5} \times 360 = 129$
3	52.4	$\frac{52.4}{221.5} \times 360 = 85$
4	23.5	$\frac{23.5}{221.5} \times 360 = 38$
5	9.1	$\frac{9.1}{221.5} \times 360 = 15$
6 or more	4.8	$\frac{4.8}{221.5} \times 360 = 8$
TOTAL	221.5	

Source: *Annual Abstract of Statistics*

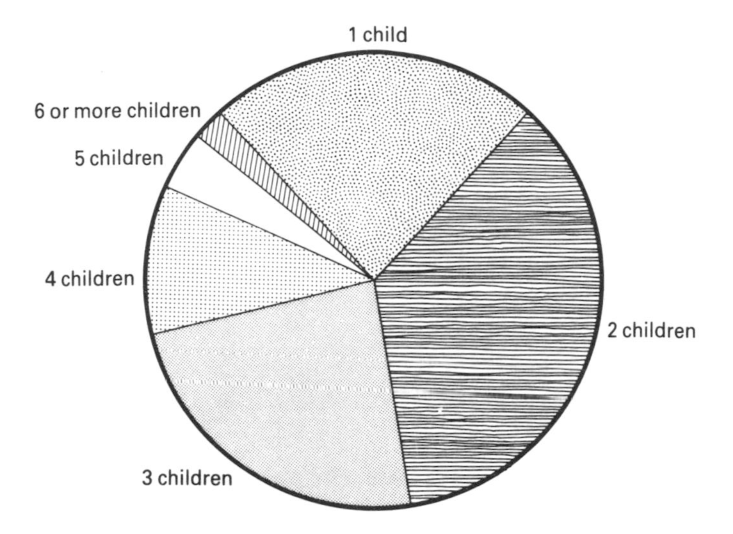

Fig. 5.9 Pie chart showing number of children in families receiving Family Credit, 1991 (Source: *Annual Abstract of Statistics*)

A pie chart is more difficult to draw than a bar chart: the angles for the sectors must be calculated and without a pair of compasses and a protractor it is difficult to draw the angles at the centre accurately.

We also run into problems if we wish to draw a series of pie charts, for example, to illustrate the Family Income Supplement/Family Credit data (Table 5.5) for each year given rather than for 1991 alone. We can follow the approach of the *percentage* component bar chart (Fig. 5.5) using the *same* size of circle in each year and dividing each circle into sectors using the method shown in Table 5.9. However, if we wish, in addition, to show the change in the total numbers of families each year, as is done in the component bar chart (Fig. 5.4), then we have to draw circles of *differing* size. The *areas* of the circles must be proportional to the total number of families each represents. Since the area of a circle depends on its radius *squared*, the radii of the circles we draw must be proportional to the *square roots* of the respective total numbers of families. In 1991 the total number of families was 220 thousands and in 1990 the total was 201 thousands. $\sqrt{201}/\sqrt{222} = 0.95$ (to 2 decimal places). Thus whatever radius we chose to draw the 1991 pie chart we would have to multiply by 0.95 to get the appropriate radius for the 1990 chart. We would then repeat the procedure for 1989, 1988 and 1987. Comparison between such charts is difficult as we are faced with sectors of differing radii and this form of presentation has little to recommend it – a component bar chart would be much clearer. Unfortunately, examiners have been known to set questions on this topic!

Pie charts are more suitable when we want to present data that falls into

Table 5.10 Types of accommodation of 500 students

Type of accommodation	Number of students
Living at home	57
Hall of residence	156
Sharing house or flat	160
Lodgings	84
Own property	43
TOTAL	500

categories. For example, 500 students were surveyed and asked what sort of accommodation they lived in. Table 5.10 shows the results of the survey.

Figure 5.10 is a pie chart representing that data. You can see that this is a good way of presenting data categorized in this way.

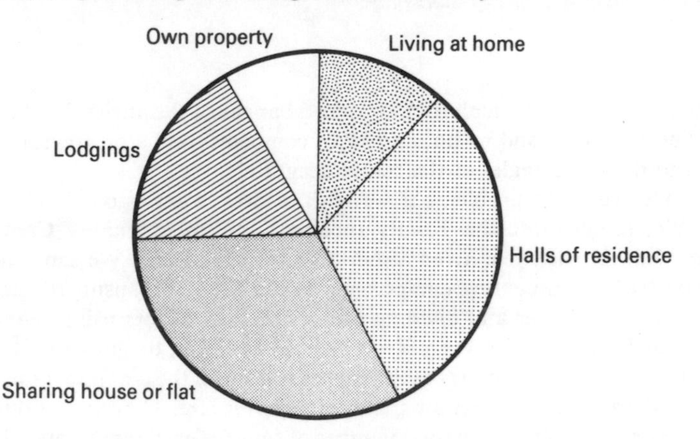

Fig. 5.10 Types of accommodation of 500 students

(c) Pictograms

A method of presentation that magazines and newspapers are very fond of is the *pictogram*, also known as an *isotype diagram* or *symbol chart*; from the point of view of the student, however, it is an impractical method. One type of pictogram is similar to a horizontal bar chart, but instead of a bar a row of symbols is used, each symbol representing a particular amount of the variable (Fig. 5.11).

Such a pictogram is extremely imprecise. How many employees are represented by the part symbols, for example? Without specialist equipment

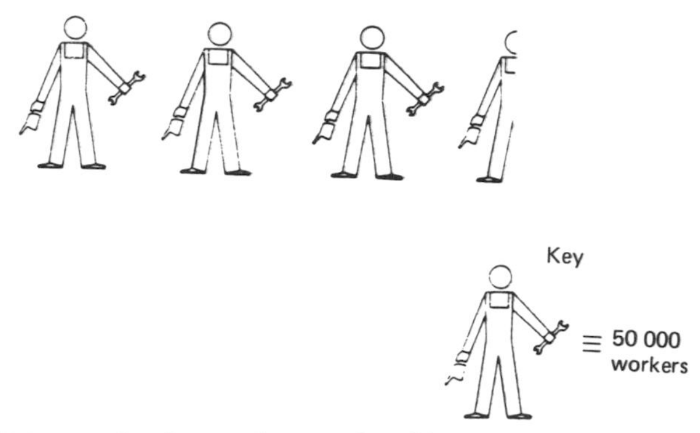

Fig. 5.11 Pictogram showing numbers employed in motor vehicle and parts manufacture, June 1990 (Source: *Annual Abstract of Statistics*)

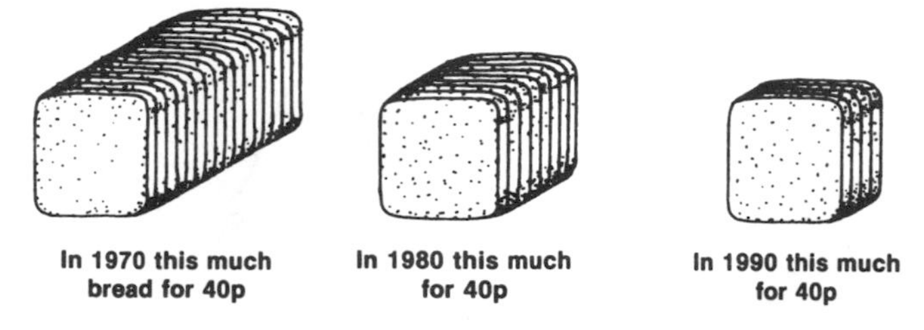

Fig. 5.12 How much bread for your money?

a pictogram is also very difficult to draw—almost impossible to draw freehand. If the numbers represented are large so that there are many symbols in each row, the diagram is difficult to read accurately.

Another form of pictogram is one where a single symbol is used and it is shown to be expanding or shrinking according to the growth or decline of the variable concerned. Suppose a commentator wanted to illustrate the increasing price of bread by showing how much bread could be bought for 40p in 1970, 1980 and in 1990. This information could be shown in the form of a pictogram (Fig. 5.12).

This form of pictogram *can* be very misleading. The amount of bread bought for 40p in 1980 must be depicted as being half of the volume that was shown as being bought for 40p in 1970. Similarly the amount bought in 1990 must be drawn as half the volume of that bought in 1980. It is easy, and tempting, to exaggerate using this type of diagram. For example, the size as well as the number of slices of bread could be reduced slightly each time to heighten the impact of the data.

(d) Cartograms

The cartogram is a map with differently shaded or coloured areas depicting statistics relating to the type of shading or colouring. The shadings or colours

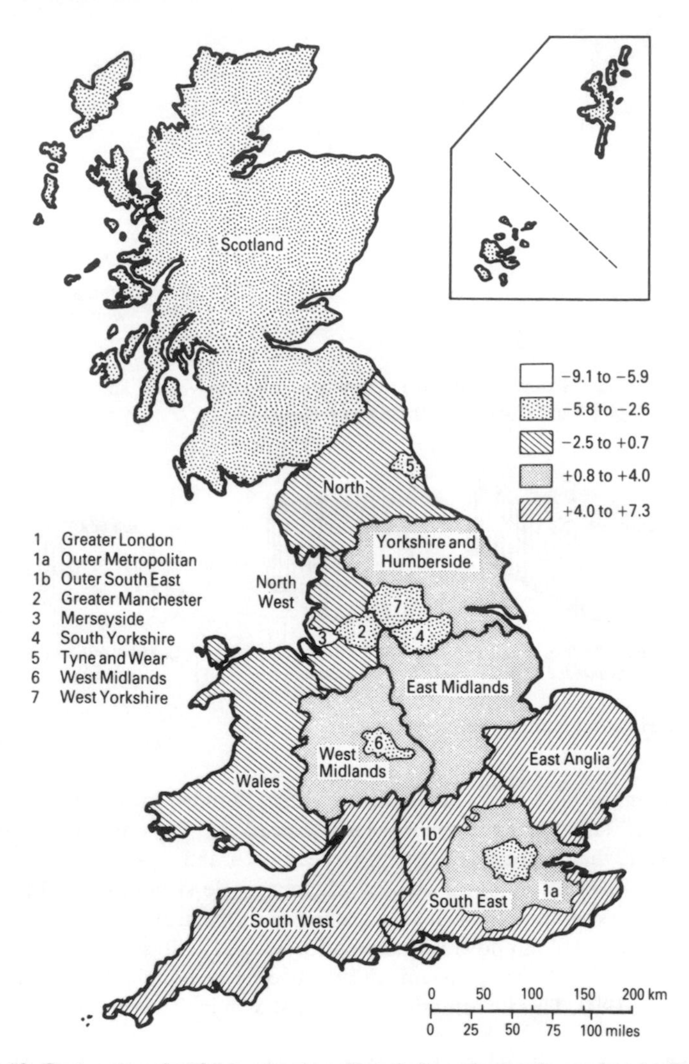

Fig. 5.13 Cartogram showing percentage population change by region in England, Scotland and Wales, 1981–91 (Source: *1991 Census Report, volume 1*)

are explained by a key. Fig. 5.13 is an example of a cartogram; you will find other examples in most atlases.

A cartogram is not difficult to draw. You can buy templates of the maps of most countries, which make the construction of cartograms even easier.

5.6 Graphs

Charts and diagrams are primarily intended to present data, whereas graphs go further than this and help in the analysis and interpretation of data. In some cases graphical methods can be used instead of a calculation. Such methods are usually not as precise as calculations, but their speed and the overall insight they give make them a valuable tool for the researcher. The use of the term 'graph' rather than 'chart' is not always adhered to and you will find that, for instance, line graphs are called line charts by other authors.

(a) Line graphs

The line graph is drawn on squared paper to a certain scale. Two axes are needed, a horizontal (or x) axis on which the scale of values for the independent variable is recorded and a vertical (or y) axis on which the scale of the dependent variable is recorded. The *independent variable* is the variable on which predictions are based, while the *dependent variable* changes in relation to the independent. Time is an example of an independent variable: it depends upon nothing. In the graph of a time series (i.e. a series of readings taken at intervals of time), the independent variable on the horizontal axis is 'time'; the dependent variable on the vertical axis is the value of the series at the time when an observation is made.

The main difficulty you will encounter when drawing a graph is the practical one of fitting the scale of each axis on to the size of paper you are using, so be sure to plan ahead. Use a piece of rough paper as a plan to ensure that the scales will fit before you begin to draw the final version. Use the available space fully; a cramped graph becomes unreadable.

Look at the figures for the population of England and Wales in Table 5.11. In a graph of these data, the time scale will be shown on the horizontal axis and the numbers in the population on the vertical axis. The vertical axis scale should always start at zero, although it is permissible to 'break' the scale on this axis if the values being shown are very large. Each reading is plotted on the

Table 5.11 Population of England and Wales, 1801–1991

Year	Population (thousands)
1801	8892
1851	17928
1901	32528
1911	36070
1921	37887
1931	39952
1951	43758
1961	46105
1971	48750
1981	49155
1991	49890

Source: *Annual Abstract of Statistics*

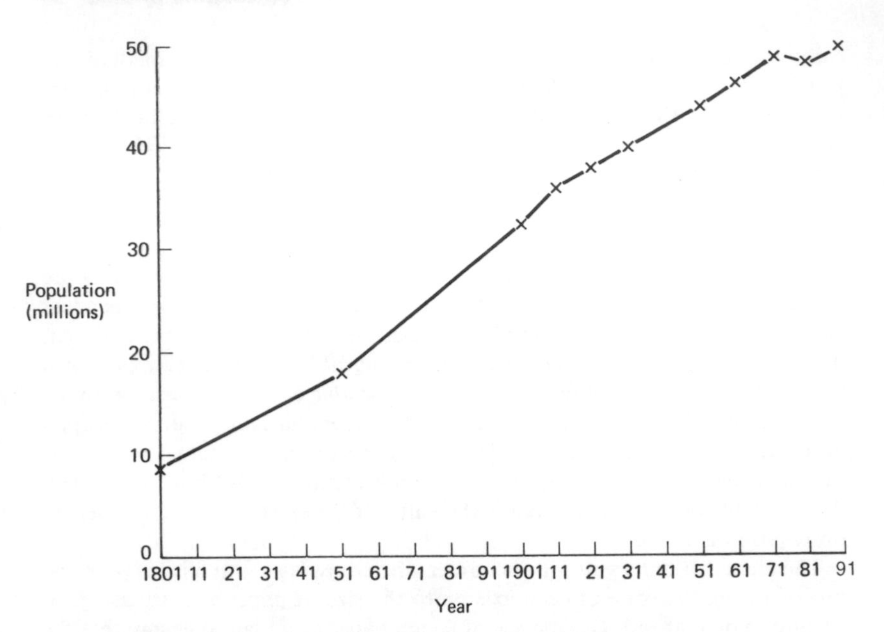

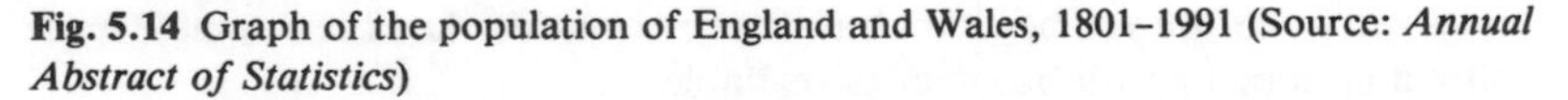

Fig. 5.14 Graph of the population of England and Wales, 1801–1991 (Source: *Annual Abstract of Statistics*)

graph and the points are joined with ruled lines. The line graph of the population series is shown in Fig. 5.14.

You can see that the graph can be used to make estimates of population for other years. What, for example, was the population in 1941? In what year was the population 35 million? The estimates of 42 million and 1908, respectively, are far from precise, but the graph does give a guide.

A graph that shows a time series such as this is known as a *historigram*—not to be confused with the *histogram* described in Unit 4.6.

Notice in Fig. 5.14 that the axes of the graph are clearly scaled and labelled, the units of the variables are shown where necessary, the graph has been given a descriptive title and the source of the data is included. (In the interest of clarity the grid of the graph paper was left out when the diagram was redrawn for inclusion in this book.) As far as possible the labelling on graphs should be horizontal; when space is limited, vertical axes can be labelled as in (i), but not as in (ii), which is difficult to read:

Price	P r i c e
(i)	(ii)

(b) Layer graphs or band charts

It may be desirable sometimes to show on a graph the parts that make up the whole: a *layer graph* or *band chart* can be used for this purpose. Consider the data on road accidents in Table 5.12. The layer graph of these values (Fig. 5.15) gives a picture of the relationship between the component parts of the total, the top line on the graph being the total number of fatalities plus serious casualties, measured on the *y* axis. The vertical distance between each line measured on the *y* axis represents the number of casualties for the type of traveller annotated between the lines. it is similar to a component bar chart and suffers from the same disadvantages.

Table 5.12 Fatal road accidents, 1984–91

	1984	1985	1986	1987	1988	1989	1990	1991
Pedestrians	1 868	1 789	1 841	1 703	1 753	1 706	1 694	1 496
Pedal cycles	345	286	271	280	227	294	256	242
All two-wheeled motor vehicles	967	796	762	723	670	683	659	548
Cars and taxis	2 179	2 061	2 231	2 206	2 142	2 426	2 371	2 053
Others	240	233	277	213	260	264	237	229

Source: *Annual Abstract of Statistics*

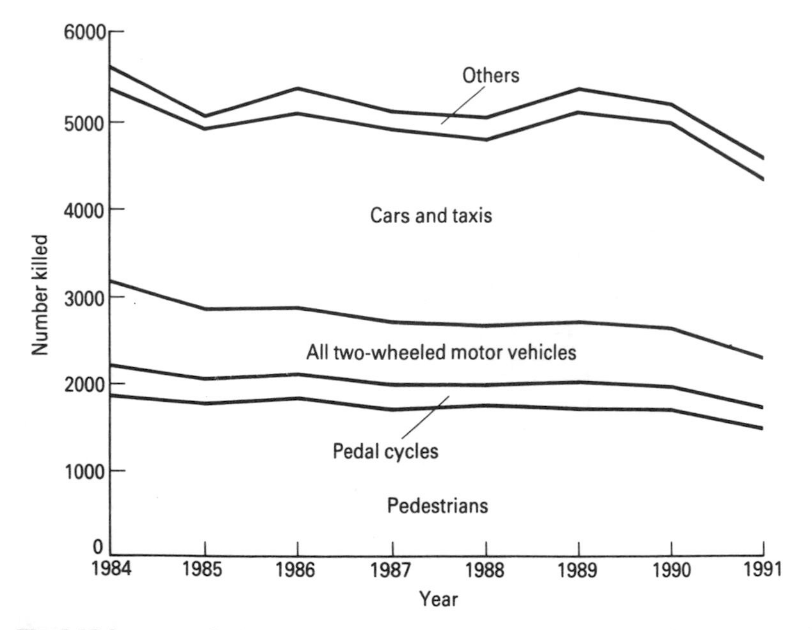

Fig. 5.15 Layer graph (band chart) of fatal road casualties, 1984–91 (Source: *Annual Abstract of Statistics*)

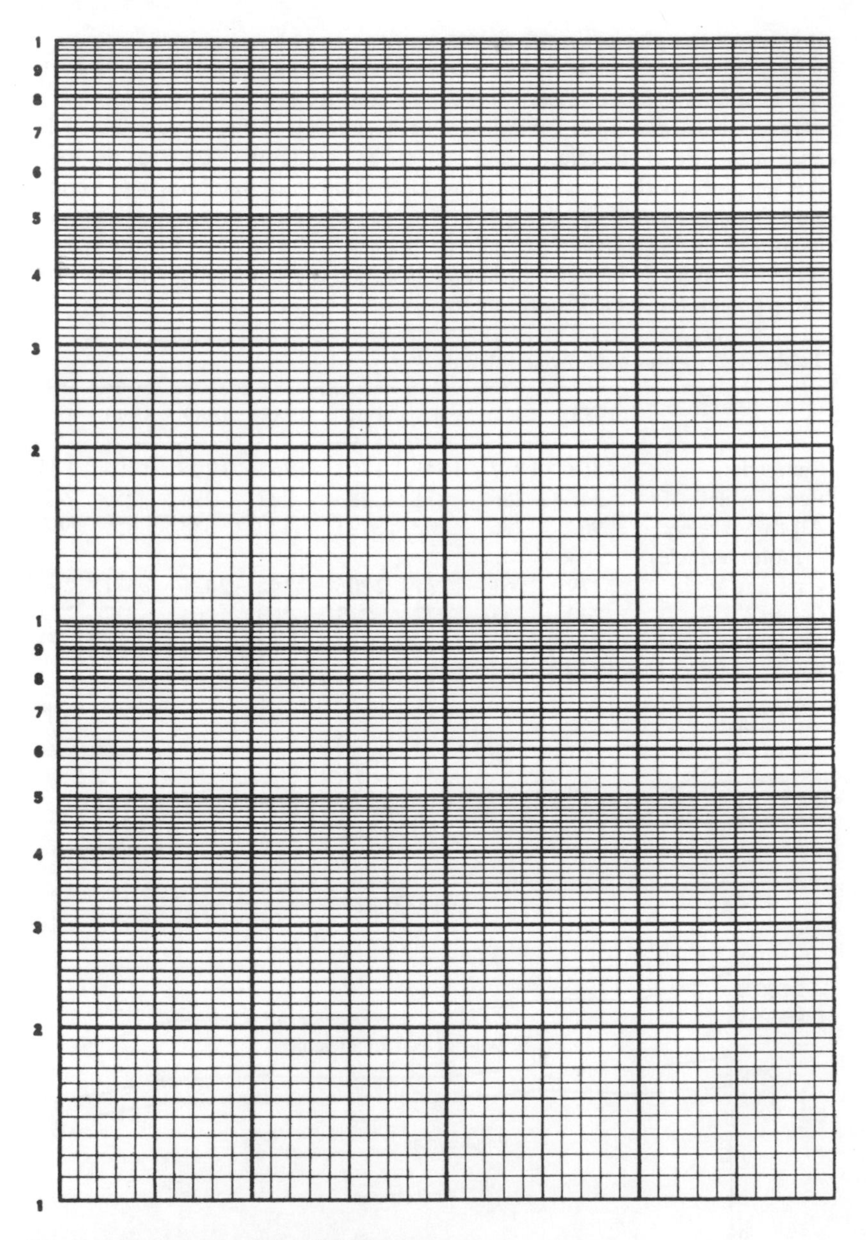

Fig. 5.16 Two-cycle semi-logarithmic paper

(c) Semi-logarithmic (ratio scale) graphs

The graphs we have looked at so far have used *arithmetic* or natural scales on each axis. On occasion it is desirable to use a *logarithmic scale*, sometimes called a *ratio scale*, on one or both of the axes.

On a logarithmic scale the distances between the values of the variable are measured according to the common logarithms of the numbers. There is thus the same distance between 10 and 100 as there is between 100 and 1 000 and as there is between 1 000 and 10 000. In fact there is the same distance between any pairs of values that are in the same ratio to each other.

You can buy graph paper with logarithmic scales, although usually in statistics you will only find a use for *semi-logarithmic* paper, that is, paper with an ordinary arithmetic scale on one axis and a logarithmic scale on the other (Fig. 5.16). You can obtain one-cycle, two-cycle, three-cycle paper etc., where the number of cycles denotes the number of repetitions of the basic grid pattern 1, 2, 3, . . . 1. On semi-logarithmic paper the vertical scale is already marked. You have to decide by considering your data whether the scale is to represent 1–10 and 10–100 or 0.1–1 and 1–10, etc. Zero and negative values cannot be represented on semi-logarithmic graphs, since their logarithms are undefined. A ratio scale graph breaks the usual rule of natural scale graphs that vertical scales start at zero. A wide range of values can be plotted clearly on a semi-logarithmic graph without obscuring small readings, since, for example, as much detail is given of the 1–10 range as of the 10–100.

If, however, you do not have logarithmic graph paper and you wish to draw up some paper with a logarithmic scale for yourself, you can do this with the help of a table of logarithms, Table A. Using logarithmic tables, take your ordinary graph paper and start the scale at 1. For subsequent numbers on the scale, write the actual number but make the distance along the scale the logarithm of the number as in Fig. 5.17. You can see that the distance between 1 and 10 on the number scale is the same as the distance between 0 and 1 on the logarithmic scale. Alternatively, to obtain a ratio scale graph you can look up the logarithms of the values you wish to record and then plot these logarithms on ordinary graph paper.

The principal use of a semi-logarithmic graph is to show proportional changes in a variable over a period of time, rather than the actual changes. A straight line sloping upwards from left to right on an ordinary graph implies a constant increase in the value of the variable, i.e. the same absolute change in the same time interval, so that if the value increased by 100 in one year it will also increase by 100 in the next year. The steeper the slope the greater the

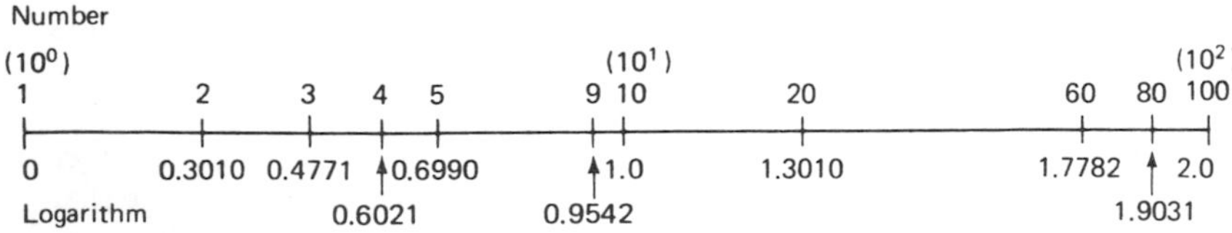

Fig. 5.17 A logarithmic scale

constant increase each year. A straight line sloping upwards from left to right on a ratio scale graph implies a constant rate of increase in the value of the variable, i.e. the same relative change in the same time interval, so that if the value doubled in one year, it will double again in the next year. A steeper line implies a greater constant rate of increase. For example, suppose sales of a new product made by a company take off in the following way:

	Sales (£)	Logarithm of sales (to 2 decimal places)
Year 1	1 000	3.00
Year 2	1 200	3.30
Year 3	4 000	3.60
Year 4	7 000	3.85

Figs. 5.18(*a*) and (*b*) show the sales figures recorded on ordinary graph paper and semi-logarithmic paper respectively. Fig. 5.18(*c*) is a semi-logarithmic graph drawn by plotting the logarithms of sales on ordinary graph paper. You can see that the use of an ordinary scale gives the impression that the sales are increasing very rapidly. Both the semi-logarithmic graphs, however, show that the *rate* of growth has not increased. In fact, the rate was constant between years 1 and 3; between years 3 and 4 the rate has fallen slightly, since the slope of the graphs is less steep.

If two line graphs of a time series are parallel on arithmetic graph paper, there is the same absolute change in the same time interval. If two line graphs are parallel on semi-logarithmic graph paper, there is the same relative change in the same time interval so the two variables are changing at the same rate. The actual position of the points on a semi-logarithmic graph is not of importance. It is their *relative* position that matters. This means that we do not need to worry about units of measurement. We come to the same conclusion in Figs. 5.18(*b*) and (*c*) about the rate of increase of sales, whether we work in units of pounds or thousands of pounds. This has the advantage that we are able to compare two or more variables with different units on the same graph. We can also divide any one set of readings by the same constant amount to enable us to detect differences in rates of change more easily. We shall illustrate these points by analysing the Road Transport Statistics given in Table 5.13 using a ratio scale graph.

Although the use of semi-logarithmic paper is easier than plotting the logarithms of the statistics, in examinations such paper is not always provided, so we shall use the second method. The logarithms of the road transport statistics are given to two decimal places in Table 5.14. A comparison of the slopes of the line graphs on the semi-logarithmic graph, Fig. 5.19(*a*), shows that casualties and the numbers of goods vehicles have increased at a slower rate than expenditure on highways and the numbers of cars. We get a clearer picture, however, if we adjust the given data before plotting the graph so that each line graph covers a greater vertical distance and slight differences in slopes are more readily detectable. Thus, before drawing Fig. 5.19(*b*) we have divided

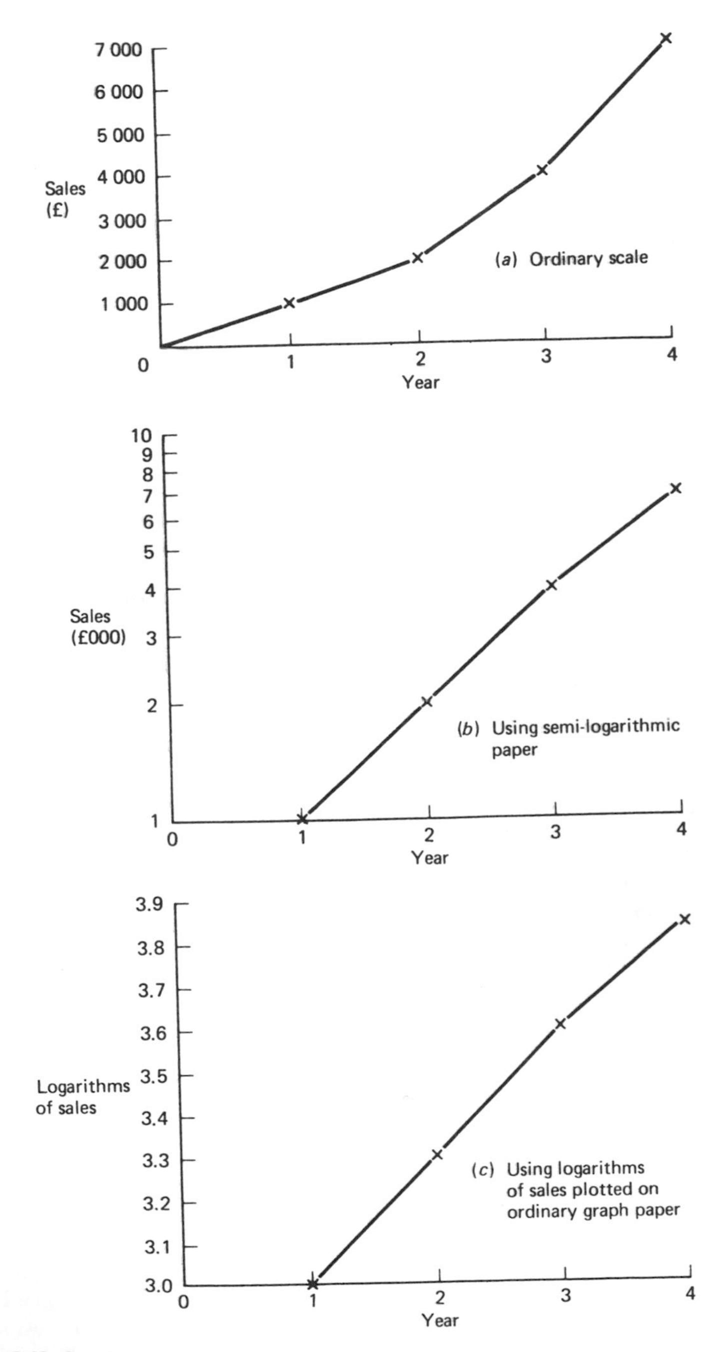

Fig. 5.18 Graphs of sales figures

Table 5.13 Road transport statistics in Great Britain

	Total expenditure on highways (£ million)	Total number of cars licensed (thousands)	Total number of goods vehicles licensed (thousands)	Total casualties in road accidents (thousands)
Year 1	228.0	4972	1378	333
Year 2	238.0	5532	1448	348
Year 3	270.7	5983	1503	350
Year 4	301.2	6560	1522	342
Year 5	342.4	7380	1582	356
Year 6	405.8	8252	1633	385
Year 7	421.2	8922	1661	397

Table 5.14 Logarithms of road transport statistics

	Total expenditure on highways (£ million)	Total number of cars licensed (thousands)	Total number of goods vehicles licensed (thousands)	Total casualties in road accidents (thousands)
Year 1	8.36	3.70	3.14	2.52
Year 2	2.38	3.74	3.16	2.54
Year 3	2.43	3.78	3.18	2.54
Year 4	2.48	3.82	3.18	2.53
Year 5	2.53	3.87	3.20	2.55
Year 6	2.61	3.92	3.21	2.59
Year 7	2.62	3.95	3.22	2.60

the data for cars and for goods vehicles by 10, equivalent in both cases to subtracting 1.00 from each of the corresponding entries in Table 5.14. Fig. 5.19(*b*) demonstrates more clearly the relatively slow rates of increase in the numbers of casualties and in the numbers of goods vehicles. The numbers of cars have risen at an almost constant rate over the seven years, whereas the increase in expenditure was greater in Years 2–6 than in Years 1–2 and 6–7. If you place a ruler between the starting- and finishing-points of each line graph in turn and slide your ruler up and down the page, you will notice that over the complete period the rate of increase in casualties is almost exactly the same as the rate of increase in the number of goods vehicles. The rate of increase in expenditure on highways and the rate of increase in the numbers of cars are almost identical, with the rate of increase in expenditure being just fractionally higher.

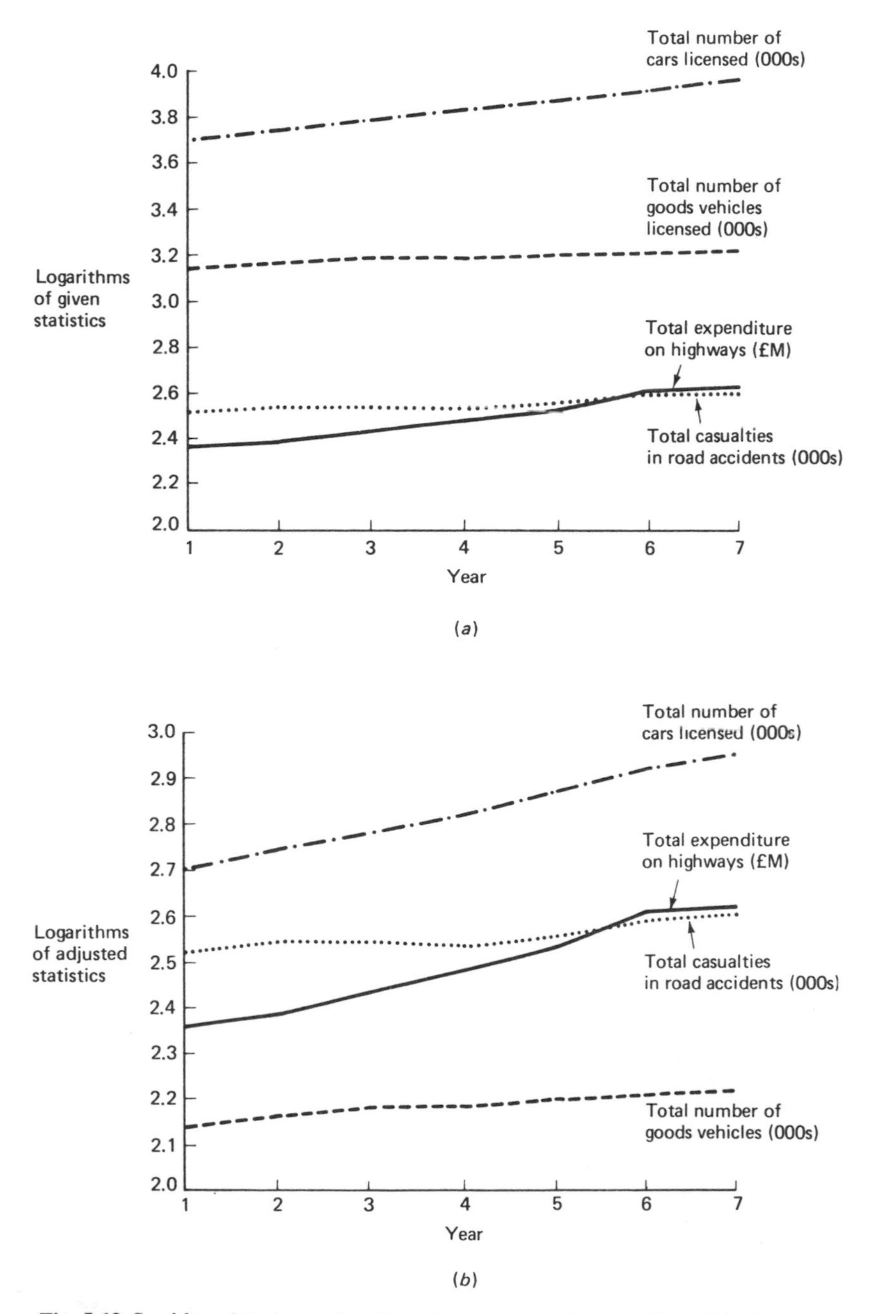

Fig. 5.19 Semi-logarithmic graphs of road transport statistics in Great Britain

(d) The misuse of graphs

As with bar charts, it is easy to give a misleading impression when using graphs to display data. One of the easiest ways of doing this is by exaggerating the vertical scale: if you draw Fig. 5.18(*a*) with 10 mm representing £500 instead of £1 000, you can see the misleading impression this gives of the sales figures.

An important point to remember is that the vertical scale on a non-logarithmic graph should start at zero. When the numbers being displayed are very large and are not changing to any great extent, it is permissible to 'break' the scale on the vertical axis as in Fig. 5.20, but it must be clearly indicated that the scale has been broken in this way.

The effect of starting the scale on the vertical axis at a figure other than zero can be extremely misleading. Consider, for example, the publishers of a magazine who wish to show the increase in circulation after an advertising campaign. Circulation before the campaign was 100000 copies and it grew as follows over the next five months:

	Number of copies
Month 1	103000
Month 2	105000
Month 3	108000
Month 4	110000
Month 5	112000

The publishers might try to display the data as in Fig. 5.21(*a*) which gives a very enhanced and incorrect picture of the increase in circulation. Compare that picture with the true position (Fig. 5.21(*b*)).

Consider another example. In the mid-1970s the value of the pound sterling fell rapidly and at one time was almost at $1.50. The fall in the value of the

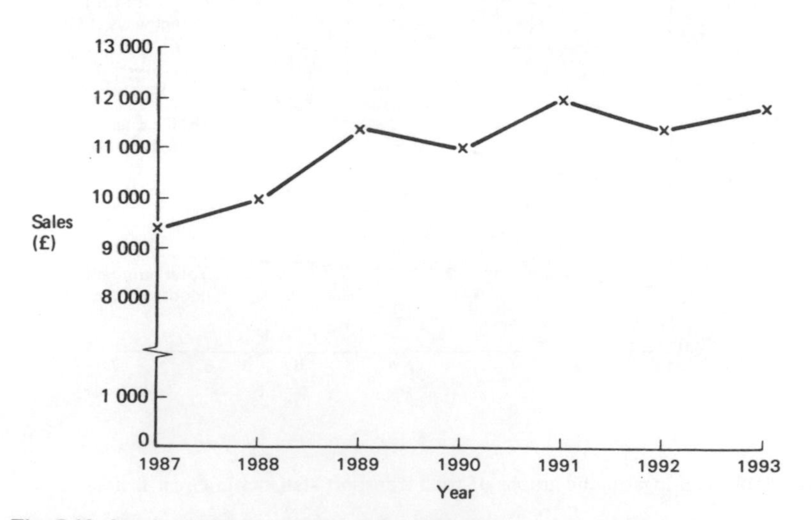

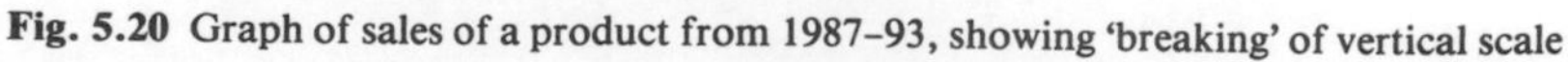
Fig. 5.20 Graph of sales of a product from 1987–93, showing 'breaking' of vertical scale

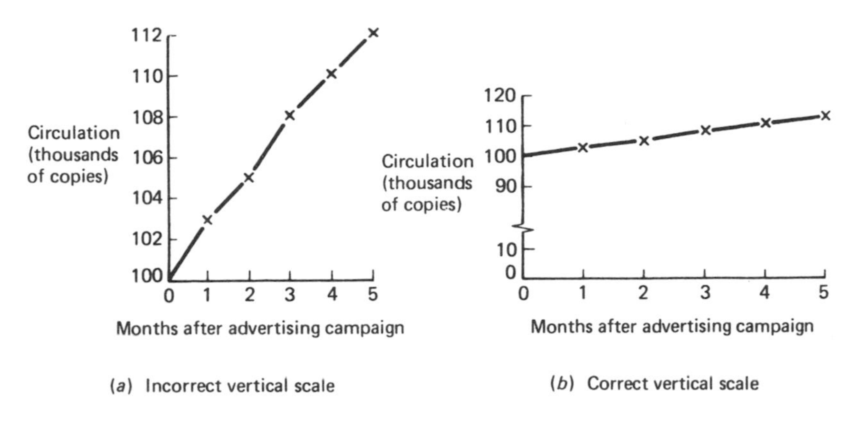

Fig. 5.21 Graphs of increase in circulation of a magazine

pound can be shown on a graph to give two completely opposing views of the situation (Figs. 5.22(*a*) and (*b*)). Because the scale on the vertical axis in Fig. 5.22(*b*) starts at $1.50 it looks as though the value of the pound dropped nearly to zero, whereas Fig. 5.22(*a*) makes it look as though there is nothing to worry about. This is a good example of how pictorial presentation can be used for effect—indeed, one of these examples was used in a party political broadcast.

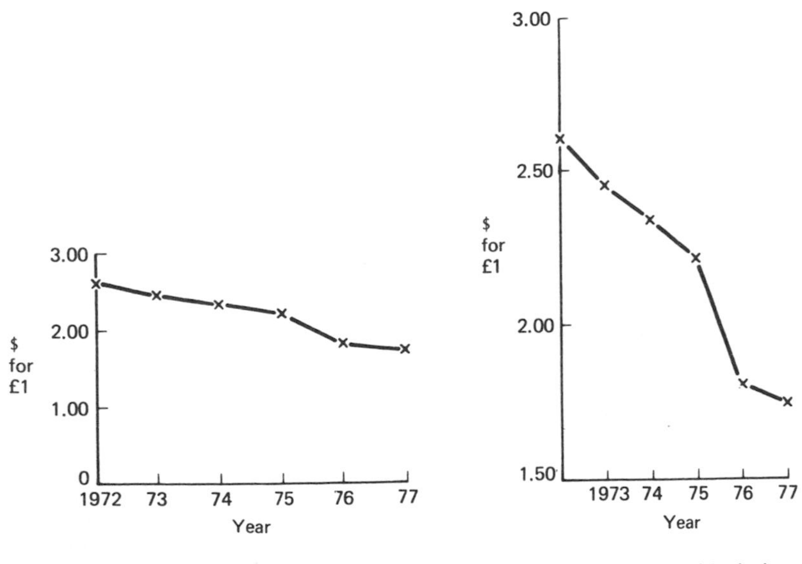

Fig. 5.22 Two interpretations of the falling value of the pound

(e) The Lorenz curve

This is a special graphical technique that shows the degree of inequality there is in the distribution of a variable among a population. For instance, economists can use this type of presentation to illustrate the uneven distribution of wealth. The *Lorenz curve* plots cumulative percentages of the variable and of the population possessing the variable on the two axes of the same graph. For example, suppose the land in a village, a total area of 662 hectares, was divided among the population of 80 families as in Table 5.15.

To draw a Lorenz curve, the number of families holding each plot size is expressed as a percentage of the total number of families and the area held for each plot size is expressed as a percentage of the total area. The cumulative percentages can then be calculated: the information is best arranged in a table (see Table 5.16).

The values of the cumulative percentages for families and land held are plotted on the vertical and horizontal axes, respectively, and the points joined by a curved line, as in Fig. 5.23. The axes should be drawn to the same scale, i.e. 0–100% both on the horizontal and vertical axes, which should be the same length. You can see from Table 5.16 that 50% of the families own 2.27% of the land, and, at the opposite end of the scale, only 3.75% of the families own 45.31% of the land. It is these inequalities that are shown by the Lorenz curve.

Table 5.15 Distribution of land in a village

Plot size (hectares)	Number of families	Hectares held (hectares)
0–1	40	15
1–5	15	34
5–10	10	80
10–20	7	103
20–30	5	130
30 and over	3	300
TOTALS	80	662

Table 5.16 Distribution of land in a village: cumulative percentages

Number of families	Percentage of total families	Cumulative percentages	Hectares held	Percentage of total area	Cumulative percentage
40	50.00	50.00	15	2.27	2.27
15	18.75	68.75	34	5.14	7.41
10	12.50	81.25	80	12.08	19.49
7	8.75	90.00	103	15.56	35.05
5	6.25	96.25	130	19.64	54.69
3	3.75	100.00	300	45.31	100.00

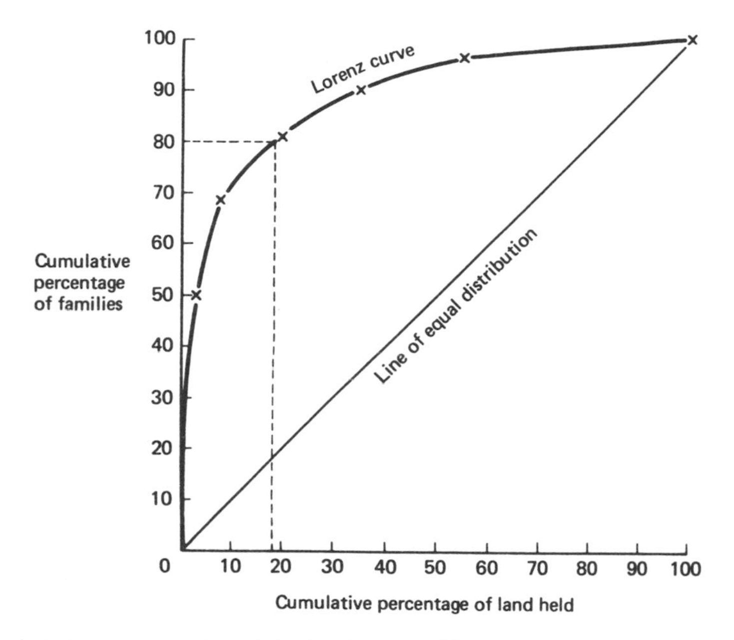

Fig. 5.23 Lorenz curve of land distribution in a village

The *line of equal distribution* shown in Fig. 5.23 running from the zero corner of the graph to the 100% corner would be the Lorenz curve if the land were exactly evenly distributed among the population—if, for example, 50% of the families owned 50% of the land. The line of equal distribution should be drawn on every Lorenz curve to complete the graph. The farther the Lorenz curve is from the line of equal distribution, the greater the degree of inequality of the variable among the population. Fig. 5.23 shows that in this village the land is very unequally distributed. We can estimate the amount of land held by any given percentage of families. For instance, 80% of the families can be seen to hold approximately 18% of the land, so the remaining 82% of the land is concentrated in the hands of 20% of the families.

5.7 Exercises

1. A survey of 200 workers was taken to find out how they travelled to their place of employment. Of the 200, 52 lived within 3 miles of their places of employment, 83 between 3 and 6 miles away, and 45 between 6 and 10 miles away. Of those living within 3 miles, 15 used public transport, 8 cycled and 16 walked to work. Of those living between 3 and 6 miles away, 23 used public transport, 1 walked and 6 cycled. Of those living between 6 and 10 miles away, only one cycled, none walked and 7 used public transport. Of those living over 10 miles away, 5 used public transport and

none walked or cycled. All workers not using the above-mentioned methods of transport used their own cars. Display these data in a table.

2. Show the following data on the percentages of persons in Great Britain aged 25–59 not in full-time education according to the socio-economic group of father using an appropriate method of pictorial presentation. Describe the method you have used and say why you think it is appropriate.

Highes qualification held[1]: by socio-economic group of father, 1990–1991

Great Britain *Percentages*

	Profes-sional	Employers and managers	Inter-mediate and junior non-manual	Skilled manual and own account non-pro-fessional	Semi-skilled manual and personal service	Unskilled manual	All Persons
Degree	32	17	17	6	4	3	10
Higher education	19	15	18	10	7	5	11
GCE A level[2]	15	13	12	8	6	4	9
GCSE, grades A-C[2]	19	24	25	21	19	15	21
GCSE, grades D-G[2,3]	4	9	7	12	12	10	10
Foreign	4	4	4	3	2	2	3
No qualifi-cations	7	19	18	40	50	60	35

[1] Persons aged 25–59 not in full-time education.
[2] Or equivalent.
[3] Includes commercial qualifications and apprenticeships.
Source: *Social Trends*

3. The following data show the individuals or heads of families receiving Income Support in the UK (in thousands) in 1991:

Retirement pensioners, 60 years and over	1272
Others over 60	254
Unemployed	335
Disabled	375
One-parent families	871
Others	331
Total recipients	4438

Source: *Social Trends*

Display the data in the form of a pie chart.

4. A company sells products in 3 product groups—A, B and C. The company's customers are located in 5 areas—north-east, north-west,

south-east, south-west and central. The sales manager requires a tabulation of sales showing sales by product group in each sales area, total sales by product group and total sales for each area. Also, the table is required to show the percentage contribution to total sales made by each area. Draw up the outline of a table that will meet these requirements.

5. Discuss the methods of pictorial presentation of data that are available and outline the advantages and disadvantages of each from the point of view of the person who has to write reports and display data without the aid of specialist printing equipment.

6. Show the following figures for imports and exports (in £ million) for 1986–93 using: (*a*) a dual bar chart; (*b*) a graph.

	1986	1987	1988	1989	1990	1991	1992	1993
Imports	9163	9980	11301	16067	23513	24431	31584	36978
Exports	8170	9290	9906	12657	16910	20198	26162	33331

7. The following figures are the population of a town:

Year	Population (hundreds)
1851	30
1901	61
1911	75
1921	79
1931	86
1941	89
1951	97
1961	105
1971	126
1981	136
1991	148

Draw a graph of the population and from that graph estimate:

(*a*) the population in 1936;
(*b*) the population in 2001;
(*c*) the year in which the population was 83000.

8. A computer manufacturer has the following record of sales:

Year	Computers sold
1975	2000
1978	4000
1981	10000
1984	18000
1987	28000
1990	42000
1993	60000

Draw graphs for these sales figures showing:

(*a*) the *actual* increase in sales over the period;
(*b*) the *rate* of increase of sales over the period.

9. The table shows the sales of electricity in a certain country, both the national total and for the Northern Region separately.

Year	National sales kilowatts (billions)	Northern Region sales kilowatts (billions)
1983	7500	840
1984	8024	908
1985	8613	943
1986	8936	966
1987	9714	987
1988	10152	1015
1989	10879	1052
1990	11261	1090
1991	12348	1148
1992	12960	1205
1993	13415	1276

(*a*) Plot these data on the same semi-logarithmic graph;
(*b*) State what your graph indicates about the comparative electricity sales of the Northern Region and the country as a whole, and explain the advantages of using a semi-logarithmic scale.

10. Use the following income statistics for the UK to draw two Lorenz curves on the same axes.

Income range (£)	Number of taxpayers (millions)	Pre-tax income (£ millions)	Post-tax income (£ millions)
3445–4999	2.2	9400	9110
5000–7499	5.5	24700	23060
7500–9999	7.6	34900	31310
10000–14999	12.2	75900	65800
15000–19999	7.8	68000	57500
20000–29999	6.5	77400	64000
30000–39999	2.9	33600	26540
40000 and over	2.4	61500	43200
TOTALS	47.1	385400	320520

Comment on what the curves show.

UNIT SIX

Approximation and error

6.1 Introduction

Statistics is not an exact science. Often statistics can only provide a guide, and sometimes a rough guide at that. It is not necessary when calculating statistics to try to attain the very finest degree of accuracy. Statisticians are often dealing with the actions of people or the consequences of their actions, and since people can be vague or imprecise it is unrealistic to expect absolute accuracy from statistics.

If you read official statistics you will see that figures are never expressed precisely. It is rarely necessary to show the size of the population of Great Britain down to single units; the size changes all the time. Who would be interested in the Gross National Product expressed in pounds and pence? An approximation to within a few million pounds is quite acceptable. Nevertheless, although final results of statistical calculations should never be stated to an unreasonable degree of precision, it may be worth while working to a finer degree of precision in the body of a calculation, since otherwise the final result may turn out incorrect at the desired level of precision.

6.2 Rounding and significant figures

Most published statistics you will see will be *rounded*, that is, adjusted to some degree. The degree of rounding will depend upon the figures being displayed, and common sense must be your guide to the degree of rounding required for your statistics if instructions are not given. The *Annual Abstract of Statistics* shows the UK population in 1991 as being 56467 thousand. With so large a number, rounding to the nearest thousand is quite adequate. In fact, an approximation of 56 million would be acceptable for most purposes. Many people would find it annoying to see some statistics shown without rounding. Imagine reports including numbers such as 247832233: it would not be long before the figures confused the eye. On the other hand, if a company's personnel officer wanted to find out the average wage within the company and found that it was £225.43, to round that figure to £230 would be carrying things too far, although rounding to £225 might be acceptable. At the same time if details of annual salaries were required, £12500 would probably be an acceptable approximation for an average salary of £12462.

As a general rule, the smaller the number the finer the degree of rounding that should be applied. Suppose we found that the average amount the adult male population spends on entertainment per week per head was £4.52, it would obviously be misleading to round that to £5. If we found that the average age of that same population was 43.6 years, it is equally obvious that the 0.6 years are trivial and that 44 years is the realistic approximation.

The mechanics of rounding are straightforward. If we round a number such as 27.2 to the nearest whole number, we obtain 27, as 27.2 is nearer to 27 than it is to 28. Similarly, 27.234 rounded to two decimal places is 27.23 since 27.234 is nearer to 27.23 than to 27.24. If we consider 27.235, however, and we wish to round this to two decimal places, we have a problem since 27.235 is equidistant from 27.23 and 27.24. You will find in practice that some people always round up in these cases (i.e. they give the answer here as 27.24) but it is preferable, particularly if you are dealing with many numbers of this type, always to round so that the digit preceding the final 5 is *even*. The reason for this is that some numbers will be rounded up and others will be rounded down and thus cumulative rounding errors will be minimized.

Examples of rounding are given below:

234 567.5345 to 3 places of decimals is 234 567.534
234 567.534 to 2 places of decimals is 234 567.53
234 567.534 to 1 place of decimals is 234 567.5
234 567.534 to the nearest whole number is 234 568
234 567.534 to the nearest ten is 234 570
234 567.534 to the nearest hundred is 234 600
234 567.534 to the nearest thousand is 235 000
234 567.534 to the nearest ten thousand is 230 000

When we measure quantities such as lengths, heights or weights, our precision is restricted by our measuring instruments. We might record the length of a box as 56 mm if our measuring apparatus can give us an answer correct to the nearest millimetre. This tells us that the *true* length of the box, if it could be measured, is somewhere between 55.5 mm and 56.5 mm. The length could perhaps be 55.8 mm since this quantity when rounded to the nearest millimetre is 56 mm. Alternatively, the true length might be 56.3 mm, but it could not be 56.7 mm since this quantity is 57 mm correct to the nearest millimetre. Nor could it be 55.4 mm since this would be 55 mm correct to the nearest millimetre. The accurate digits in such a number, apart from any zeros at the start of the number which merely position the decimal point, are called the *significant* figures of the number. In the above example 56 has two significant figures.

Numbers arising from enumeration or counting are *exact*, i.e. there is no error of measurement, and these numbers are sometimes said to have an unlimited number of significant figures.

When you are performing calculations, always give your answer to the correct number of significant figures. It is easy, particularly with a calculator, to write down many digits but these give only spurious accuracy. When multiplication, division and extraction of roots are involved, the final result can have no more significant figures than the quantity with the fewest significant figures.

Examples
(i) 6.31×22.54
If you work out the product in full, you obtain 142.2274, but, assuming that 6.31 is correct to 3 significant figures since no more than three digits are quoted, the answer can be correct only to 3 significant figures.

$$6.31 \times 22.54 = 142 \text{ to 3 significant figures.}$$

Writing down any more of the digits above gives an impression of greater precision than is warranted.

When you are performing a calculation in several stages, work with as many digits as is practicable during the calculation and then round off your calculation at the end to the appropriate number of significant figures.

When you do addition and subtraction, you need to consider each case individually to decide how many significant figures are appropriate.

(ii) $12.01 + 4.5$
The sum appears to be 16.51 but, as 4.5 is correct only to 1 decimal place, our final answer can be correct only to 1 decimal place, i.e. it is 16.5.

(iii) $1.450 - 1.203$
Assuming that the original numbers are correct to three decimal places, as is implied by the explicit 0 in the first number, the final answer can be given correct to three decimal places as 0.247. Note that, although we started with two numbers both correct to four significant figures, we ended with an answer correct to three significant figures. (Remember that a zero at the *start* of a number is not a significant figure.) It is easy to 'lose' significant figures in this way when subtracting numbers of similar amount.

(iv) $12 + 22.22$
If we know that 12 is an *exact* number (i.e. it has an unlimited number of significant figures), then the answer is 34.22. You have to consider the context of the numbers in an example such as this to determine whether or not such a number is exact.

6.3 Truncation

When a number is *truncated* it is shortened to an approximation simply by dropping the required number of digits from the right of the number. The effect is the same as rounding when the digit which is being rounded is less than 5; for example, if 146.4 is rounded to the nearest whole number, to give 146, the effect is to truncate the number. It is not as accurate to truncate as it is to round and, if the digits being dropped are all higher than 5, results can be very inaccurate, especially if a series of truncated numbers is being totalled in some form of statistical presentation.

6.4 Unbiased or compensating errors

If we are totalling a large number of rounded figures, some of the figures will be too low and some too high; but the excesses and deficits will tend to cancel each other out so that the error in the final total will be small. The larger the number of rounded figures being totalled the less error there is likely to be in the final total. Because the differences are not all in the same direction the error is said to be *unbiased*.

Consider the numbers in Table 6.1.

Table 6.1 Errors and rounding

Number	Rounded to nearest hundred	Error
6863	6900	+37
5942	5900	−42
4324	4300	−24
3658	3700	+42
3763	3800	+37
3448	3400	−48
5555	5600	+45
7327	7300	−27
7016	7000	−16
6992	7000	+ 8
54888	54900	+12

In a total of 54900 there is only an error of 12 because the pluses have tended to cancel out the minuses. The error is only 0.02% of the total.

6.5 Biased or cumulating errors

Consider the numbers in Table 6.1 after they have been approximated by truncation (see Table 6.2).

In this case we have *biased* errors because all the differences are in the same direction. The error is now 0.9% of the total. We must consider the effects of biased error. It is unlikely in the interests of accuracy that we would produce a report with numbers truncated but it is common practice to truncate under particular circumstances. For example, ages are usually truncated. When people are asked their age, they usually state it in a completed number of years and ignore the months: you would say you were eighteen until you had reached your nineteenth birthday. If you totalled the ages of a group of people in this way in order to find the average age, your final answer would under-estimate the true average age.

Occasionally you will meet a series of numbers which have all been rounded up rather than truncated or rounded to the nearest appropriate decimal place.

Table 6.2 Errors and truncation

Number	Truncated to hundreds	Error
6863	6800	−63
5942	5900	−42
4324	4300	−24
3658	3600	−58
3763	3700	−63
3448	3400	−48
5555	5500	−55
7327	7300	−27
7016	7000	−16
6992	6900	−92
54888	54400	−488

For instance, when individual departments within a local authority are submitting their estimated expenditure budgets for the following financial year, they will each tend to state an amount higher than their expected need, 'to be on the safe side'. The financial controller considering the sum of these estimates will know that it is an over-estimate of the likely total expenditure, although, of course, with the passage of time, unforeseen circumstances and more rapidly rising inflation may make even this estimate too small.

6.6 Absolute and relative errors

An *absolute error* is the difference between the true value and its approximate value. In the example of Table 6.1, the absolute error in the sum is 12. We consider errors positive when the approximation is above the true value. Often the absolute error on its own does not tell us much about the precision of our calculations. To be 12 out in a total which amounts to 100 is much more serious than to be 12 out in a total of 54888. The absolute error expressed as a proportion of the true value is termed the *relative error*. Thus to be 12 out in 100 is a relative error of 0.12 and to be 12 out in 54888 is a relative error of 0.0002 to 4 decimal places. Relative errors are usually expressed as percentages, i.e. 12% and 0.02% respectively. Often the true value is not known so the relative error is then given as a percentage of the approximate value.

6.7 Calculating with approximations

We are told that the population of a town is 36000 to the nearest thousand. This means that the population is somewhere between 35500 and 36500. To state this concisely we write that the population is 36000 plus or minus 500, i.e. 36000 ± 500. +500 and −500 are referred to as the *limits of error* of the estimate. If we use such approximate values in calculations, the total error in the final result will be greater than the errors in the individual items. We now consider how these errors accumulate under the basic arithmetical operations.

(a) Addition and subtraction

Suppose we wish to add one number which we know is 1000 correct to the nearest 100 and a second number which is 500 correct to the nearest 10. The first number must lie somewhere between 950 and 1050 so we can write it as 1000 ± 50. The second number lies between 495 and 505 and can be written as 500 ± 5. The largest value the sum of these two numbers could be is $1050 + 505$ which is 1555. The smallest value the sum can take is $950 + 495$ which is 1445. Thus the sum must lie between 1445 and 1555 and can be written as 1500 ± 55. We have shown that:

$$(1000 \pm 50) + (500 \pm 5) = (1500 \pm 55)$$

This illustrates the general rule that when approximated values are added, the absolute error in their sum is the sum of the absolute errors in each of the values.

If we subtracted (500 ± 5) from (1000 ± 50), the largest value the difference can be is 1050–495 which is 555 and the smallest value is 950–505 which is 445. Thus, the difference lies between 445 and 555 and can be written as 500 ± 55. Our result is:

$$(1000 \pm 50) - (500 \pm 5) = (1500 \pm 55)$$

Thus again when approximated values are subtracted, the absolute error in the difference is the *sum* of the absolute errors in each of the values.

(b) Multiplication and division

We can work out the error in a product or quotient using a similar procedure to (*a*), that is, working out the largest and smallest values the result could be but there is no simple rule that tells us what the absolute error is. However, it is usually easier to apply the following approximate result: when approximated values are multiplied or divided the total *relative error* is the sum of the *relative errors* in each of the values. If our numbers are stated in absolute error form, we must first convert the absolute errors to relative errors. To multiply 1000 ± 50 and 500 ± 5, we first express 50 as a percentage of 1000, i.e. 5% and 5 as a percentage of 500, i.e. 1%. Using the approximate rule given above, the multiplication can be set out as:

$$\begin{array}{r} 1000 \pm 5\% \\ \times \quad 500 \pm 1\% \\ \hline 500000 \pm 6\% \\ \hline\hline \end{array}$$

This relative error can then be reconverted to an absolute error, if desired, by calculating 6% of the approximated product of 500000, that is:

$$500000 \pm 30000$$

To divide 1000 to the nearest 100 by 500 to the nearest 10, the relative errors are again summed to find the total relative error:

$$\begin{array}{r} 1000 \pm 5\% \\ \div \quad 500 \pm 1\% \\ \hline 2 \pm 6\% \\ \hline\hline \end{array}$$

Converted to an absolute error this becomes:

$$2 \pm 0.12$$

(c) Worked examples

(i) The total weekly wages bill in a clothing factory is £186 300 correct to the nearest hundred pounds and the number of employees is 960 correct to the nearest ten. Find the limits within which the average wage will lie.

$$\begin{aligned} \text{Average wage} &= \frac{\text{Total wages paid to all employees}}{\text{Number of employees}} \\ &= \frac{£186\,300 \pm £50}{960 \pm 5} \\ &= \frac{£186\,300 \pm 0.027\%}{960 \pm 0.521\%} \\ &= £194 \pm 0.548\% \\ &= £194 \pm £1.06 \end{aligned}$$

Therefore the average wage lies between £192.94 and £195.06 or £192.90 and £195.10 correct to the nearest 10p.

(ii) A builder has given a quotation of £5 000 for the construction of an out-building. He estimates that the work will take 120 hours correct to the nearest 10 hours. The labour costs will be £6.10 per hour correct to the nearest 10p. He estimates that materials will cost £1 800 correct to the nearest £50. Estimate his profit and state the limits of error in your estimate.

$$\begin{aligned} \text{Time taken} &= 120 \pm 5\text{ hr} \\ &= 120\text{ hr} \pm 4.17\% \\ \text{Labour cost} &= £6.10 \pm £0.05 \text{ per hr} \\ &= £6.10 \pm 0.82\% \text{ per hr} \\ \text{Total labour cost} &= (120 \pm 4.17\%)(£6.10 \pm 0.82\%) \\ &= £732 \pm 4.99\% \\ &= £732 \pm £37.52 \\ \text{Material costs} &= £1\,800 \pm £25 \\ \text{Total costs} &= (£732 \pm £37.52) + (£1\,800 \pm £25) \\ &= £2\,532 \pm £62.52 \\ \text{Revenue} &= £5\,000 \\ \text{Profit} &= £5\,000 - (£532 \pm £62.52) \\ &= £2\,468 \pm £62.52 \end{aligned}$$

Therefore, the estimated profit is £2468 and the limits of error are ±£63, correct to the nearest £.

6.8 Exercises

1. Round the following:

(*a*) 8.6363 to 2 places of decimals;
(*b*) 96.56 to the nearest whole number;
(*c*) 66.6666 to 3 places of decimals;
(*d*) 86023676 to the nearest thousand;
(*e*) 25.8763 to 3 significant figures;
(*f*) 90.8656 to the nearest one-tenth.

2. Add the following figures, which are all rounded, giving the answers as accurately as possible showing the limits of error and the percentage relative error: 8.6; 13.56; 4.573; 5.007

3. A manufacturer plans to make 200 machines for which the estimated costs are:

Materials	£240000 (to the nearest £10000)
Labour	£130000 (±2%)
Overheads	£80000 (±4%)

The machines will be sold at a price of £3000 ± £100 each. If all the machines are sold, show the estimated profits with the limits of error.

4. Use logarithm tables or a calculator to find the following:

(*a*) $\sqrt{243 \times 457 \times 363 \times 540}$ to 3 significant figures;

(*b*) $30 + \left(\frac{248}{256} \times 10\right)$ to 1 place of decimals;

(*c*) $\sqrt{\frac{210}{12} - \frac{40}{12}}$ to 2 places of decimals;

(*d*) $\frac{93919 - (1657 \times 1105/20)}{140199 - (1657)^2/20}$ to 2 places of decimals.

5. For three different areas, the number of households and the average consumption of fruit per week are as follows:

Area	Number of households	Kilos per household
A	435000	1.56
B	273000	1.29
C	527000	2.17

If the number of households is rounded to the nearest thousand and the average consumption to the nearest one-hundredth of a kilo, estimate (*a*) the total consumption for each separate area, and (*b*) the total consumption for the three areas combined, showing the limits of error in each case.

6. Explain what is meant by: (*a*) relative error, (*b*) absolute error, (*c*) biased error, and (*d*) unbiased error.

7. (i) A factory has an output of 28 million tonnes and employs 206000 workers. What is the average output per worker with the possible absolute error if the figure for output is stated to the nearest million and the number of workers to the nearest thousand?
   (ii) The above factory produces talcum powder in two varieties, 'Rose of the West' and 'Oriental Orchid'. It produces per day 1250 tins, to the nearest 10 tins, of 'Rose of the West' and 2700 tins, to the nearest 10 tins, of 'Oriental Orchid'. Each tin contains between 80 and 100 grams of powder. Calculate the total weight of two days' output of 'Rose of the West' and one day's output of 'Oriental Orchid' in kilograms.

8. A firm produces 50 ± 1 articles per hour working a 40 hour week. With overtime and short working the actual working week varies by up to half an hour from the standard 40 hours. The production costs are £2 per article and the selling price is £3 per article, each figure being correct to the nearest 10 pence. Estimate the weekly profit, showing the possible limits of error.

(Answers at the end of the book.)

UNIT SEVEN

Measures of central tendency or location: averages

7.1 The purpose of averages

A frequency distribution (see Unit 4.2) gives us concise information about the frequency of occurrence of the values of the variable we are interested in. This may be extremely useful, as we often need to have just this kind of information.

Usually, though, we need to be able to condense the information from the frequency distribution so that we can make comparisons more readily. We might want to compare two or more frequency distributions obtained from different populations to see if there are differences between the populations – for example, whether the cost of living is higher in Exeter than in Norwich. You might want to look at a frequency distribution to see how you compare with other members of your profession, and whether, for example, your salary is comparable with other people of your age.

When making these comparisons it is obviously useful to have a single measure that is representative or typical of the distribution. Such measures are referred to as *measures of central tendency, measures of location* or simply as *averages*. There are several different measures of central tendency which we shall now define. Each has its advantages and disadvantages. It is important to choose the appropriate measure for the particular comparison we wish to make.

7.2 The arithmetic mean

In the goal averages shown in football league tables, what is the type of average used? It is the arithmetic mean and this measure of location is the one with which the layman is familiar and he refers to it as *the* average.

We have already looked at the numbers of visits to a local doctor's surgery made by 10 mothers (see Unit 4.2). The numbers of visits made by the mothers were:

$$8 \quad 6 \quad 5 \quad 5 \quad 7 \quad 4 \quad 5 \quad 9 \quad 7 \quad 4$$

If you were asked to calculate the average number of visits, what would you do? Probably you would add together the numbers of visits, making a total of 60, and divide by the number of mothers, giving 6 as the average number of visits. This particular type of average is known as the *arithmetic mean*. It is calculated by totalling the values of the observations and dividing that total by the number of observations.

7.3 Introduction to formulae

At this point you need to be introduced to some symbols and formulae. Do not let these worry you: they are merely a shorthand method of telling you how to calculate results, and are meant to help, not confuse you. With modern pocket calculators and with the many computing tools that are now available (see unit 18) those who are purely *using* statistical methods need have little knowledge of these formulae. They are mostly required for examination purposes. In most statistics examinations a list of formulae is provided with the question papers: so long as you understand the formulae you should have no trouble with the questions. Let us look at a simple formula:

$$\bar{x} = \frac{\Sigma x}{n}$$

This is the formula for calculating the arithmetic mean. We will now go through the formula and see exactly what it means.

The symbol $\bar{x}$ (pronounce it *x-bar*) stands for the arithmetic mean. Whenever you see it in a formula it represents the arithmetic mean of the variable we are interested in. It occurs in many formulae.

The symbol Σ stands for *summation*: whenever you see this symbol in a formula it means 'the total of'. (Σ is a Greek capital letter which is pronounced *sigma*.) The letter x refers to the value of an observation, and n is the number of observations.

Relating this formula to the example of the numbers of visits made by mothers to the doctor's surgery, x refers to the value of an observation, that is, the number of visits made by one mother. Σx means the total of all these numbers of visits, that is, the total number of visits made by all the mothers:

$$8 + 6 + 5 + 5 + 7 + 4 + 5 + 9 + 7 + 4 = 60$$

The number of mothers, n, is 10. If we substitute the values for Σx and n in the formula, we get a value of 6, ($60 \div 10$), for $\bar{x}$, the arithmetic mean of the numbers of visits to the surgery. Although there is a geometric mean and a harmonic mean (see Units 7.12 and 7.13), it is usual to refer to the arithmetic mean as *the mean* when no confusion can arise. We would thus state that the mean number of visits to the surgery is 6.

7.4 The arithmetic mean of a frequency distribution

Now we will go a stage further and consider how to find the arithmetic mean of a frequency distribution. We will use our example of Table 4.1, the numbers of visits made to the doctor's surgery by 100 mothers (see Table 7.1), and find the mean number of visits to the surgery.

Each value of the variable occurs more than once. To find the sum of the

Table 7.1 Numbers of visits made to doctor's surgery by 100 mothers

Number of visits (variable)	Number of mothers (frequency)
4	8
5	12
6	15
7	25
8	17
9	13
10	10
TOTAL	100

Table 7.2 Frequency distribution and sum of visits made by 100 mothers to doctor's surgery

Number of visits (variable) x	Number of mothers (frequency) f	Sum of visits fx
4	8	32
5	12	60
6	15	90
7	25	175
8	17	136
9	13	117
10	10	100
TOTALS	100	710

numbers of visits made by all the mothers we multiply each value of the variable by the frequency with which it occurs. It is best to set out the calculation in the form of a table (see Table 7.2).

The arithmetic mean is found by dividing the total number of visits made by all the mothers (710) by the number of mothers (100): this gives 7.1.

The formula for the arithmetic mean calculated from a frequency distribution has to be amended to include the frequency. It becomes:

$$\bar{x} = \frac{\Sigma fx}{\Sigma f}$$

This shows you exactly what to do. To find Σfx, multiply each observed value of the variable by its frequency and total the results: this gives 710. Σf stands for the sum of the frequencies, that is, the total number of observations: this is 100, the total number of mothers. Now we can substitute the values in the formula:

$$\bar{x} = \frac{\Sigma fx}{\Sigma f} = \frac{710}{100} = 7.1$$

Thus, the mean number of visits made to the doctor's surgery by 100 mothers is 7.1. Time spent in making sure that you can relate this numerical example to the formula will help you with the formulae you will meet in later units.

7.5 The arithmetic mean of a grouped frequency distribution

So far we have seen how to calculate the arithmetic mean for a discrete variable. The numbers of visits to the surgery were whole numbers and the range of observations was small. What about continuous variables covering a large range of values for which we drew up a frequency distribution by grouping the observations in classes? We called that a *grouped frequency distribution* (see Unit 4.3).

To show how we can calculate the arithmetic mean of a grouped frequency distribution we shall use the example of the weights of 75 pigs, from Unit 4. The classes and frequencies are given in Table 7.3.

Table 7.3 Weights of 75 pigs

Weight (kg) (variable)	Number of pigs (frequency)
Under 20	1
20 and under 30	7
30 and under 40	8
40 and under 50	50
50 and under 60	19
60 and under 70	10
70 and under 80	7
80 and under 90	5
90 and under 100	4
Over 100	3
TOTAL	75

With such a frequency distribution we have a range of values of the variable comprising each group. As our values for x in the formula for the arithmetic mean we use the midpoints of the classes. (We again make the assumption that the open-ended class intervals are of the same length as the adjoining classes.) This assumes that within each class the arithmetic mean value of the observations is at the class midpoint, an assumption that may not be completely sound, but it is the best that we can do. Having made that assumption, we can calculate the arithmetic mean as in Table 7.4. We have used the rounded values of the midpoints (25, 35 etc.) rather than the more cumbersome precise values (24.995, 34.995 etc.) (see Unit 4.4).

Substituting in the formula $\bar{x} = \Sigma fx/\Sigma f$, we find that the arithmetic mean of the weights is 4305 kg ÷ 75, that is, 57.4 kg.

7.6 The assumed mean or short-cut method

The calculation of the arithmetic mean using the formula $\bar{x} = \Sigma fx/\Sigma f$ is easy if we use an electronic calculator. If you do not possess one, however, or, perish the thought, yours breaks down half-way through an examination, the calculation can be simplified using a procedure known as either the *assumed mean* method or the *short-cut* method. In this method we *assume* that one of the values of x is the arithmetic mean and then find the deviations of our observations from that assumed mean. If we happen to choose as our assumed mean the true value of the arithmetic mean, then the sum of these deviations is zero. To illustrate this consider again our 10 mothers who visited the doctor's surgery the following numbers of times:

8 6 5 5 7 4 5 9 7 4

We found the true mean to be 6. The deviations of the observations from 6 are:

2 0 −1 −1 1 −2 −1 3 1 −2

Table 7.4 Grouped frequency distribution of weights of 75 pigs

Weight (kg) (variable)	Midpoint of class x	Number of pigs (frequency) f	fx
Under 20	15	1	15
20 and under 30	25	7	175
30 and under 40	35	8	280
40 and under 50	45	11	495
50 and under 60	55	19	1045
60 and under 70	65	10	650
70 and under 80	75	7	525
80 and under 90	85	5	425
90 and under 100	95	4	380
Over 100	105	3	315
TOTALS		75	4305

(Observations *below* 6 have *negative* deviations.) If we total all these deviations, we find their sum is zero. If we worked out the deviations from any number other than the true arithmetic mean, the sum of the deviations would not be zero. For example, the deviations of the observations from an assumed mean of 7 are:

$$1 \quad -1 \quad -2 \quad -2 \quad 0 \quad -3 \quad -2 \quad 2 \quad 0 \quad -3$$

The sum of these deviations is -10. The minus sign tells us we have over-estimated with our assumed mean of 7 since we have too little positive contribution to the sum of the deviations. To adjust our assumed mean to obtain the true mean, we notice that in 10 observations we have over-estimated by a total amount of 10 so, on average, we have over-estimated by 10/10 for each observation, so the true mean must be

$$7 - 10/10 = 7 - 1 = 6$$

This adjustment procedure works with whatever value we choose for the assumed mean but the arithmetic can be very greatly simplified by the appropriate choice of assumed mean. The method can be extended to the calculation of the standard deviation (see Unit 8.6). It is particularly advantageous to use the assumed mean method with a grouped frequency distribution when you wish to calculate both the arithmetic mean and the standard deviation. Even with a reliable calculator it is recommended that you use this method as the numbers you have to key-in are smaller, so fewer slips are likely.

Let us recalculate the arithmetic mean of the pig weights using this method, taking an assumed mean of 55 kg. To ensure simplification of the arithmetic we must choose one of the class midpoints as the assumed mean. It is also helpful to choose a midpoint in the centre of the distribution and one with a high frequency. The calculation is set out as in Table 7.5.

The deviations in column (4) are found by subtracting the assumed mean from each value of x. To simplify the calculation further, in column (5) each deviation has been divided by its highest common factor, in this instance 10. Although taking out a common factor of 10 does not simplify this particular calculation a great deal, the procedure can often make a calculation much easier, so always look for a common factor.

In the final column of Table 7.5 each deviation has been multiplied by its frequency to give the total deviation for each class; for instance, there were 7 readings in the 20 and under 30 kg class so these 7 readings give a contribution of $7 \times (-3)$ to the total deviation. The sum of the deviations is divided by the total number of observations to find the mean of the deviations but we must remember to multiply by the common factor of 10 that we took out earlier. Adding this mean of the deviations to the assumed mean gives us the true arithmetic mean of the weights:

$$55 + \frac{18}{75} \times 10 = 55 + 2.4 = 57.4 \text{ kg}$$

Table 7.5 Assumed mean method of calculating the arithmetic mean of the weights of 75 pigs

(1) Weight (kg) (variable)	(2) Midpoint x (kg)	(3) Number of pigs (frequency) f	(4) Deviation of x from assumed mean (kg)	(5) Deviation in units of 10 kg d	(6) fd
Under 20	15	1	−40	−4	− 4
20 and under 30	25	7	−30	−3	−21
30 and under 40	35	8	−20	−2	−16
40 and under 50	45	11	−10	−1	−11
50 and under 60	55	19	0	0	0
60 and under 70	65	10	10	1	10
70 and under 80	75	7	20	2	14
80 and under 90	85	5	30	3	15
90 and under 100	95	4	40	4	16
Over 100	105	3	50	5	15
TOTALS		75			18

The formula for performing this calculation is:

$$\bar{x} = \text{Assumed mean} + \left(\frac{\Sigma fd}{\Sigma f} \times i\right)$$

where i is the common factor that was taken out of the deviations to simplify the arithmetic and d is the difference between each value of x and the assumed mean in units of i. If the class intervals are of equal length, then the length of the class interval will always be a common factor of the deviations. Some formula booklets use the symbol c in place of i and use the symbol x_0 or A to stand for the assumed mean.

7.7 Characteristics of the arithmetic mean

Each measure of central tendency has its own particular characteristics. The arithmetic mean is fully representative of a frequency distribution as it is based on all, and not merely some, of the observations. It is not necessary, however, to know the value of each individual observation in order to calculate the arithmetic mean. Only the total of the observations and the number of observations are required. For example, to work out the arithmetic mean wage paid by a company all we need to know are the total wages bill and the number of employees. The arithmetic mean is readily understood and is straightforward to calculate. It has the added advantage in more advanced statistical work of being amenable to algebraic manipulation. The main disadvantage of the arithmetic mean is that its value is distorted by one or two extreme values. For instance, suppose that our 10 mothers had visited the doctor's surgery the following numbers of times:

8 6 5 5 7 4 5 9 7 44

(perhaps the tenth mother had a chronic illness). The arithmetic mean number of visits is now 100/10, that is, 10. This gives a false picture of the distribution, since all but the tenth mother visited the surgery less than ten times in the year. Another disadvantage is that the arithmetic mean is not necessarily a physically possible value of the variable. Although in the above example the arithmetic mean number of visits was exactly 10, the arithmetic mean number of visits made by 100 mothers (see Unit 7.4) was 7.1, a value which cannot actually occur.

7.8 The mode

The *mode*, as its name might suggest, is the value of the variable which is most 'fashionable', the one with the highest frequency. It is at the highest peak of the frequency curve. The mode is easy to find in a discrete frequency distribution such as the numbers of visits made by 100 mothers to the doctor's surgery (see Table 7.1), where the mode is 7. Twenty-five mothers made 7 visits to the surgery whereas fewer mothers made any other number of visits.

The mode, however, is difficult to determine for a grouped frequency distribution with a continuous variable, such as the weights of pigs (see Table 7.3). Its value can be estimated using a histogram (Fig. 7.1). To do this,

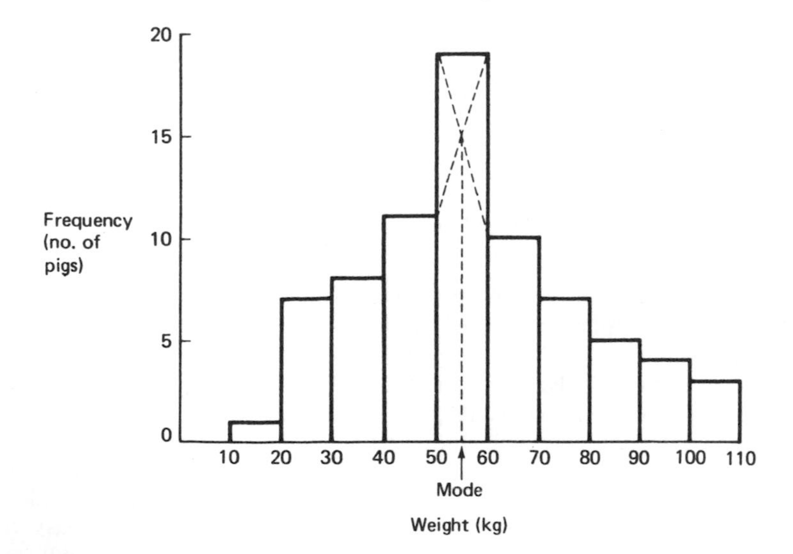

Fig. 7.1 Finding the mode in a grouped frequency distribution with a continuous variable

identify the class with the highest frequency (the *modal* class). Draw two lines at the top of this class, one from the top right-hand corner of the column before the modal class to the top right-hand corner of the modal class, and one from the top left-hand corner of the column after the modal class to the top left-hand corner of the modal class. From the point of their intersection drop a vertical line to the horizontal axis of the histogram and that value of the variable gives the estimate of the mode. From Fig. 7.1 we see that the mode is just less than 55 kg.

The graphical procedure can be replaced by the following formula:

$$\text{Mode} = L + \frac{(f_z - f_l)}{(f_z - f_l) + (f_z - f_h)} \times i$$

where

L = lower class boundary of the modal class
i = class interval
f_z = frequency of the modal class
f_l = frequency in the adjacent *lower* class
f_h = frequency in the adjacent *higher* class

Substituting in the formula we find:

$$\text{Mode} = 50 + \frac{(19 - 11)}{(19 - 11) + (19 - 10)} \times 10$$

$$= 50 + \frac{8}{8 + 9} \times 10$$

$$= 50 + \frac{8}{17} \times 10$$

$$= 50 + 80/17$$

$$= 54.7 \text{ kg to 1 decimal place}$$

This agrees with the graphical result.

Any obsession with the precise estimation of the mode for a continuous variable is somewhat misplaced although examiners do request such calculations.

The mode has the advantage over the arithmetic mean that for a discrete variable its value is one that actually can occur: the modal number of visits to the doctor's surgery made by the 100 mothers was 7 whereas the mean was 7.1. The mode is not affected by the occurrence of a few extreme values. In Unit 7.7 the modal number of visits made by the 10 mothers is 5, the same value as the mode for the original 10 mothers in Unit 7.2. The mode is very easy to find for a discrete variable since no calculation is required, but it can only be estimated for a continuous distribution. The mode is not necessarily unique, that is, there can be more than one mode. For example, in the set of numbers 5, 6, 6, 6, 7, 7, 8, 9, 9, 9, 11, 12 there are two modes: 6 and 9. Distributions with *two* modes are referred to as *bimodal* and their histograms have two distinct peaks. Not every set of numbers has a mode. For example, in the set 5, 6, 7, 8, 9, 11, 12 each number occurs once only, so we say its mode does not exist. Because of these characteristics and because it is not based on all the observations, the mode is not used much in advanced statistical work.

7.9 The median

We would expect a measure of central tendency to be near the middle of the distribution to which it refers. The value of the variable which divides the distribution so that exactly half of the distribution has the same or larger values and exactly half has the same or lower values is called the *median*.

To find the median we need to arrange our observations in ascending sequence and then locate the middle value. We shall again look at the numbers of visits to the doctor's surgery made by ten mothers (see Unit 7.2):

8 6 5 5 7 4 5 9 7 4

For a start we shall consider only the first nine observations. Arranged in order of magnitude these are

4 5 5 5 6 7 7 8 9

There are nine readings so the middle one is the fifth one in the sequence. The value of the fifth observation is 6. The median number of visits made by these nine mothers is 6. When we consider all ten mothers, the sequence is

4 4 5 5 5 6 7 7 8 9

As we have an even number of observations, there are two 'middle' readings: 5 and 6. The median in this case is defined as the arithmetic mean of the two middle observations so the median number of visits made by the ten mothers is 5.5.

Calculating the median for a simple frequency distribution of a discrete variable such as the number of visits made to the doctor's surgery by 100 mothers (see Table 7.1) is straightforward because we only have to identify the value of the variable possessed by the middle item (i.e. mother) in the distribution with the variable arranged in ascending sequence. This can readily be found from a cumulative frequency table (Table 7.6).

There are 100 mothers altogether and we wish to know the number of visits made by the mother in the middle of the distribution. As 100 is an even number we need to take the arithmetic mean of the numbers of visits made by the 50th and the 51st mothers. From the table we see that 35 mothers had made less than 7 visits whereas 60 mothers had made less than 8 visits. The 50th and 51st observations both have the value 7 so the median number of visits made by the 100 mothers is 7. This tells us that half the mothers made 7 or fewer visits to the surgery and half the mothers made 7 or more visits to the surgery.

Table 7.6 Numbers of visits made by 100 mothers to a local doctor's surgery

Number of visits (variable) x	Number of mothers (frequency) f	'Less than' cumulative (frequency)
4	8	–
5	12	8
6	15	20
7	25	35
8	17	60
9	13	77
10	10	90
11	–	100

7.10 The median of a grouped frequency distribution

When we have a grouped frequency distribution of a continuous variable, we have to adopt another technique for finding the median, although we still need to use the cumulative frequencies. Take as an example the weights of 75 pigs (see Table 7.7).

We want to find the weight of the pig half-way through the distribution, that is, the weight of the 38th pig. This is the weight of the $(n + 1)/2$th pig (where n is the total number of pigs). The 38th pig is somewhere in the '50 and under 60 kg' class, since 27 pigs weighed less than 50 kg and 46 pigs weighed less than

Table 7.7 Weights of 75 pigs

Weight (kg) (variable)	Number of pigs (frequency) f	'Less than' cumulative frequency
Under 20	1	1
20 and under 30	7	8
30 and under 40	8	16
40 and under 50	11	27
50 and under 60	19	46
60 and under 70	10	56
70 and under 80	7	63
80 and under 90	5	68
90 and under 100	4	72
Over 100	3	75
TOTAL		75

60 kg. We assume that the weights of the pigs in the '50 and under 60 kg' class are evenly spread across the class. We have accounted for 27 of the 38 pigs before we reached this class, so we need to consider 11 out of the 19 pigs in the class to arrive at the 38th pig. The median weight will thus be 11/19ths of the way across the '50 and under 60 kg' class interval:

$$50 + \frac{11}{19} \times 10 = 50 + 5.79 = 55.79$$

The median weight is 55.8 kg to 1 decimal place. We conclude that half the pigs weighed less than 55.8 kg and half weighed 55.8 kg or more.

It is common practice when dealing with grouped data to calculate the median as the value of the $n/2$th item in the distribution, in this case the weight of the $37\frac{1}{2}$th pig, although strictly speaking it should be the value of the $(n + 1)/2$th item. For large values of n the difference between the two results is very small. The formula for the median of a grouped frequency distribution calculated in this way is:

$$\text{Median} = L_M + \left(\frac{\frac{1}{2}n - F_{M-1}}{f_M}\right) \times i$$

where

L_M = lower class boundary of the median class
n = total number of observations, i.e. Σf
F_{M-l} = cumulative frequency below the median class
f_M = median class frequency
i = class interval

The symbol M or, for preference, Q_2 may be used to denote the median.

Substituting in the formula, we find:

$$\text{Median} = 50 + \frac{(\frac{75}{2} - 27)}{19} \times 10$$

$$= 50 + \frac{10.5}{19} \times 10$$

$$= 55.5 \text{ kg to 1 decimal place}$$

This value corresponds to the weight of the $37\frac{1}{2}$th pig whereas our previous value was the weight of the 38th pig.

The median can also be determined by the use of the *ogive* or cumulative frequency curve (see Unit 4.6 and Figs 4.6 and 4.7). We draw a horizontal line from the midpoint of the total frequency on the vertical axis to the ogive and then drop a perpendicular line from that point on the curve to the horizontal axis. This value on the horizontal axis is the median. Reference to Fig. 4.6 at a cumulative frequency of $37\frac{1}{2}$ or to Fig. 4.6 at a cumulative percentage frequency of 50% leads to a median value of approximately $55\frac{1}{2}$ kg which agrees with our calculation.

The median is a measure of central tendency which is easily understood as being the 'half-way' point. It is less affected by extreme values than the arithmetic mean: the tenth mother who visited the surgery 44 times (see Unit 7.7) raises the median value from 5.5 to 6.5 whereas the mean went from 6 to 10. The median can be found without actually knowing the values of all the observations. If, for example, we wished to determine the median height of a class of 33 school-children, we do not need to measure the height of every single one. We line the pupils up in order of their heights and only the middle pupil, that is, the 17th in the line, needs to be measured. The median is thus of great use when we have open-ended classes at the edges of a distribution. The calculation of the median depends only on the numbers of observations in these classes whereas the arithmetic mean requires a knowledge of the values of the observations as well.

7.11 The relationship between the arithmetic mean, the median and the mode

In a symmetrical frequency distribution that is peaked in the centre, the arithmetic mean, the median and the mode coincide. Fig. 7.2(*a*) illustrates this for a normal distribution.

If the distribution of the variable is not symmetrical, that is, if we have a skew distribution, the arithmetic mean is not so typical of the distribution. Fig. 7.2(*b*) shows that in a positively skewed distribution, the mean is not at the centre. The mean is dragged to the right of centre by the few very high values of the variable that have been observed. When the arithmetic mean of salaries is taken as the average salary for comparative purposes, the majority of people will be left earning less than the average. The median salary would be more typical.

In a negatively-skewed distribution the mean is *reduced* by the few small values of the variable and hence will be left of centre (Fig. 7.2(*c*)). The few people who have their teeth knocked out at an early age make the mean age at which we lose our teeth appear low. Again, the median age would be more representative.

In a moderately skew distribution the following relationship holds approximately:

$$\text{Arithmetic mean} - \text{Mode} = 3(\text{Arithmetic mean} - \text{Median})$$

This can be used as a method of estimating the mode from the values of the mean and the median.

7.12 The geometric mean

A less commonly used measure of central tendency is the *geometric mean*. It is calculated using the formula:

$$\text{Geometric mean} = \sqrt[n]{x_1 \times x_2 \times x_3 \times \ldots \times x_n}$$

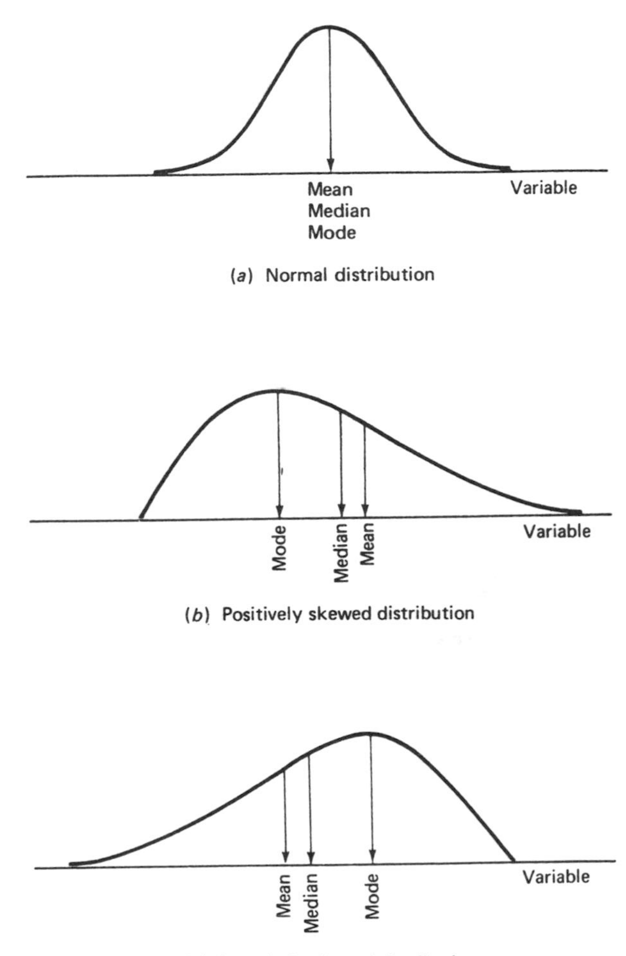

Fig. 7.2 The arithmetic mean, the median and the mode in frequency distributions

where n is the number of observations made of the variable x and $x_1, x_2, \ldots, x_n$ are the values of these observations. A simple example will explain the formula. Let us find the geometric mean of the numbers 3, 25 and 45. There are three observations, so $n = 3$.

$$\text{Geometric mean} = \sqrt[3]{3 \times 25 \times 45}$$
$$= \sqrt[3]{3375}$$
$$= 15 \text{ (by inspection, since } 15^3 = 3375\text{)}$$

The geometric mean cannot be calculated if we have negative or zero observations. The geometric mean of a set of readings is always less than the arithmetic mean (unless all the readings are identical) and is less influenced by very large items. Take, for example, a small company where the following salaries (in thousands of pounds per annum) are paid to the staff:

10 12 14 14 14 16 20

The arithmetic mean annual salary is £14286 to the nearest pound. The geometric mean of the salaries is:

$$= \sqrt[7]{10 \times 12 \times 14 \times 14 \times 14 \times 16 \times 20}$$

$$= \sqrt[7]{105\,369\,600}$$

Finding roots such as this is not an easy arithmetical operation but if you have a calculator with an x^y key, you can read off the root directly.

$$\sqrt[7]{105\,369\,600} = (105\,369\,600)^{1/7}$$

$$= (105\,369\,600)^{0.14285714}$$

$$= 13.999 \text{ to 3 decimal places (using calculator)}$$

Alternatively, if you have a spreadsheet package on your computer (see unit 18) you can enter the data and get the package to calculate the geometric mean for you.

To the nearest pound, the geometric mean annual salary is £13999.

Suppose that the owner of the company, having paid himself £20000 per annum, now decides that the success of the company warrants him trebling his salary to £60000 so that the salaries (in thousands of pounds per annum) now are:

10 12 14 14 14 16 60

The arithmetic mean salary is £20000 per annum to the nearest pound. The geometric mean is:

$$= \sqrt[7]{10 \times 12 \times 14 \times 14 \times 14 \times 16 \times 60}$$

$$= \sqrt[7]{316\,108\,800}$$

$$= 16.378 \text{ to 3 decimal places}$$

The geometric mean salary is £16378 per annum to the nearest pound. The trebling of the owner's salary has had less of an effect on the geometric mean because the geometric mean is only affected by a factor of $\sqrt[7]{3}$.

The geometric mean is useful when only a few items in a distribution are changing: it is then much more stable than the arithmetic mean. For this reason it is used in the calculation of the *Financial Times* Industrial Ordinary Share Index. The geometric mean is also useful for making estimates from data which grow or decay in geometric progression. Population grows in this way; that is,

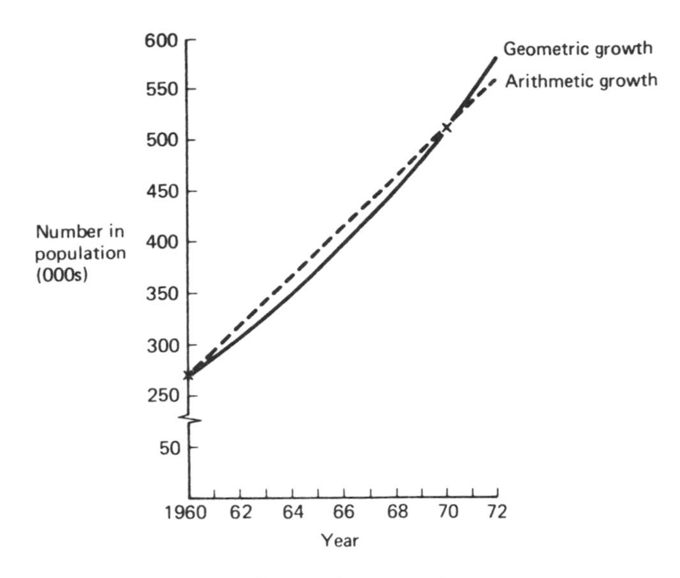

Fig. 7.3 Geometric and arithmetic growth compared

an increase in population is proportional to the number in the population at any one time, *not* proportional to the number in the population at the *start* of the time span as it would be if it grew arithmetically. Fig. 7.3 demonstrates this. If we knew that the population of a city in 1980 was 270000 and 510000 in 1990 and we wanted to estimate what the population was in 1985, we could say that it was approximately:

$$\frac{270000 + 510000}{2} = 390000$$

This assumes that the population grows by the same *number* each year. The more people there are in a city, however, the greater the number added to the population each year so it is more realistic to assume a geometric growth: the population is growing at the same *rate* each year. The geometric mean is:

$$\sqrt[2]{270000 \times 510000} = 363730$$

You will find this a useful technique when making estimates from any data that grow in this way, such as sums of money invested at compound interest.

7.13 The harmonic mean

The *harmonic mean* does not have many uses in business. It is used for averaging *rates*.

If a car travelled on an outward journey at 60 kilometres per hour and on the return journey at 40 kilometres per hour, what is its average speed? It is not 50 kilometres per hour as it might appear. That would be the average speed if the car had travelled for one hour at 60 kilometres per hour and then a further hour at 40 kilometres per hour. To find the average speed when different speeds have been attained over the same *distance*, the harmonic mean is used. The formula for this is:

$$\text{Harmonic mean} = \frac{n}{\sum\left(\frac{1}{x}\right)}$$

where n is the number of observations.

In our example, the speeds of 60 and 40 kilometres per hour are the observations. The average speed, therefore, is:

$$\frac{2}{\left(\frac{1}{60}+\frac{1}{40}\right)} = \frac{2}{\left(\frac{5}{120}\right)} = \frac{2 \times 120}{5} = 48 \text{ kilometres per hour}$$

The principle can be applied to a business situation. Suppose that a typist in an office can type invoices at a rate of 30 per hour, statements at a rate of 40 per hour and reminders at a rate of 80 per hour and that equal numbers of these documents have to be typed. The office manager is asked to report on the average output per hour of the typist: the rates are added together, $30 + 40 + 80 = 150$, and divided by three giving an average rate of 50 documents per hour. This does not take into account the fact that invoices take longer to produce than statements and statements take longer to produce than reminders. The true average rate is found by taking the harmonic mean:

$$\frac{3}{\left(\frac{1}{30}+\frac{1}{40}+\frac{1}{80}\right)} = \frac{3}{\left(\frac{17}{240}\right)} = \frac{3 \times 240}{17} = 42 \text{ (to the nearest document)}$$

7.14 Exercises

1. What is meant by a measure of central tendency? Explain the measures available and the differences between them.
2. Calculate the median of the following frequency distribution:

Hours worked per week	Number of workers
25.0–29.9	10
30.0–34.9	40
35.0–39.9	100
40.0–44.9	45
45.0–49.9	15
50.0 and over	10

Discuss the advantages of the median as an average and indicate the meaning of your answer in respect of the above frequency distribution.

3. A manufacturer of a consumer-good analyses the purchases at various outlets as follows:

Purchases per annum (£)	Number of outlets
2000 and under 4000	30
4000 and under 6000	49
6000 and under 8000	107
8000 and under 10000	62
10000 and under 12000	25
12000 and under 14000	15
14000 and under 16000	8
16000 and under 18000	4
TOTAL	300

   (*a*) What is the arithmetic mean of the purchases at these outlets?
   (*b*) What is the median purchase?
   (*c*) Is the mean a representative average for this distribution? If not, which average is, and why?

4. The following is the wage distribution of an engineering company:

Wage group (£)	Number in group
8000 and under 9000	20
9000 and under 10000	42
10000 and under 11000	65
11000 and under 12000	108
12000 and under 13000	38
13000 and under 14000	11
14000 and under 15000	7
15000 and under 16000	5
16000 and under 17000	3
17000 and under 18000	1

   (*a*) What is the mean wage?
   (*b*) If this company was involved in a wage dispute can the mean wage be used as a fair measure of the average wage? If not, say which average can, and why.

5. Using the frequency distribution you constructed for Exercise 8 in Unit 4.7 calculate:

(*a*) the mean price of the cars;
(*b*) the median price;
(*c*) the modal price.
(*d*) Say which price is the most appropriate average having regard to the distribution, and why.

6. Using the frequency distribution you constructed for Exercise 9 in Unit 4.7 calculate: (*a*) the mean expenditure on food for the week; (*b*) the median expenditure on food for the week.

7. The training manager for a large manufacturing company has the following marks which have been obtained by the apprentices in the final examinations at the end of the apprentice training scheme:

99 98 96 94 90 90 87 85 85 81 78 78 77 76 74 72 72
70 69 65 65 64 64 63 60 59 59 58 57 55 54 54 54 54
51 50 50 50 46 43 41 38 36

Which average—the mean, the median or the mode—would show these marks in the best light? Give reasons for your choice.

8. The following table gives the age distribution of the estimated population of the UK at 30 June 1993:

Age (years)	Number of persons (to nearest thousand)
0–9	7621
10–19	7132
20–29	9217
30–39	8257
40–49	7766
50–59	6002
60–69	5597
70–84	5472
85 and over	936

Source: *Monthly Digest of Statistics*

Calculate: (*a*) the mean age and (*b*) the median age of the population.
(Answers at the end of the book.)

UNIT EIGHT

Measures of dispersion and measures of skewness

8.1 Why measure dispersion?

We have now looked at the different measures of central tendency or location. A measure of central tendency is a value of the variable that we are interested in and can be said to be typical of the population which we have observed. In Unit 7 we saw that there is more than one measure of central tendency, and that sometimes one particular measure is more appropriate than another.

A measure of central tendency, however, is a single value of the variable, useful if we want to compare ourselves with the rest of the population or to compare two populations. It does not tell us a great deal about the population from which it comes. We would like to know more about a population before we can draw proper conclusions or even make fair comparisons. For example, suppose we do a survey and find that the mean height of adult males in Norwich is 180 cm and that the mean height of adult males in Coventry is also 180 cm. Does this imply that adult males in Coventry are of similar heights to those in Norwich? We should be rash in coming to such a conclusion solely from the comparison of these two means. We need to look a little further.

We might find that the frequency distributions of the heights in the two populations were quite different (Fig. 8.1). Although on average the men in both cities are of the same height, there is a greater range of heights in Coventry than in Norwich. In Norwich the majority of the heights is close to 180 cm whereas in Coventry we are more likely to find both very short and very tall men.

You can think of many more situations where a measure of central tendency on its own gives insufficient information about two populations. Suppose that you are starting a new job and find two firms that attract you, both paying average salaries of £15 000 per annum. Would you rather work for one where the lowest salary paid is £6000 but where you could one day earn £30 000, or one where the lowest salary is £10 000, so that you are assured of a more reasonable starting pay, but where you are unlikely to earn over £22 000 during your career there? Or suppose that you are the manager of a team in the Football League and you want a good experienced goal-scorer for your team. Of two players who have scored an average of 20 goals per season during their careers, would you rather have the player who consistently scores around 20 goals every season or one who can grab 40 or 50 goals in a season but is just

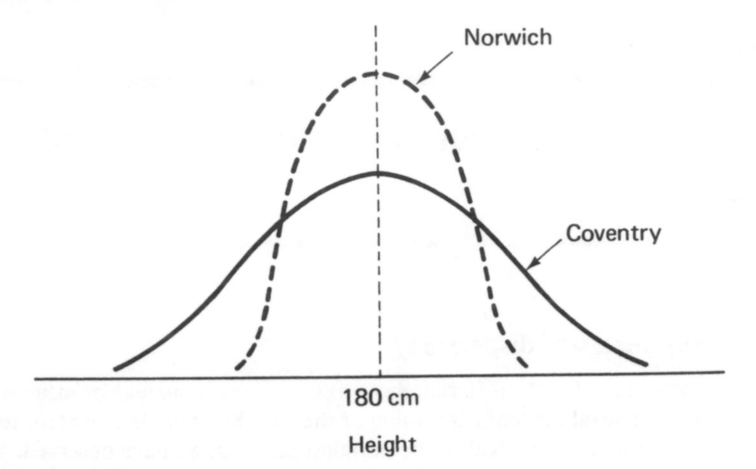

Fig. 8.1 Frequency curves of heights of adult males in Norwich and Coventry

as likely to score only 4 or 5? These examples show why the spread of data can be very important.

We are now going to look at ways of measuring and comparing the spread of data in a population. These measures are called *measures of dispersion* or *measures of variation*.

8.2 The range

The first measure of dispersion to consider is one that you may have used unknowingly at times when you have needed a rough estimate. If you are going on holiday and deciding how much money to take, you may well say 'I need at least so much, but will not need more than so much'. You are then in fact finding the *range* of your requirements. In the example we used in Unit 4.3 of the examination marks of A-level statistics candidates, the range is simply the difference between the highest mark, 96%, and the lowest, 34%, giving a range of 62%.

The range is a fairly crude measure of dispersion. This is because it looks at only two values in a distribution, the two extremes. This can be dangerous. Consider again the problem of deciding on a job. If you found that the range of salaries in a firm is £8000 to £80000, you might be highly encouraged, but could feel less so by discovering that the two partners running the firm each take salaries of £80000 and the next highest salary below theirs is £24000. Nevertheless, besides being quickly calculated, the range can still be a useful measure of dispersion. Suppose you were going on a trip abroad and found out in advance that the mean temperature where you were going was 20 °C. You might well go equipped for a temperate climate and be very distressed to find when you arrive that the temperature drops to 4 °C in the evening while by mid-afternoon it has risen to over 40 °C in the shade. Knowing the temperature range in advance would have helped you in that situation. When we are dealing with a continuous variable the range can only be found accurately from the original raw data. Its value can only be estimated from a grouped frequency distribution. Because of its simplicity it is used in quality control work (see Unit 15.6).

8.3 The quartile deviation or semi-interquartile range

A problem with using the range is that extreme values in a distribution can cause the range to be misleading as a measure of dispersion. The *quartile deviation* helps with this problem because it just looks at the middle 50% of the distribution, therefore leaving out the extreme values.

As their names suggest, quartiles are the values of the variable that belong to the members of the population that are 25%, 50% or 75% of the way through the distribution. There are therefore three quartiles, the first known as Q_1, the second as Q_2 and the third as Q_3. The second quartile we have already come across: it is the *median*, discussed in Unit 7, the value of the variable that belongs to the item half-way through the distribution. The values of the other quartiles can be calculated in exactly the same way as the median was calculated. We shall again use the example of the weights of 75 pigs (see Table 8.1).

Table 8.1 Weights of 75 pigs

Weight (kg) (variable)	Number of pigs (frequency)	'Less than' cumulative frequency
Under 20	1	1
20 and under 30	7	8
30 and under 40	8	16
40 and under 50	11	27
50 and under 60	19	46
60 and under 70	10	56
70 and under 80	7	63
80 and under 90	5	68
90 and under 100	4	72
Over 100	3	75

There are 75 pigs altogether and we wish to find the weight of the pig that is one quarter of the way through the distribution. Strictly speaking this is the weight of the 19th pig (that is the $(n + 1)/4$th pig, where n is the total number of pigs) but in practice you will often find that the weight of the $n/4$th pig is taken. We discussed this procedure when studying the median in Unit 7. The difference between the two results is small when n is large.

The 19th pig is in the '40 and under 50 kg' class. We have accounted for

16 pigs before we reach this class, so we need to include (19 − 16), i.e. 3 pigs out of the 11 in the '40 and under 50 kg' class before we reach the first quartile, so:

$$Q_1 = 40 + \frac{3}{11} \times 10$$
$$= 40 + 2.73$$
$$= 42.73 \text{ kg}$$

We conclude that one quarter of the pigs weighed less than 42.73 kg.

Similarly the third quartile is the $3(n + 1)/4$th item, that is, the weight of the 57th pig. This is in the '70 and under 80 kg' class:

$$Q_3 = 70 + \frac{(57 - 56)}{7} \times 10$$
$$= 70 + 1.43$$
$$= 71.43 \text{ kg}$$

Thus three-quarters of the pigs weighed less than 71.43 kg so that one quarter weighed 71.43 kg or more. The *interquartile range* is the difference between Q_3 and Q_1:

$$Q_3 - Q_1 = 71.43 - 42.73$$
$$= 28.70 \text{ kg}$$

This tells us that the middle 50% of the data spans 28.70 kg. The *quartile deviation* is the *semi*-interquartile range:

$$\text{Quartile deviation} = \frac{Q_3 - Q_1}{2}$$
$$= 14.35 \text{ kg}$$

Alternatively we can use the ogive (see Unit 4.6) to find approximate values of the quartiles as in Fig. 8.2.

If we calculated the quartile deviation for another group of pigs, perhaps of a different breed, we could tell by comparison of the quartile deviations whether one breed was more variable than the other over the middle range of weights. The breed with the higher quartile deviation would be of more variable weight. Since the quartile deviation looks only at the middle 50% of the data, its value is unaffected by extreme values and it can be used when there are open-ended classes at the edges of the distribution. In general, it is used as a measure of dispersion when the median is chosen as the appropriate measure of location, for instance, when dealing with income statistics.

If you compare the median pig weight, 55.8 kg, with the quartiles, 42.7 and 71.4 kg, you will notice that the median weight is *not* the arithmetic mean of

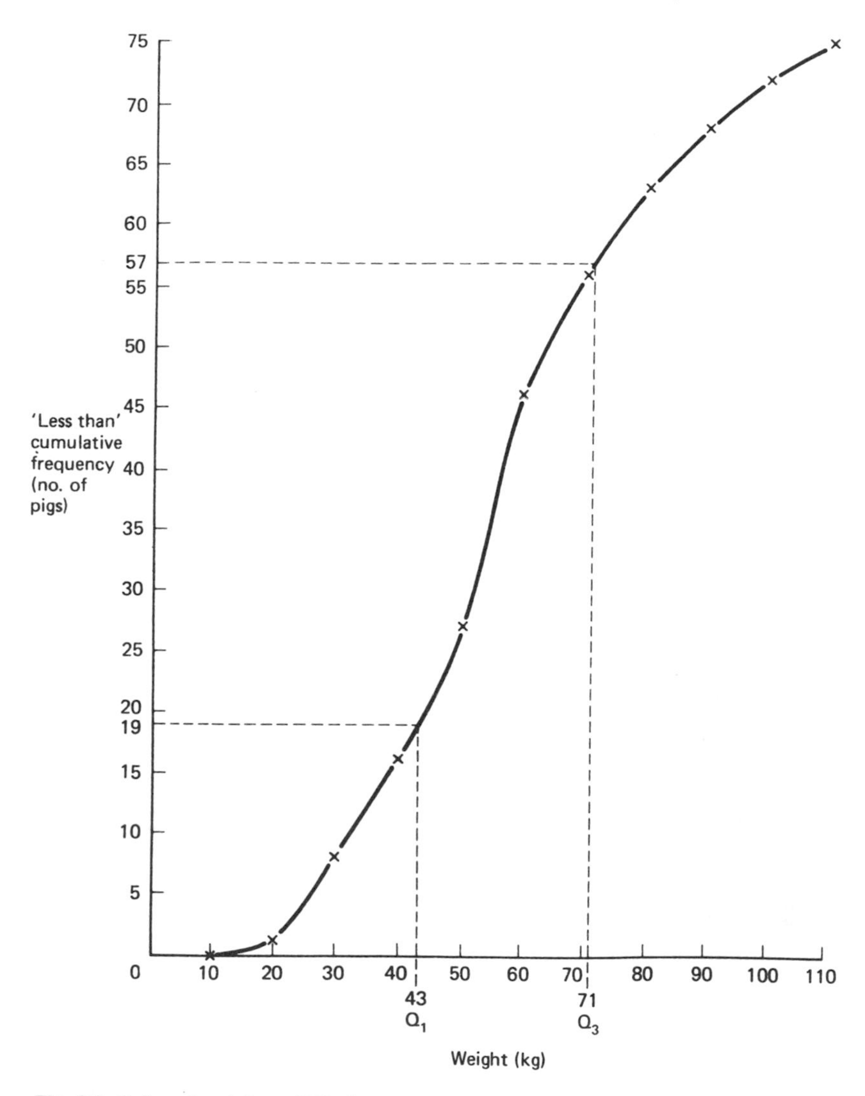

Fig. 8.2 Ogive of weights of 75 pigs

the quartile weights. It is only when the distribution is symmetrical that the median is exactly half-way between the two quartile values.

8.4 Deciles and percentiles

It is convenient, particularly when dealing with statistics of income and employment, to consider values similar to the quartiles but which subdivide the distribution more finely. Such partition values are *deciles* and *percentiles*.

The deciles are the values which divide the total frequency into tenths and the percentiles are the values which divide the total frequency into hundredths. It is only meaningful to derive such statistics when we are dealing with a very large number of observations, preferably thousands.

The deciles are denoted by $D_1, D_2, \ldots D_9$. The third decile D_3, for example, is the value below which 30% of the data lie. The method of calculation follows the same pattern as the calculation of the median and quartiles. In our pig weight example (see Table 8.1), 30% of 75 is 22.5 so D_3 is the weight of the 22.5th pig. As we assume that the pig weights are evenly spread across each group, we can calculate the weight of the 22.5th pig in exactly the same way as we would calculate the weight of the 22nd or 23rd pig, the fraction making no difference to the method. Consulting the cumulative frequency column in Table 8.1 we see that the 22.5th pig is in the '40 and under 50 kg' class so:

$$D_3 = 40 + \frac{(22.5 - 16)}{11} \times 10$$

$$= 40 + \frac{6.5}{11} \times 10$$

$$= 40 + 5.91$$

$$= 45.91 \text{ kg}$$

Thus, 30% of the pigs weighed less than 45.91 kg.

The percentiles are denoted by $P_1, P_2, \ldots P_{99}$ and, for example, P_6 is the value below which 6% of the data lie and P_{66} is the value below which 66% of the data lie. The percentiles most commonly quoted from UK income statistics are P_1, P_5, P_{95} and P_{99}. These values give useful information about the lowest and highest incomes, respectively, found in the UK.

8.5 The mean deviation

It is desirable from a mathematical point of view that a measure of dispersion should take all values of the observations into account. The first such measure that we shall look at is the *mean deviation*. The mean deviation measures how far on average the readings are from the arithmetic mean. (The median is very occasionally used instead of the arithmetic mean.) If the data have a small spread about the mean, the mean deviation has a lower value than for data which show large variations about the mean.

To calculate the mean deviation

(i) find the arithmetic mean, $\bar{x}$, of the data;
(ii) find the deviation of each reading from $\bar{x}$, i.e. work out the difference between each reading and $\bar{x}$;
(iii) find the arithmetic mean of the deviations, ignoring their signs.

As an example take the numbers of visits made by 10 mothers to the local doctor's surgery (see Unit 4.2). The numbers of visits are:

8 6 5 5 7 4 5 9 7 4

The arithmetic mean is:

$$\bar{x} = 60/10 = 6$$

It is easier to set out the deviations in columns as in Table 8.2.

Table 8.2 Calculation of the mean deviation for ungrouped data

	Number of visits x	Mean $\bar{x}$	Deviation $(x - \bar{x})$
	8	6	2
	6	6	0
	5	6	−1
	5	6	−1
	7	6	1
	4	6	−2
	5	6	−1
	9	6	3
	7	6	1
	4	6	−2
TOTALS	60		0

If we total the deviations in the final column, we find the total deviation from the mean comes to zero. This will always be so, as we demonstrated in Unit 7.6, provided we have done our arithmetic correctly. To get round this problem we simply ignore the signs as we total the deviations, so in our example the total is 14 and the mean deviation is 14 ÷ 10 = 1.4 .

For ungrouped data the formula for the mean deviation is:

$$\text{Mean deviation (MD)} = \frac{\Sigma f|x - \bar{x}|}{n}$$

where $|x - \bar{x}|$ stands for the *absolute deviation*, that is, the deviation *ignoring* the sign.

When calculating the mean deviation of a grouped frequency distribution the formula is modified to include the class frequencies:

$$\text{MD} = \frac{\Sigma f|x - \bar{x}|}{\Sigma f}$$

x in this formula is the class midpoint. Using the example of the weights of 75 pigs, we set out the calculation as in Table 8.3. We again assume that the open-ended classes are of the same length as adjacent classes.

Table 8.3 Calculation of the mean deviation

Weight (kg) (variable)	Midpoint x	Number of pigs (frequency) f	fx	Deviation $\lvert x-\bar{x}\rvert$	$f\lvert x-\bar{x}\rvert$
Under 20	15	1	15	42.4	42.4
20 and under 30	25	7	175	32.4	226.8
30 and under 40	35	8	280	22.4	179.2
40 and under 50	45	11	495	12.4	136.4
50 and under 60	55	19	1045	2.4	45.6
60 and under 70	65	10	650	7.6	76.0
70 and under 80	75	7	525	17.6	123.2
80 and under 90	85	5	425	27.6	138.0
90 and under 100	95	4	380	37.6	150.4
Over 100	105	3	315	47.6	142.8
TOTALS		75	4305		1260.8

$$\bar{x} = \frac{\Sigma fx}{\Sigma f} = \frac{4305}{75} = 57.4 \text{ kg}$$

$$\text{MD} = \frac{\Sigma f\lvert x-\bar{x}\rvert}{\Sigma f} = \frac{1260.8}{75} = 16.8 \text{ kg to 1 decimal place}$$

The mean deviation can be used to compare the variation of different distributions but it is not very useful in advanced statistical work because the negative signs in the deviations are ignored when it is calculated. The calculation is tedious for a grouped frequency distribution and it cannot be simplified by the use of the assumed mean method. For these reasons the mean deviation is little used nowadays although examiners occasionally set questions on this topic. The mean deviation suffers from the same problem encountered with the arithmetic mean, namely, that its value is distorted by extreme values of the variable.

8.6 The standard deviation

The *standard deviation* is the most important measure of dispersion as it is of great use in many of the statistical techniques you will study. It takes the value of every observation into account but does not suffer from the same arithmetical deficiencies as the mean deviation.

When we calculated the mean deviation for ungrouped data, we used the formula:

$$\text{MD} = \frac{\Sigma\,|x - \bar{x}|}{n}$$

where $\Sigma\,|x - \bar{x}|$ meant the total of the absolute deviations from the mean. We had to take the absolute deviations because the sum of the actual deviations was zero. There is another way of getting round this problem: instead of ignoring the signs we can remove them simply by squaring the deviations. From your elementary mathematics you will know that if you multiply *like* signs the product always has a *positive* sign. It does not matter therefore whether $(x - \bar{x})$ has a positive or a negative sign: squaring it will always give a positive result.

If we find the mean of the sum of the squared deviations, we obtain the *variance*:

$$\text{Variance} = \frac{\Sigma(x - \bar{x})^2}{n}$$

The variance is used in advanced statistical analysis as a measure of dispersion but for comparative purposes it suffers from the disadvantage that it has square units. For instance, if our observations are weights in kilograms (kg), the variance has units of square kilograms (kg^2). For this reason we use the square root of the variance as our measure of dispersion for comparative purposes as this has the same units as our original observations. This measure of dispersion is the *standard deviation*:

$$\text{Standard deviation (SD)} = \sqrt{\text{Variance}}$$

$$= \sqrt{\frac{\Sigma(x - \bar{x})^2}{n}}$$

To calculate the standard deviation for ungrouped data using this formula:

(i) find the arithmetic mean, $\bar{x}$, of the data;
(ii) find the deviation of each reading from $\bar{x}$;
(iii) square each of these deviations;
(iv) total the squared deviations;
(v) divide this sum by the total number of readings to obtain the variance;
(vi) find the square root of the variance to obtain the standard deviation.

As an example we shall take the numbers of visits made by 10 mothers to the local doctor's surgery (see Unit 4.2) and the complete calculation of the standard deviation is set out in Table 8.4.

$$\bar{x} = \frac{\Sigma x}{n} = \frac{60}{10} = 6$$

$$\text{SD} = \sqrt{\frac{\Sigma(x-\bar{x})^2}{n}} = \sqrt{\frac{26}{10}} = \sqrt{2.6} = 1.61 \text{ visits, to 2 decimal places}$$

In this particular example, $\bar{x}$ happened to be a convenient whole number and this made the arithmetic easy. In general, however, $\bar{x}$ can be any number and the arithmetic using the method of Table 8.4 becomes very messy. In such cases it is more convenient to use the alternative formula:

$$\text{SD} = \sqrt{\frac{\Sigma x^2}{n} - \left(\frac{\Sigma x}{n}\right)^2} \quad \text{or} \quad \sqrt{\frac{\Sigma x^2}{n} - \bar{x}^2} \quad \text{since} \quad \bar{x} = \frac{\Sigma x}{n}$$

This formula is obtained from the original one by algebraic manipulation which it is unnecessary for you to know at this stage. The arithmetic using this formula is usually much simpler, particularly if you have a calculator with a memory. Remember also that if you have a computer available with a modern spread sheet package (see Unit 18), once you have entered the columns of data into the spread sheet the package will calculate the standard deviation for you.

The stages in the calculation of the standard deviation for ungrouped data using the alternative formula are:

Table 8.4 Calculation of the standard deviation of ungrouped data using the formula $\text{SD} = \sqrt{\frac{\Sigma(x-\bar{x})^2}{n}}$

	Number of visits (variable) x	Deviation $(x - \bar{x})$	Squared deviation $(x - \bar{x})^2$
	8	2	4
	6	0	0
	5	−1	1
	5	−1	1
	7	1	1
	4	−2	4
	5	−1	1
	9	3	9
	7	1	1
	4	−2	4
TOTALS	60	(Check) 0	26

(i) sum the readings;
(ii) find the arithmetic mean, taking care not to round off too much;
(iii) square each reading;
(iv) find the sum of the squares;
(v) divide the total sum of squares by the number of readings;
(vi) square the arithmetic mean and subtract it from the result of (v) to obtain the variance (if this number turns out to be negative you have made a mistake and you should check your arithmetic);
(vii) find the square root of the variance to obtain the standard deviation. The calculation of the standard deviation of the numbers of visits made by 10 mothers to the local doctor's surgery using the alternative formula is set out in Table 8.5.

$$SD = \sqrt{\frac{\Sigma x^2}{n} - \left(\frac{\Sigma x}{n}\right)^2} = \sqrt{\frac{386}{10} - \left(\frac{60}{10}\right)^2}$$

$$= \sqrt{38.6 - 6^2} = \sqrt{38.6 - 36}$$

$$= \sqrt{2.6} = 1.61 \text{ visits, to 2 decimal places}$$

This agrces with our previous result.

Table 8.5 Calculation of the standard deviation of ungrouped data using the formula $SD = \sqrt{\frac{\Sigma x^2}{n} - \left(\frac{\Sigma x}{n}\right)^2}$

	Number of visits (variable) x	x^2
	8	64
	6	36
	5	25
	5	25
	7	49
	4	16
	5	25
	9	81
	7	49
	4	16
TOTALS	60	386

8.7 The standard deviation of a frequency distribution

When calculating the standard deviation of a frequency distribution, the formula for the standard deviation has to be modified to include the frequencies as follows:

$$\text{SD} = \sqrt{\frac{\Sigma f(x - \bar{x})^2}{\Sigma f}}$$

where f is the number of readings having the value x and $\bar{x} = \Sigma fx/\Sigma f$. Again this formula can be rewritten in the more convenient form:

$$\text{SD} = \sqrt{\frac{\Sigma fx^2}{\Sigma f} - \left(\frac{\Sigma fx}{\Sigma f}\right)^2} \quad \text{or} \quad \sqrt{\frac{\Sigma fx^2}{\Sigma f} - \bar{x}^2}$$

It is almost invariably quicker to use this alternative formula and we shall use it to calculate the standard deviation of the numbers of visits made by 100 mothers to the local doctor's surgery (see Unit 4.2, Table 4.1). The calculation is set out in Table 8.6.

$$\begin{aligned} \text{SD} &= \sqrt{\frac{\Sigma fx^2}{\Sigma f} - \left(\frac{\Sigma fx}{\Sigma f}\right)^2} \\ &= \sqrt{\frac{5334}{100} - \left(\frac{710}{100}\right)^2} \\ &= \sqrt{53.34 - 50.41} \\ &= \sqrt{2.93} \\ &= 1.71 \text{ visits, to 2 decimal places} \end{aligned}$$

Always set out your calculation in the form of a table, even when using a calculator, as you can check your arithmetic more readily. Note particularly that fx^2 stands for $f \times x \times x$ and *not* for $(fx)^2$—this is a common error made by beginners. You should check for yourself that the same answer is given by the original formula:

$$\text{SD} = \sqrt{\frac{\Sigma f(x - \bar{x})^2}{\Sigma f}}$$

You can see that the arithmetic involved in calculating the standard deviation is fairly complicated. With high values of the variable and large frequencies it can be rather forbidding, but fortunately the arithmetic can be simplified using an extension of the assumed mean method of calculating the arithmetic mean (see Unit 7.6). The formula for the standard deviation can be written as:

$$\text{SD} = \sqrt{\frac{\Sigma fd^2}{\Sigma f} - \left(\frac{\Sigma fd}{\Sigma f}\right)^2}$$

Table 8.6 Calculation of the standard deviation of a frequency distribution

Number of visits (variable) x	Number of mothers (frequency) f	fx	x^2	fx^2
4	8	32	16	128
5	12	60	25	300
6	15	90	36	540
7	25	175	49	1225
8	17	136	64	1088
9	13	117	81	1053
10	10	100	100	1000
TOTALS	100	710		5334

where d is the deviation of each value of the variable from an assumed mean. Using our example of the numbers of visits made by 100 mothers to the local doctor's surgery, with an assumed mean of 7, the calculation is set out as in Table 8.7.

$$\text{SD} = \sqrt{\frac{\Sigma fd^2}{\Sigma f} - \left(\frac{\Sigma fd}{\Sigma f}\right)^2}$$

$$= \sqrt{\frac{294}{100} - \left(\frac{10}{100}\right)^2}$$

$$= \sqrt{2.94 - 0.1^2}$$

$$= \sqrt{2.94 - 0.01}$$

$$= \sqrt{2.93}$$

$$= 1.71 \text{ visits, to 2 decimal places}$$

We have obtained the same answer with smaller numbers throughout although we do have the additional complication of negative signs appearing in the fd column. This method is to be particularly recommended when you do not have a calculator available.

8.8 The standard deviation of a grouped frequency distribution

Calculating the standard deviation of a grouped frequency distribution uses the same formulae as in Unit 8.7 but x now refers to the midpoint of the class

Table 8.7 Calculation of the standard deviation of a frequency distribution by the assumed mean method

Number of visits (variable) x	Number of mothers (frequency) f	Deviation from assumed mean d	fd	fd^2
4	8	−3	−24	72
5	12	−1	−24	48
6	15	−1	−15	15
7 (assumed mean)	25	0	0	0
8	17	1	17	17
9	13	2	26	52
10	10	3	30	90
TOTALS	100		10	294

which has f readings in it. For a grouped frequency distribution it is almost invariably quicker to use the assumed mean method, as is shown in Table 8.8, to calculate the standard deviation of the weights of 75 pigs. It is important that you master this procedure as almost every examination paper requires the calculation of the arithmetic mean and the standard deviation of a grouped frequency distribution.

$$\text{SD} = \sqrt{\frac{\Sigma fd^2}{\Sigma f} - \left(\frac{\Sigma fd}{\Sigma f}\right)^2}$$

$$= \sqrt{\frac{344}{75} - \left(\frac{18}{75}\right)^2} \text{ (in tens of kg)}$$

$$= \sqrt{4.5867 - 0.0576} \text{ (in tens of kg)}$$

$$= \sqrt{4.52911} \text{ (in tens of kg)}$$

$$= 2.13 \text{ (in tens of kg) to 2 decimal places}$$

$$= 21.3 \text{ kg}$$

Note that the calculation was further simplified by taking out the highest common factor (10) of each of the deviations. This is always worth doing but remember that allowance must be made for this at the end of the calculation by multiplying by the common factor *after* taking the square root.

In the next section we shall look at one of the important uses of the standard deviation. You must not think, however, that the standard deviation is a 'perfect' measure of dispersion. It gives even more weight to extreme observations than does the mean deviation and thus its value can be distorted by a few unusually high or low values. In calculating the standard deviation of the

Table 8.8 Calculation of the standard deviation of a grouped frequency distribution using the assumed mean method (Assumed mean = 55 kg)

Weight (kg)	Midpoint (kg) x	Number of pigs (frequency) f	Deviation of x from assumed mean (in tens of kg) d	fd	fd^2
Under 20	15	1	−4	−4	16
20 and under 30	25	7	−3	−21	63
30 and under 40	35	8	−2	−16	32
40 and under 50	45	11	−1	−11	11
50 and under 60	55 (assumed mean)	19	0	0	0
60 and under 70	65	10	1	10	10
70 and under 80	75	7	2	14	28
80 and under 90	85	5	3	15	45
90 and under 100	95	4	4	16	64
100 and under 110	105	3	5	15	75
TOTALS		75		18	344

grouped frequency distribution, we had to make two assumptions; firstly, that the open-ended classes were of the same length as adjacent classes and, secondly, that the observations in each group had their mean at the midpoint of the group.

The standard deviation has the same units as the variable. Occasionally when we are comparing distributions with different units, for example, UK wages given in £ and US wages given in $, it is convenient to have a dimensionless measure, i.e. one which does not depend on the units of measurement. The *coefficient of variation* is such a measure and is the standard deviation expressed as a percentage of the mean:

$$\text{Coefficient of variation} = \frac{\text{SD}}{\bar{x}} \times 100\%$$

8.9 The standard deviation and the normal distribution

In Unit 4.6 we identified the normal distribution as having a bell-shaped curve that is symmetrical about the arithmetic mean of the distribution. The two ends of the curve do not touch the horizontal axis although they approach it as the values of the variable become extremely small or extremely large.

The normal distribution curve has two *parameters*—the arithmetic mean and the standard deviation—that determine the position and spread of the distribution. There is a relationship between the normal curve and its parameters which enables the proportion of a population that lies between any two values of the variable in which we are interested to be found. There are tables available that give the required proportions so that all you need to do is to learn the relationship between the curve and its parameters and the tables.

It is convenient at this stage to introduce some more notation. We shall use the Greek letter μ (pronounced 'mew') to denote the arithmetic mean of a *population* and the Greek letter σ (small sigma) to denote the standard deviation of a population. It is conventional to use Greek letters to stand for population parameters, that is, fixed quantities which determine the particular population of interest, and to reserve ordinary letters, for example $\bar{x}$, to stand for statistics calculated from a *sample* of readings taken from a population. These statistics will have different values for the same population depending on which particular samples we have taken.

In a normal distribution almost all of the population lies between the values of the variable that are four standard deviations (4σ) either side of the arithmetic mean (μ) of the distribution. In fact the range $\mu - 4\sigma$ to $\mu + 4\sigma$ spans 0.9999 or 99.99% of the population. There will only be a tiny proportion of the population (0.01%) outside this range. For example, if the mean height of an adult male population is 180 cm with a standard deviation of 5 cm, then 0.9999 of that population will have heights between $(180 \pm 4 \times 5)$ cm, that is, between 160 and 200 cm. This proportion is the area under the normal curve between these two values (Fig. 8.3). The total area under a normal curve is always one, for, in this example, we know that everyone in the population has a height of some value. As a normal curve is symmetrical, the mean, the median and the mode coincide at the centre of the distribution and half the population has a value of the variable above the mean and half lies below the mean.

Other proportions of the area under the curve that can be found are:

$$0.6826 \text{ between } (\mu - 1\sigma) \text{ and } (\mu + 1\sigma)$$
$$0.9545 \text{ between } (\mu - 1\sigma) \text{ and } (\mu + 2\sigma)$$
$$0.9973 \text{ between } (\mu - 3\sigma) \text{ and } (\mu + 3\sigma)$$

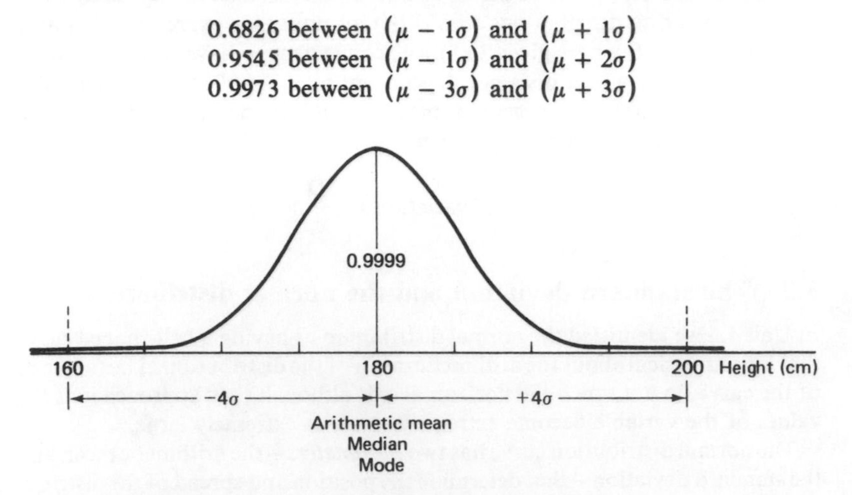

Fig. 8.3 Area under the normal curve between $\mu \pm 4\sigma$

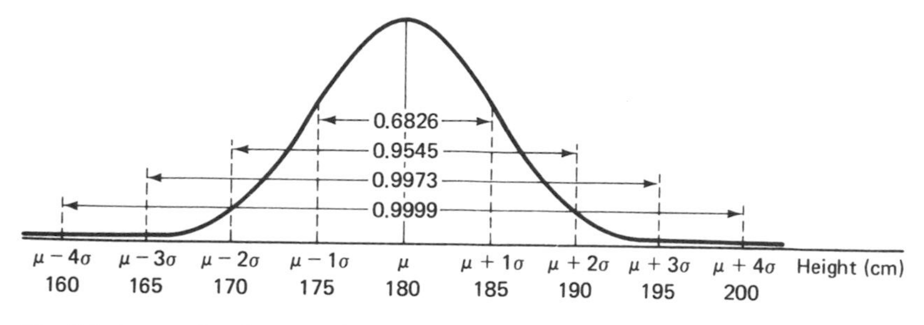

Fig. 8.4 Area under the normal curve between $\mu \pm n\sigma$

Using again our example of the heights of an adult male population, we can state that 0.6826 or 68.26% of that population will be between the heights (180 − 5) cm and (180 + 5) cm, that is, between 175 and 185 cm. We can also state that 0.9545 or 95.45% will be between 170 and 190 cm in height and that 0.9973 or 99.73% will be between the heights of 165 cm and 195 cm. Fig. 8.4 shows these areas under the normal curve.

8.10 Z scores

Table D at the end of the book enables us to work out any required areas under a normal curve. The first column of the table, headed Z, contains what are known as *Z scores*. A Z score is the difference between any value of the variable (x) in a normal distribution and the arithmetic mean (μ) of that distribution, divided by the standard deviation (σ) of the distribution, that is:

$$Z = \frac{x - \mu}{\sigma}$$

A Z score can therefore be thought of as the difference between x and μ measured in units of the standard deviation. Once a Z score has been calculated, Table D shows us the proportion of the population that will be between the Z score and the mean of the distribution and also the proportion that will be between the Z score and the edge of the distribution, or beyond Z. (Not all tables of areas under the normal curve are displayed in this form but any tables that you may need to use in examinations will give equivalent information and sufficient explanation of which particular areas are tabulated will be given.)

You can use Table D to check the proportions of the area under a normal curve that are within one, two and three standard deviations of the mean as given in Unit 8.9. The Z score for a height of 175 cm, for example, is:

$$\frac{x - \mu}{\sigma} = \frac{175 - 180}{5} = -1$$

The minus sign indicates that we are dealing with a value of the variable that is below the mean. Looking in Table D for a Z score of 1 we find that 0.3413

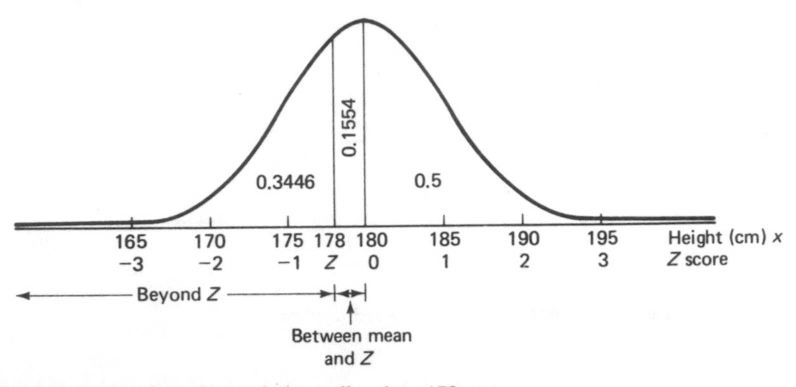

Fig. 8.5 Proportion of population taller than 178 cm

of the population is between the mean and Z, or between the mean and one standard deviation below the mean. It follows that twice that, 0.6826 of the population, is in the range $(\mu - 1\sigma)$ to $(\mu - 1\sigma)$, that is, within one standard deviation of the mean.

We can now look at some more examples. Suppose we want to find the proportion of the population of adult males that has heights taller than 178 cm. Fig. 8.5 illustrates this example. The value of the variable we are interested in is 178 cm and the mean is 180 cm. The difference between x and μ is -2, and the standard deviation is 5 cm, so:

$$Z = \frac{178 - 180}{5} = \frac{-2}{5} = -0.4$$

If we look at Table D in the Z column for 0.4 we see that 0.1554 of the population is between the mean and Z, that is, between 178 cm and 180 cm. As our Z score is negative we know that the value of the variable we are interested in is less than the mean. We also know that half the population lies above the mean, that is, has a height greater than 180 cm. It follows that 0.1554 + 0.5 or 0.6554 of the population is taller than 178 cm.

It also follows from the above example that (1 − 0.6554), that is, 0.3446 of the population is shorter than 178 cm since the total area under the normal curve is one. We can confirm this by looking again at Table D for the Z score of 0.4. The table tells us that 0.3446 is the proportion of the area under the normal curve beyond Z, that is, between the Z score for 178 cm and the lower edge of the distribution.

As a final illustration let us work out the proportion of the population between 178 cm and 188 cm in height. We already know from the previous example that 0.1554 of the population is between 178 cm and 180 cm. We also now need to know the proportion between 180 cm and 188 cm (Fig. 8.6). The Z score for 188 cm is:

$$Z = \frac{x - \mu}{\sigma}$$

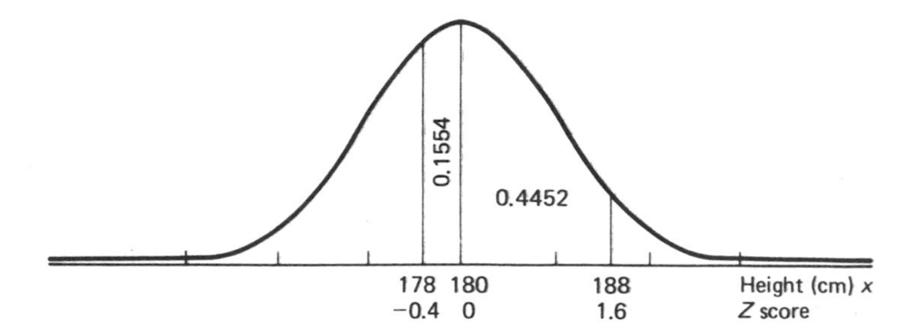

Fig. 8.6 Proportion of population between 178 cm and 188 cm

$$= \frac{188 - 180}{5}$$

$$= 1.6$$

Looking at Table D for the Z score of 1.6 we find that 0.4452 of the population is between the mean and Z. By adding this proportion to that between 178 cm and 180 cm we find that 0.4452 + 0.1554 or 0.6006 of the population is between the heights of 178 cm and 188 cm.

8.11 Measures of skewness

You now know how to compare the positions and the spreads of distributions using measures of central tendency and measures of dispersion. We can also compare the *shapes* of distributions using *measures of skewness*. In Fig. 7.2 we showed the positions of the mean, median and mode for symmetrical, positive skew and negative skew distributions. In a peaked symmetrical distribution the mean, the median and the mode coincide. In a skew distribution the mean and the median lie on the side of the distribution which has the longer tail although the mode is still at the point where the curve is highest. The more skew the distribution, the greater the distance from the mode to the mean or the median and we can use the amount of this spread to measure the degree of skewness. The most usual way of doing this is to calculate:

$$\text{Pearson's first coefficient of skewness} = \frac{\text{Mean} - \text{Mode}}{\text{Standard deviation}}$$

As we have seen, the mode is not always easy to find in a grouped frequency distribution and so the following formula can be used instead:

$$\text{Pearson's second coefficient of skewness} = \frac{3(\text{Mean} - \text{Median})}{\text{Standard deviation}}$$

You are expected to use one of these formulae when an examiner asks for the skewness or coefficient of skewness of a distribution. When you do the calculation, take care with the sign when subtracting the mode or median from the mean. You will get negative answers for negatively skew distributions, positive answers for positively skew distributions and zero for symmetrical distributions. The value of the coefficient of skewness is between −3 and +3 although values below −1 and above +1 are rare and indicate very skew distributions.

8.12 Exercises

1. 'The concept of dispersion is essential to the interpretation of statistical averages.' Explain this statement and describe three measures of dispersion and their merits.

2. A survey of motorists reveals the following distances covered by them in a year:

Kilometres per annum	Number of motorists
2000 and under 3000	6
3000 and under 4000	12
4000 and under 5000	16
5000 and under 6000	25
6000 and under 7000	14
7000 and under 8000	12
8000 and under 9000	9
9000 and under 10000	6
TOTAL	100

   What is the standard deviation of these distances?

3. An employer finds the following figures from the personnel department:

Number of days absent in 1993	Number of employees
0	3
1	6
2	15
3	26
4	40
5	52
6	42
7	27
8	14
9	7
10	2
TOTAL	234

Find (*a*) the standard deviation for this distribution, and (*b*) the mean deviation.

4. A shopper looks at the price of minced beef in various butchers' shops and finds the following prices (in pence) per half-kg:

100	130	96	146	118	144	126	130	120	126
88	158	126	122	132	138	128	142	116	126

Find: (*a*) the mean price; (*b*) the range, the quartile deviation and the mean deviation of the prices; (*c*) the variance and the standard deviation of the prices.

5. Two batsmen playing for a first-class cricket team have the following scores over 15 matches:

Batsman A	44	58	67	34	75	101	34	45	32	28	55	56	48	67	56
Batsman B	16	120	8	0	108	86	2	0	0	93	145	5	96	0	130

Discuss the merits of the two batsmen, using appropriate statistics. Say which player you would prefer to have in your team and why. What do the statistics *fail* to show you?

6. Using the frequency distribution you constructed for Exercise 8 in Unit 4.7 calculate:

(*a*) the mean deviation of the price of cars;
(*b*) the standard deviation of the price of cars;
(*c*) the quartile deviation of the price of cars.

7. Using the frequency distribution you constructed for Exercise 9 in Unit 4.7 calculate:

(*a*) the mean deviation of the expenditure on food for the week;
(*b*) the standard deviation of the expenditure on food for the week;
(*c*) the quartile deviation of the expenditure on food for the week.

8. Find the proportion of the population above Z, below Z and between Z and the mean for the following Z scores:

(*a*) -1.16	(*d*) $+0.96$	(*g*) -0.78
(*b*) $+1.72$	(*e*) -2.90	(*h*) $+1.96$
(*c*) -2.01	(*f*) $+3.08$	(*i*) $+2.64$

9. A store-keeper sells the following number of bottles of lemonade during June:

Date	*Number sold*	*Date*	*Number sold*	*Date*	*Number sold*
1	20	11	15	21	25
2	15	12	20	22	15
3	25	13	30	23	30
4	35	14	25	24	0
5	30	15	20	25	10
6	20	16	40	26	15
7	15	17	20	27	20
8	25	18	5	28	25
9	20	19	20	29	10
10	35	20	10	30	5

Find the mean number of bottles sold per day together with the standard deviation. Using these values as estimates of the mean and standard deviation of the population (assumed normal) of the number of bottles sold daily, find the Z scores for the numbers of bottles sold on the following dates:

(*a*) 1; (*b*) 10; (*c*) 20; (*d*) 30; (*e*) 15; (*f*) 16.

10. The mean weight of the male staff of a company is 75 kg with a standard deviation of 2 kg. The distribution of weights is normal. What proportion of the staff weighs more than 72 kg?

11. The staff of the above company are given a standard IQ test. The score on the test is 120 with a standard deviation of 5. What proportion of the staff scored:

(*a*) less than 110;
(*b*) more than 135;
(*c*) between 110 and 130;
(*d*) less than 118.

12. (*a*) Explain and illustrate graphically the relationship between the mean, median and mode in:

(i) a positively skewed distribution;
(ii) a negatively skewed distribution.

(*b*) Calculate the coefficient of skewness for a distribution with the following values: mean = 20, median = 21, standard deviation = 5. What does your answer tell you about this distribution?

(Answers at the end of the book.)

UNIT NINE

The relationship between two variables

9.1 Introduction

We have so far looked at statistics that are concerned with the values of one variable. We are often, however, interested in looking at the relationship between two variables, particularly to see if one variable depends on the other. We can ascertain if this is so by seeing if the values of one of the variables, the *dependent variable*, change in response to fluctuations in the other, the *independent variable*. If changes in the values of the independent variable are matched by similar proportional changes in the values of the dependent variable, there may well be a *causal relationship* between the variables. All kinds of people are interested in finding causal relationships. Scientists are often looking for them in their controlled experiments: the effect of varying temperature on the expansion or contraction of metal, for example, or the effect of the application of nutrients on the rate of plant growth.

In the world of business, causal relationships can also be important. The production engineer may be looking for relationships similar to those which interest the scientist. The personnel and training manager may attempt to see if money spent on training is reflected by better performance. The accountant may be trying to assess if heavy investment in, say, a computerized system has resulted in lower stock-holding costs. The sales manager may be interested in seeing if advertising has a causal effect on sales.

It is possible for a relationship to exist between two variables without this being a *causal* relationship. Looking through tables of official statistics one may discover that when ice-cream sales increase in Great Britain, so do road accidents in France. We cannot conclude that the increase in sales of ice-cream in Great Britain *causes* more road accidents in France. It could be that both are dependent on a third variable, namely, time of year. It is quite likely that at holiday times sales of ice-cream in Great Britain and road accidents in France both increase, although further investigation would be necessary to establish this. Alternatively, it could be that the official statistics just happen to show that these variables increase and decrease together over the time-span considered but over the longer term there is no reason to suppose that this relationship will be maintained—we have obtained what is called a *spurious correlation*.

9.2 Scatter diagrams

A *scatter diagram* or *scattergraph* shows related figures as single points on a diagram that has two axes, a vertical axis for the dependent variable and a horizontal axis for the independent variable (exactly the same as those for a graph). Take for example a firm's advertising and sales figures over a period of time (see Table 9.1). The scatter diagram of these values is shown in Fig. 9.1.

The resulting cluster of points on the scatter diagram can often indicate immediately if there is any relationship between the two variables. In Fig. 9.1 it can be seen that higher values of advertising expenditure tend to be associated with higher sales. (In Fig. 9.1 we did not start the scales on the axes

Table 9.1 Advertising and sales figures

Advertising expenditure (£) (x)	Sales (£) (y)
1 000	22 000
1 200	25 000
1 800	26 000
1 500	30 000
800	23 000
1 700	27 000
2 000	32 000
1 500	27 000
1 100	25 000
1 900	29 000

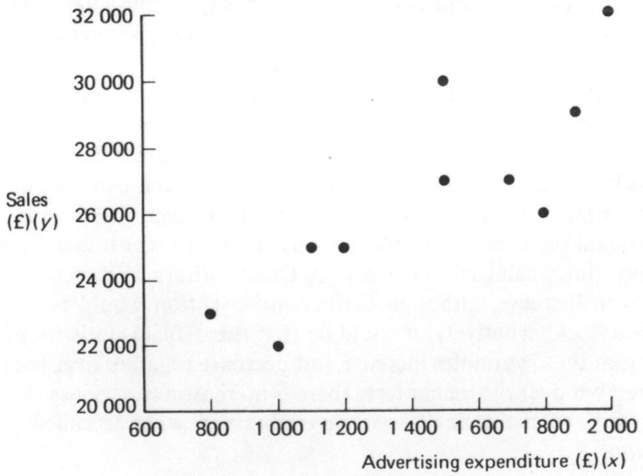

Fig. 9.1 Scatter diagram showing sales and advertising figures

at zero. This is acceptable in a scatter diagram as we want to be able to see as clearly as possible the relative changes in the two variables rather than being concerned with the absolute changes in a single variable as compared with zero.)

It may be that a straight line can be drawn that will pass very close to every point on a scatter diagram, as in Figs 9.2(*a*) and (*b*). If all the points on the diagram lie very near to a straight line then there is said to be a *strong linear correlation* between the two variables. (We will be looking at the method of measuring that correlation in Unit 9.5.) Fig. 9.2(*a*) shows strong *positive linear correlation*, which means that the value of the dependent variable *increases* as the value of the independent variable *increases*, as may well happen with advertising and sales. Fig. 9.2(*b*), on the other hand, shows strong *negative linear correlation*, which means that the dependent variable *decreases* as the independent variable *increases*, for example, as temperature increases so the weight of clothing worn decreases.

A scatter diagram may however show no evidence of correlation at all (Fig. 9.3(*a*)) or there may appear to be a *curvilinear* or *non-linear* relationship (Fig. 9.3(*b*)) rather than a linear relationship.

If there is some degree of linear relationship between the two variables, a

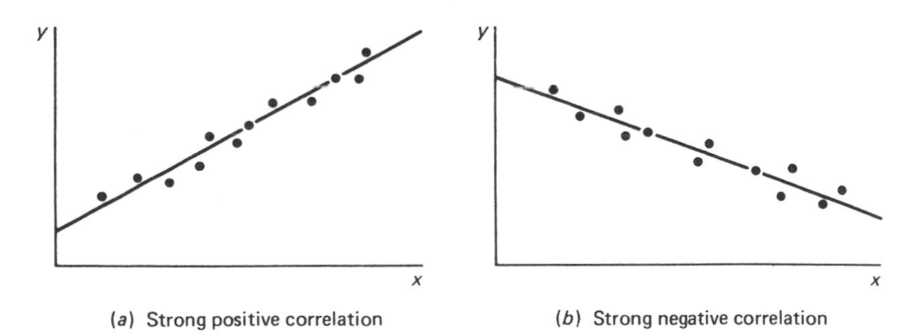

Fig. 9.2 Scatter diagrams showing positive and negative correlations

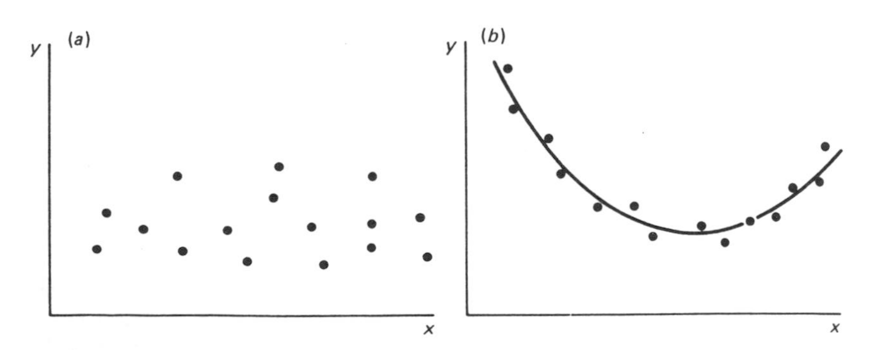

Fig. 9.3 Scatter diagrams showing (a) no correlation and (b) curvilinear correlation

scatter diagram can be used for making estimates of the dependent variable by using a straight line that passes as close as possible to all points on the diagram. Lines like this are known as *lines of best fit*, for obvious reasons. If the points all lie fairly near to a straight line, it may be quite easy to use a ruler to draw a line that passes as closely as you can judge to every point. In Unit 9.3 we will look at methods of drawing a line of best fit that will be more reliable than one that depends on the eye alone.

In the examples we will look at, we are concerned only with relationships measured from a few values of each variable. Unfortunately, for the purposes of examples and examinations it is only possible to consider just a few items. In practice a scatter diagram should be constructed from many more observations of the two variables concerned, if possible. As with all our statistics we must always make sure that we use enough observations of our variables to eliminate as much bias as possible from our data.

9.3 Regression analysis

By *regression analysis* we mean the use of the observations of the variables we are considering to calculate a curve of best fit so that we can make estimates and predictions about the behaviour of the variables. Because of the difficulty of the analysis we shall restrict ourselves to fitting straight lines only. The calculated line of best fit is called a *regression line*.

(a) The three-point method

The simplest way of calculating a regression line is by the *three-point method*: all that it involves is the calculation of arithmetic means. As an example we shall use the advertising and sales figures given in Table 9.1. The first of the three points is $(\bar{x}, \bar{y})$, that is, we find the arithmetic mean, $\bar{x}$, of all the x values, the arithmetic mean, $\bar{y}$, of all the y values and plot a point on the scatter diagram at $x = \bar{x}$ and $y = \bar{y}$ (point A in Fig. 9.4). We have:

$$\bar{x} = \frac{\Sigma x}{n} = \frac{£14\,500}{10} = £1\,450$$

$$\bar{y} = \frac{\Sigma y}{n} = \frac{£266\,000}{10} = £26\,600$$

Next we rewrite the entries in Table 9.1 starting with the pair of readings that has the lowest x-value and then continuing with the x-values in order of magnitude. If two x-values are the same, the one with the lower y-value has precedence in the table. The redrawn table is Table 9.2.

Now we divide this table into two with an equal number of pairs of readings in each group—one group corresponds to the lower values of x and one to the higher values of x. If there is an odd number of pairs we simply discard the pair of readings in the middle of the table. We then work out $(\bar{x}, \bar{y})$ for each of the two groups of points and plot these two points on the scatter diagram.

Table 9.2 Advertising and sales figures in order of advertising expenditure

Advertising expenditure (£) (x)	Sales (£) (y)
800	23000
1000	22000
1100	25000
1200	25000
1500	27000
1500	30000
1700	27000
1800	26000
1900	29000
2000	32000

1st group:

$$\bar{x} = \frac{\Sigma x}{n} = \frac{£5600}{5} = £1120$$

$$\bar{y} = \frac{\Sigma y}{n} = \frac{£122000}{5} = £24400$$

$$(\bar{x}, \bar{y}) = (£1120, £24400)\ (\text{Point B in Fig. 9.4})$$

2nd group:

$$\bar{x} = \frac{\Sigma x}{n} = \frac{£8900}{5} = £1780$$

$$\bar{y} = \frac{\Sigma y}{n} = \frac{£144000}{5} = £28800$$

$$(\bar{x}, \bar{y}) = (£1780, £28800)\ (\text{Point C in Fig. 9.4})$$

The regression line is drawn through the three points A, B and C on the scatter diagram (Fig. 9.4). The regression line goes exactly through these three points if there is an even number of pairs of readings. It goes through point A and approximately through points B and C if there is an odd number of pairs of readings. Using this method reduces the amount of personal judgment involved when drawing a line of best fit by eye but it is still not a very reliable method as a lot of information has been condensed in a crude manner.

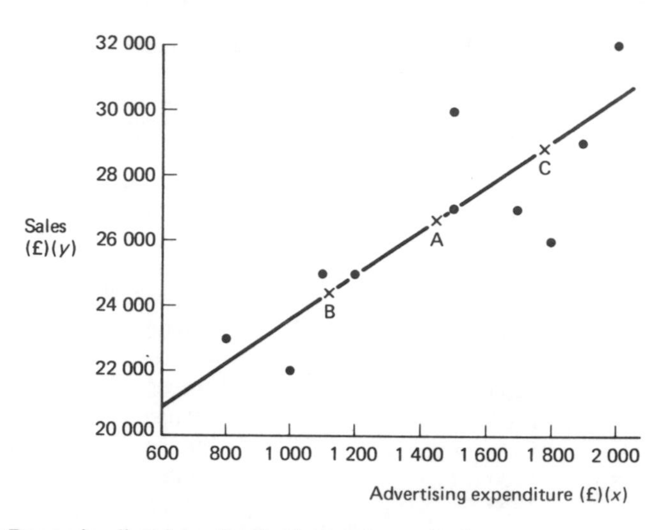

Fig. 9.4 Regression line drawn by the three-point method

(b) The least squares method

A more exact method of finding the line of best fit is to use the *least squares method*. This gives an algebraic equation connecting x and y. The type of equation that gives a straight line relationship between x and y is always of the form:

$$y = a + bx$$

where y is the value of the dependent variable, x is the value of the independent variable, and a and b are constants. Each different straight line has different values of a and b. a is the value of y, the dependent variable, when x equals zero. a is termed the *intercept* on the y-axis. b is the *slope or gradient* of the line. The gradient of a line is the change in y for unit increase in the value of x. For a line sloping upwards from left to right b is positive. For a line sloping downwards from left to right b is negative.

A simple example will help you understand the concept of a straight-line equation. Suppose a taxi hire company charges a basic rate of £1.00 plus 80p a kilometre for its contracts. You can quite quickly calculate the cost of any journey according to the length of the journey: it is $y = 100 + 80x$, where y is the cost of the journey in pence, x is the number of kilometres travelled, a is the constant basic rate in pence and b the cost per kilometre in pence. You can draw a graph of this equation as in Fig. 9.5.

To draw the line all you need to do is calculate the value of y for each of two values of x and join these points on the graph with a straight line, continuing the line in each direction. For a journey of 1 km the cost is $100 + (80 \times 1) = 180$p; for a journey of 3 km the cost is $100 + (80 \times 3) = 340$p. You can see that the line meets the vertical axis where $y = 100$p, the value of a. You can also see that the slope of the line is 80 p

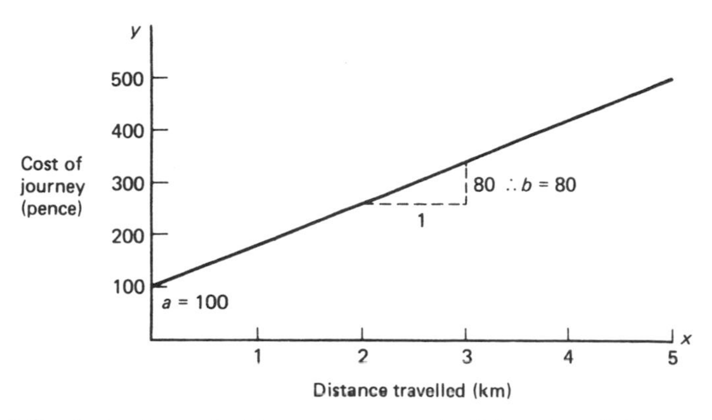

Fig. 9.5 Graph showing cost of journey according to distance travelled

because as x increases by 1 unit, y increases by 80 units. This equation is in exactly the same form as for a regression line and tells you the value of y for any value of x. The line could be used to estimate graphically the cost of any journey according to the distance travelled; alternatively the algebraic formula corresponding to this line:

$$y = 100 + 80x$$

could be used to calculate the cost without needing to draw a graph. You can probably think of other systems of charging that use the same principles as the imaginary taxi company.

The least squares method enables us to find the equation of the straight line which mathematically is the 'best fit' to the points in the scatter diagram. We assume that values of x, the independent variable, can be specified exactly and that when finding the line of best fit we are trying to minimize the variation about the line of the observed values of y. Thus we are interested in the *vertical* deviations (d) between the observed points and the line (Fig. 9.6). The line of best fit will pass somewhere through the middle of the observed points so some of these vertical deviations will be positive and some negative. In order to remove the negative signs we consider the *squares* of these vertical deviations and the *least squares regression* line is the line that minimizes the sum of the squares of these vertical deviations. It can be shown by mathematics which you are not required to know at this stage that this is achieved when the values of a and b in the equation $y = bx$, are given by:

$$b = \frac{n\Sigma xy - \Sigma x \Sigma y}{n\Sigma x^2 - (\Sigma x)^2}$$

and

$$a = \bar{y} - b\bar{x}$$

where y denotes the observed y value corresponding to the value, x, of the

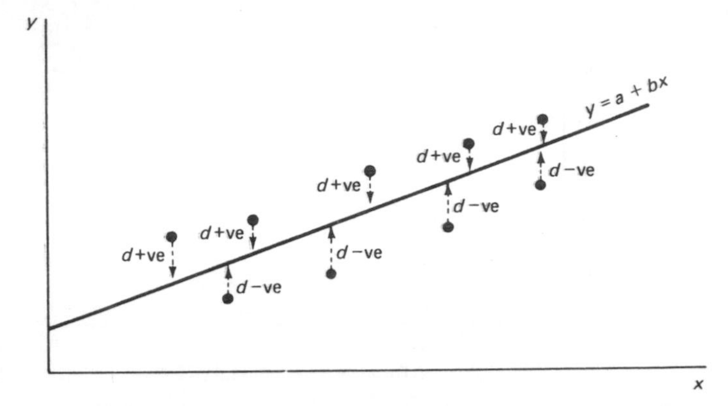

Fig. 9.6 Relationship between points on a scatter diagram and the least squares regression line of y upon x

independent variable and n is the number of pairs of values of x and y that have been observed.

Let us work out the least squares regression line for the advertising and sales data of Table 9.1. The sums required are set out in Table 9.3. In order to reduce the arithmetic the advertising expenditure, x, has now been expressed in hundreds of pounds and the resulting sales, y, are now in thousands of pounds.

$$n = 10 \qquad \bar{x} = 14.5 \qquad \bar{y} = 26.6$$

Substituting in the formulae for b and a, we find:

$$b = \frac{10 \times 3952 - 145 \times 266}{10 \times 2253 - (145)^2}$$

$$= \frac{39520 - 38570}{22530 - 21025}$$

$$= \frac{950}{1505}$$

$$= 0.631 \text{ to 3 significant figures}$$

$$a = 26.6 - \left(\frac{950}{1505}\right) 14.5$$

$$= 26.6 - 9.1528$$

$$= 17.447$$

$$= 17.4 \text{ to 3 significant figures}$$

The equation of the regression line is:

$$y = 17.4 + 0.631x$$

Table 9.3 Values for the least squares regression calculation

	Advertising expenditure (£ hundred) x	Sales (£ thousand) y	xy	x^2
	10	22	220	100
	12	25	300	144
	18	26	468	324
	15	30	450	225
	8	23	184	64
	17	27	459	289
	20	32	640	400
	15	27	405	225
	11	25	275	121
	19	29	551	361
TOTALS	145	266	3952	2253

where x is the advertising expenditure in hundreds of pounds and y is the sales in thousands of pounds.

In all regression (and correlation) calculations it is important to keep in as many significant figures as practicable until the end of the calculation because the subtractions involved in the calculation can lead to loss of significant figures. Note that the exact form for b of 950/1 505 was used in the calculation of a rather than the rounded value of 0.631 which is quoted in the final regression equation.

To draw the least squares regression line on the scatter diagram, choose two values of x, one towards the left of the diagram and one towards the right. Calculate y from the regression equation for each of these values of x, plot the two points on the diagram and join them up with a straight line. If you have done the calculations correctly, the line will pass through the point $(\bar{x}, \bar{y})$ so this will provide a check of your arithmetic. In the advertising and sales example, when advertising expenditure is £1 000, ($x = 10$), the sales are given by:

$$y = 17.4 + (0.631 \times 10)$$

$$= 23.7 \text{ to 3 significant figures}$$

y is in thousands of pounds so this corresponds to sales of £23 700. Similarly, when advertising expenditure is £2000, ($x = 20$), the sales are £30000 to three significant figures. The scatter diagram and least squares regression line are shown in Fig. 9.7.

There are other ways of finding the values of a and b in the least squares regression equation $y = a + bx$. These are equivalent to the procedure adopted above.

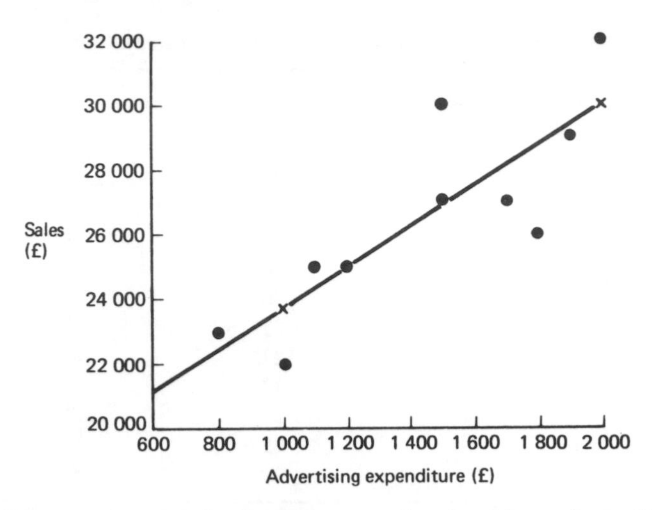

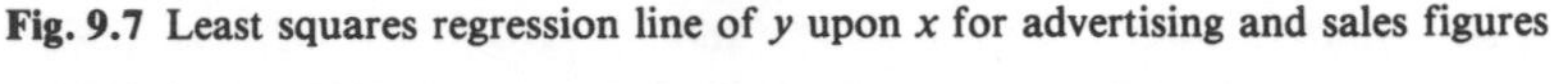
Fig. 9.7 Least squares regression line of y upon x for advertising and sales figures

(i) The formula for b can be written as:

$$b = \frac{\dfrac{\Sigma xy}{n} - \left(\dfrac{\Sigma x}{n}\right)\left(\dfrac{\Sigma y}{n}\right)}{\dfrac{\Sigma x^2}{n} - \left(\dfrac{\Sigma x}{n}\right)^2} = \frac{\dfrac{\Sigma xy}{n} - \bar{x}\bar{y}}{\dfrac{\Sigma x^2}{n} - \bar{x}^2}$$

As before, a is given by:

$$a = \bar{y} - b\bar{x}$$

(ii) a and b can be found by solving the following simultaneous equation which are sometimes referred to as the *normal* equations for a and b:

$$\Sigma y = na + b\Sigma x$$

$$\Sigma xy = a\Sigma x + b\Sigma x^2$$

You should check for yourself that the methods (i) and (ii) above give the same results for a and b as our original method. Formula sheets issued in examinations usually quote one of the three forms mentioned above although you may find the letter c used in place of a and m in place of b. Although we have seen that there are methods of finding a regression line other than the least squares method, when an examination question asks you to calculate *the* regression line, you should assume that the least squares method is to be used. If a question requires the three point method or freehand method to be used, this will be specifically stated.

9.4 Estimating with regression lines

Once we have a regression line we can use it for estimating. Suppose, for example, we wish to estimate the sales that will result from an advertising expenditure of £1 600. We can do this graphically by drawing a vertical line from that value on the horizontal axis up to the regression line; from that point we draw a horizontal line to the vertical axis and read off the corresponding sales figure. Using the three point regression line (Fig. 9.4), we obtain an estimated sales of £27 600. From the least squares regression line (Fig. 9.7), the estimated sales are £27 500. Alternatively we can use the least squares regression equation:

$$y = 17.4 + 0.631x$$

When $x = 16$, $y = 17.4 + 10.1 = 27.5$ to 3 significant figures. Thus the estimated sales are £27 500.

We are estimating within the range of our sample data, a technique known as *interpolation*. Our resulting estimates should be good provided we know that the observed points lie fairly close to a straight line on the scatter diagram. There is little point in attempting to fit a regression line when the scatter of points appears non-linear. Estimating outside the range of the observed data is known as *extrapolation*. For example, we may estimate the sales that will result from an advertising budget of £2 500. Using our regression equation:

$$\begin{aligned} y &= 17.4 + (0.631 \times 25) \\ &= 33.2 \text{ to 3 significant figures} \end{aligned}$$

which gives a sales figure of £33 200. It is extremely dangerous to extrapolate too far outside the range of available data, however, since we have no guarantee that the same linear relationship will hold outside the sample range. For instance, a massive increase in advertising expenditure will not automatically produce a corresponding dramatic increase in sales once the market for the product is saturated.

The least squares regression line which we have drawn in Fig. 9.7 and whose equation we have determined is referred to as *the regression line of y upon x*, that is, in this case, the regression line of sales upon advertising expenditure. An examination question will not always tell you which is the independent variable (x) and which is the dependent variable (y). You must decide for yourself which is the appropriate labelling. In the advertising and sales example, we assumed that a change in advertising expenditure brought about a change in sales so that sales was the dependent variable. It could be, however, that changes in sales produced a corresponding change in advertising expenditure—if a company has higher sales, it presumably has higher profits and can spend more on advertising. In order to predict the level of advertising expenditure from a given sales figure, we would need to calculate the regression line of advertising expenditure on sales, that is, *the regression line of x upon y*. This does not produce the same least squares regression line as before because this time we are minimizing the sum of the squares of the *horizontal* deviations of the observed points from the line (Fig. 9.8). It is only if all the

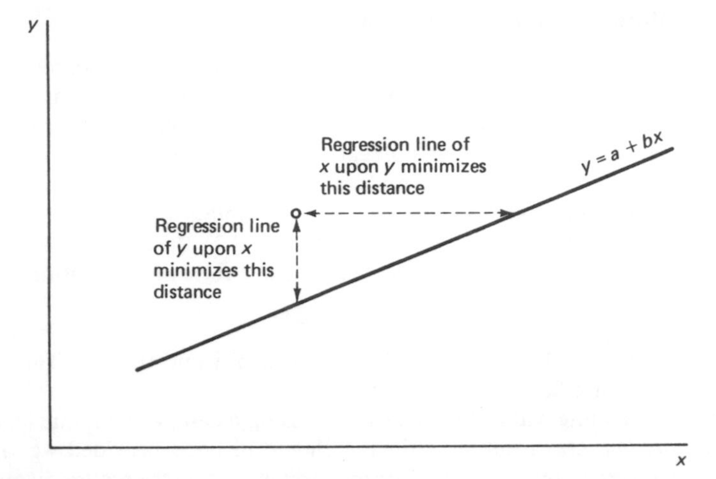

Fig. 9.8 Difference between the regression lines of y upon x and of x upon y

observed data lie exactly on a straight line in the scatter diagram that the two regression lines coincide.

9.5 Correlation

The techniques of correlation enable us to measure the *degree of linear relationship* between two variables. If we look at a scatter diagram we can often detect whether there is any linear relationship between the two variables by the pattern of points on the diagram. We saw in Unit 9.2 that sometimes there is strong positive correlation (Fig. 9.2(*a*)) and sometimes strong negative correlation (Fig. 9.2(*b*)). We also saw scatter diagrams which gave no evidence of linear correlation (Fig. 9.3).

It is easy to see why we refer to the correlation as positive or negative if we look at the way in which the degree of linear correlation is calculated. A scatter diagram can be divided into four quarters or *quadrants* (Fig. 9.9) by drawing a vertical line from the mean of the x series and a horizontal line from the mean of the y series. Fig. 9.9 shows that for any value of x, the sign of $(x - \bar{x})$ depends on the quadrant in which x lies. The same applies to the values of y. For points in quadrant 1, for instance, $(x - \bar{x})$ and $(y - \bar{y})$ are positive. In quadrant 2 the values of x are all below the mean value x, so $(x - \bar{x})$ is negative although $(y - \bar{y})$ is still positive. You can similarly work out the signs of $(x - \bar{x})$ and $(y - \bar{y})$ for points in the other quadrants. The product $(x - \bar{x})\,(y - \bar{y})$ is positive in quadrants 1 and 3 and negative in quadrants 2 and 4.

Suppose that we total all the individual products $(x - \bar{x})\,(y - \bar{y})$. If most or

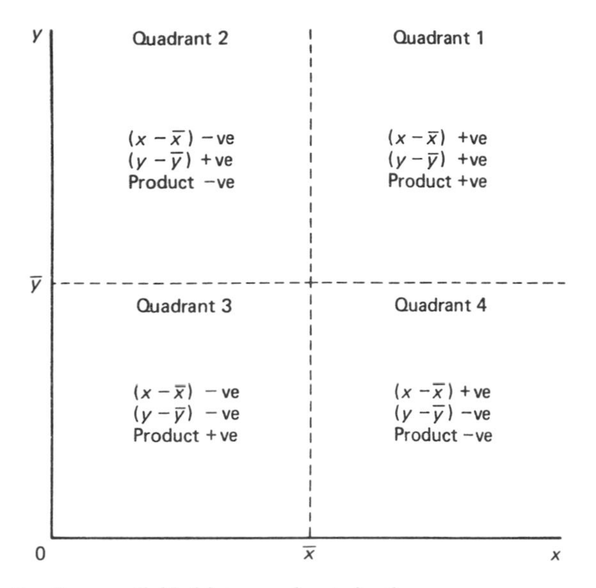

Fig. 9.9 Scatter diagram divided into quadrants by the means

all of the points on the scatter diagram arc in quadrants 1 and 3, as in Fig. 9.2(*a*), the result will be a large positive number. However, if all or most of the points are in quadrants 2 and 4 as in Fig. 9.2(*b*), the result will be a high negative number. Obviously if some of the points are in quadrants 1 and 3 and some are in quadrants 2 and 4 as in Fig. 9.3, then the positives will tend to cancel out the negatives and as a result there will be a small positive or negative number, in other words, the result will be nearer to zero.

As a measure of the degree of correlation between two variables we can use the arithmetic mean of the sum of these products. This is known as the *covariance* of x and y and is $\frac{\Sigma(x - \bar{x})(y - \bar{y})}{n}$. However, this measure of correlation depends on the units of measurement. For instance, if x is measured in pounds (£) and y is measured in pounds (£), the covariance has units of square pounds (£^{2}) but if we decided to measure x in hundreds of pounds and y in thousands of pounds we would obtain a different value for the covariance although the original data are the same. We would prefer to have a dimensionless measure, that is, one which does not depend on the units of measurement. Because of this and also to allow for the spread of data about the means of the x and the y series respectively, it is convenient to divide the covariance by the standard deviation of the x values and the standard deviation of the y values. The result of this calculation gives us a number which is called *Pearson's product moment correlation coefficient* or *the correlation coefficient* for short. It is denoted by the symbol, r, for a calculation based on a sample of paired observations. The Greek letter rho, ρ, is used to denote a population correlation coefficient.

Referring to Unit 8.6 where we discussed the standard deviation, we see that the formula for r can be written as:

$$r = \frac{\text{Covariance of } x \text{ and } y}{\text{Standard deviation of } x \times \text{Standard deviation of } y}$$

$$= \frac{\dfrac{\Sigma(x - \bar{x})(y - \bar{y})}{n}}{\sqrt{\dfrac{\Sigma(x - \bar{x})^2}{n}}\sqrt{\dfrac{\Sigma(y - \bar{y})^2}{n}}}$$

$$= \frac{\Sigma(x - \bar{x})(y - \bar{y})}{\sqrt{\Sigma(x - \bar{x})^2}\sqrt{\Sigma(y - \bar{y})^2}}$$

This is a complex expression and it is often easier to use one of the following rearrangements of the formula:

$$r = \frac{n\Sigma xy - \Sigma x \Sigma y}{\sqrt{n\Sigma x^2 - (\Sigma x)^2}\sqrt{n\Sigma y^2 - (\Sigma y)^2}}$$

or

$$r = \frac{\dfrac{\Sigma xy}{n} - \left(\dfrac{\Sigma x}{n}\right)\left(\dfrac{\Sigma y}{n}\right)}{\sqrt{\dfrac{\Sigma x^2}{n} - \left(\dfrac{\Sigma x}{n}\right)^2}\sqrt{\dfrac{\Sigma y^2}{n} - \left(\dfrac{\Sigma y}{n}\right)^2}}$$

$$= \frac{\dfrac{\Sigma xy}{n} - \bar{x}\bar{y}}{\sqrt{\dfrac{\Sigma x^2}{n} - \bar{x}^2}\sqrt{\dfrac{\Sigma y^2}{n} - \bar{y}^2}}$$

It is one of the above forms that you will usually find in a formula list in an examination.

The formula for r looks rather formidable, but, as long as you are methodical when making the calculation, it is quite straightforward. You will notice that many of the sums required are the same ones we needed when calculating b in Unit 9.3 on regression.

As an example let us calculate r for our sales and advertising figures (see Table 9.4).

Setting out the figures as in the table has given us the values to substitute in the formula:

$$r = \frac{n\Sigma xy - \Sigma x \Sigma y}{\sqrt{n\Sigma x^2 - (\Sigma x)^2}\sqrt{n\Sigma y^2 - (\Sigma y)^2}}$$

$$= \frac{(10 \times 3952) - (145 \times 266)}{\sqrt{(10 \times 2253) - (145)^2}\sqrt{(10 \times 7162) - (266)^2}}$$

$$= \frac{39\,520 - 38\,570}{\sqrt{1\,505}\sqrt{864}}$$

$$= \frac{950}{1\,140.316}$$

$= 0.83$ to 2 decimal places

You should check for yourself that using any of the other formulae leads to the same result. It is worth noting the following points which can help to simplify the arithmetic involved when calculating r:

(i) we can take any constant amount off every value of x;
(ii) we can take any constant amount off every value of y;
(iii) we can divide or multiply every value of x by a constant amount; and
(iv) we can divide or multiply every value of y by a constant amount;

all without altering the value of r. Check that if in Table 9.4 you subtract 8 from all the x values and 20 from all the y values, you still obtain a correlation coefficient of 0.83.

Table 9.4 Values for calculating Pearson's product moment coefficient of correlation

	Advertising expenditure (£ hundred) x	Sales (£ thousand) y	xy	x^2	y^2
	10	22	220	100	484
	12	25	300	144	625
	18	26	468	324	676
	15	30	450	225	900
	8	23	184	64	529
	17	27	459	289	729
	20	32	640	400	1024
	15	27	405	225	729
	11	25	275	121	625
	19	29	551	361	841
TOTALS	145	266	3952	2253	7162

9.6 The interpretation of the correlation coefficient

The correlation coefficient is always between -1 and $+1$ inclusive. If you obtain a numerical value greater than 1, then you have made a mistake and you should check your arithmetic. A correlation coefficient of -1 occurs when there is *perfect negative linear correlation*, that is, all the points in the scatter diagram lie *exactly* on a straight line sloping down from left to right. A correlation coefficient of 0 occurs when there is *no linear correlation*. This does not necessarily imply that there is no relationship between x and y—it could be that there is a relationship but that it is non-linear as in Fig. 9.3(*b*). A correlation coefficient of $+1$ occurs when there is *perfect positive linear correlation*, that is, all the points in the scatter diagram lie *exactly* on a straight line sloping upwards from left to right. A correlation coefficient between -1 and 0 indicates that the variables are *partly negatively correlated*. This implies that as one variable increases the other tends to decrease but the relationship is not exactly linear. A correlation coefficient between 0 and 1 implies some degree of *positive correlation*, that is, as one variable increases so does the other but again, the relationship is not exactly linear. The closer the coefficient is to -1 or to $+1$ the stronger the degree of correlation and the nearer the relationship is to being truly linear.

The conclusions that we can draw from a correlation coefficient about the degree of linear relationship between two variables depend on the number of pairs of observations that has been made. It is obviously easier for 10 points to lie close to a straight line in a scatter diagram than for 100 points to do so. Thus, we would automatically expect a lower value for a correlation coefficient based on a larger number of observations, even though the actual degree of relationship between the two variables was the same. In our example relating advertising expenditure and sales we assessed the degree of relationship based on a sample of 10 pairs of observations. We would like to assess whether the value of the correlation coefficient, $r = 0.83$, implies that there is always likely to be a linear relationship between advertising expenditure and sales revenue for any values of these two variables, that is, we want to be able to test to see whether ρ, the *population* correlation coefficient, is what is termed 'significantly different from zero'. We shall see in Unit 15.4(*d*) how to to do this provided we assume that x and y have what is called a bivariate normal distribution. You do not need to know the mathematical form of this distribution but you should be aware that if x and y follow this distribution then the distribution of the x variable is normal when the value of the y variable is fixed and similarly, the distribution of the y variable is normal when the value of the x variable is fixed.

9.7 Rank correlation

We do not always have a list of numerical values of the two variables whose relationship we are investigating. Suppose a company recruits personnel for its internal secretarial training course by means of an aptitude test. At the end of the course the trainees are ranked in order of merit because a variety of methods of assessment is used on the course and giving an overall course mark would be impractical. Performance on the aptitude test is also ranked in order of merit and the personnel officer wants to see if there is any relationship between the performance of trainees in the aptitude test and their performance on the training course.

In this situation another measure of correlation is appropriate. This measure is known as *Spearman's rank correlation coefficient* (R). The formula for this coefficient is:

$$R = 1 - \frac{6\Sigma d^2}{n(n^2 - 1)}$$

where d is the difference between the rankings of each item observed and n is the number of paired observations.

Suppose the trainees were ranked in the aptitude test and on the course as in Table 9.5.

Substituting in the formula we have:

$$R = 1 - \frac{6 \times 34}{8(64 - 1)} = 1 - 0.4047$$

$$= 0.60 \text{ to 2 decimal places.}$$

The value of R is always between -1 and $+1$, both inclusive. The closer the agreement between the rankings, the nearer R is to 1. In this example, with $R = 0.60$, there appears to be some positive relationship between the two variables but it is not very strong, so there does not appear to be much agreement between the performance of trainees in the aptitude test and their performance on the training course.

A negative value for R implies an inverse relationship and in this example

Table 9.5 Calculation of Spearman's correlation coefficient with ranked data

Trainee	Ranking in aptitude	Ranking on course	Difference in ranking (d)	d^2
Beryl	6	5	1	1
Agnes	8	6	2	4
Laura	1	1	0	0
Edith	3	2	1	1
Ann	2	3	−1	1
James	5	8	−3	9
Sylvia	4	7	−3	9
Elizabeth	7	4	3	9
TOTALS		(Check)	0	34

Table 9.6 Calculation of Spearman's rank correlation coefficient for advertising and sales figures

Advertising expenditure (£ hundred) x	Sales (£ thousand) y	Ranking of x	Ranking of y	Difference in ranking (d)	d^2
10	22	2	1	1	1
12	25	4	3.5	0.5	0.25
18	26	8	5	3	9
15	30	5.5	9	−3.5	12.25
8	23	1	2	−1	1
17	27	7	6.5	0.5	0.25
20	32	10	10	0	0
15	27	5.5	6.5	−1	1
11	25	3	3.5	−0.5	0.25
19	29	9	8	1	1
TOTALS			(Check)	0	26.0

would have suggested that those who did well in the aptitude test performed badly on the training course and vice versa. A zero value for R implies no relationship between the variables.

Besides being used when rankings only of the variables are available, Spearman's rank correlation coefficient can also be used with any paired observations such as the advertising and sales figures (see Table 9.6) provided we rank the data first. Spearman's coefficient is quicker to calculate than Pearson's product moment correlation coefficient and, although the latter gives a more exact result since it is based on actual values rather than ranks, Spearman's coefficient is a close approximation, in general, and gives a good indication of the amount of correlation present.

Sometimes it is not possible to distinguish between the ranks of two or more observations. In such a case you should give each of the tied ranks the arithmetic mean of the ranks they jointly occupy. Strictly speaking the given formula for R does not hold when there are tied ranks, but unless there are many equal observations this need not worry you. The calculation of Spearman's coefficient for the advertising and sales figures is set out in Table 9.6.

Substituting in the formula we have:

$$R = 1 - \frac{6 \times 26}{10(100 - 1)}$$

$$= 1 - 0.1576$$

$$= 0.84 \text{ to 2 decimal places}$$

This is in close agreement with the value of Pearson's correlation coefficient ($r = 0.83$) found in Unit 9.5.

An examination question will specify if a rank correlation coefficient is to be used—if the word 'rank' or 'Spearman's' is not mentioned, then you should assume that Pearson's product moment correlation coefficient is required.

9.8 Exercises

1. Two sets of variables are observed and the following random sets of observations are made:

Set 1:	x	1	2	3	4	5	6
	y	7	5	4	3	2	1
Set 2:	x	1	2	3	4	5	6
	y	2	3	4	5	6	7

For each set of observations calculate Pearson's coefficient of correlation and interpret your result.

2. The manager of a large branch of a national chain of women's clothing stores does a small survey and finds the following:

Woman's age (years)	18	21	36	45	23	53	25	37	30	32
Annual expenditure on clothes (£)	330	300	180	120	310	200	200	150	250	190

(*a*) Calculate Pearson's product moment coefficient of correlation for these data.

(*b*) Interpret your result.

3. A small-holder is interested in increasing the growth rate of a particular crop during the critical spring growing period so that it may be harvested ahead of that of competitors. Experiments in growing several plots of the crop over some years have produced the following results:

Water fed to plot during critical period (thousand litres)	Growth during period (cm)
2	4
2	5
1	3
3	9
4	8
5	11
3	7
7	13
6	10
7	15
5	9

(*a*) Use Pearson's product moment coefficient of correlation to measure the association between the amount of water fed and the growth; interpret your results.
(*b*) Calculate the regression line formula for y upon x and estimate with it the growth during the period if 9000 litres of water are fed to the plants.
(*c*) Draw the regression line on a scatter diagram of the data.

4. A sociologist observes ten children and measures the cost of keeping each child, with the following results:

Age of child (years) (x)	1	2	3	4	2	5	7	8	5	2
Annual cost (£000) (y)	2.5	0.5	2.0	1.0	1.0	2.5	3.0	3.5	2.0	1.0

(*a*) Estimate the cost of keeping a 9-year-old child.
(*b*) Draw a scatter diagram from the data and impose the y upon x regression line on that graph.

5. (*a*) On graph paper, draw a scatter diagram of the following data and, by the method of least squares, draw a straight line which fits the data.

Year	Output (tonnes)	Cost (£)
1986	10000	32000
1987	20000	39000
1988	40000	58000
1989	25000	44000
1990	30000	52000
1991	40000	61000
1992	50000	70000
1993	45000	64000

(*b*) Estimate the costs likely to be incurred at output levels of 26000 and 48750 tonnes.
(*c*) Comment briefly on the reliability of your estimates.

6. A mail-order warehouse employs casual labour on a day-to-day basis to deal with orders received. Each morning the manager needs to be able to assess very quickly the number of orders there are in the incoming mail so that the right number of staff can be taken on to deal with them. To help with this it is decided to find out if the weight of mail gives an indication of the number of orders received. The following records of the weight in kilograms and the number of orders in the mail are produced over 20 days:

Weight	Number of orders	Weight	Number of orders
20	5400	24	5400
15	4200	16	4300
23	5800	28	6100
17	5000	15	3600
12	3500	30	6200
35	6400	18	5300
29	6000	27	5800
21	5200	30	3000
10	4000	20	5200
13	3800	24	5000

(*a*) Measure the degree of correlation between weight and number of orders.
(*b*) On the basis of this degree of correlation do you think that weight is a fair indication of the number of orders?
(*c*) If ten people must be engaged to deal with 1000 orders how many should be employed to handle: (i) 22 kilograms of mail; (ii) 18 kilograms of mail; (iii) 35 kilograms of mail?

7. Two TV critics were asked to rank in order of preference 10 television series. They did so as follows:

TV series	Critic I	Critic II
A	4	7
B	3	1
C	6	3
D	9	8
E	2	2
F	1	6
G	7	5
H	5	4
J	10	10
K	8	9

Are the views of these two critics consistent?

8. The following data show the latitude and the mean high and mean low temperatures of 20 cities in the Northern Hemisphere:

City	Latitude (to nearest degree)	Mean high temperature (°C)	Mean low temperature (°C)
London	52	14.4	6.7
Paris	49	15.0	6.1
Rome	42	21.7	10.6
Amsterdam	52	12.2	10.6
Lisbon	39	19.4	12.8
Berlin	53	12.8	4.4
Copenhagen	56	11.1	5.0
Oslo	60	10.0	2.2
Montreal	46	10.0	1.7
Prague	50	12.2	5.6
Calcutta	22	31.7	21.1
Dublin	53	13.3	5.6
Bucharest	44	16.7	5.6
Ottawa	45	10.6	0.0
Rangoon	17	31.7	22.8
Saigon	11	32.2	23.3
Helsinki	60	7.8	1.7
Hong Kong	22	25.0	20.0
Belgrade	45	16.7	7.2
Bogota	5	18.9	10.0

(*a*) Measure the degree of correlation between latitude and mean high temperature.

(*b*) Measure the degree of correlation between latitude and mean low temperature.

(*c*) Obtain rank correlation coefficients between latitude and mean low and mean high temperatures and compare them with the Pearsonian correlation coefficients.

9. The quality inspector in a factory suspects that the number, of defective items produced by a machine is linearly related to the speed at which the machine runs. The machine is run for half-hour intervals at various speeds and the number of defective items produced in each interval is measured. The results are as follows:

Speed (rpm) x	12.6	10.2	12.7	12.4	10.9	13.2	10.0	11.5	12.0	10.8
No of items defective y	81	70	86	90	68	84	62	76	70	10.8

Plot the data on graph paper and comment on the inspector's suspicion.
Estimate the line of best fit for predicting defectives from running speed and draw it in your graph.

(Answers at the end of the book.)

UNIT TEN

Index numbers

10.1 Introduction

The use of index numbers in statistical work has become extremely important. It is a convenient method of showing comparisons of a variable or variables over periods of time such as the quarterly sales of a product from 1990–93 (see Table 10.1). Instead of looking at the *actual* sales in hundreds of pounds, an index number measures the *relative* changes so that the sales in each quarter are expressed as a percentage of the sales in one particular quarter, termed the *base period*. In this way we can clearly see whether sales have increased or decreased. If sales have risen compared with the base period, the percentage will be greater than 100. If sales have fallen with time, this will be shown by a percentage value of less than 100.

The final column in Table 10.1 gives an index number of sales for each

Table 10.1 Sales of a product over four years

Year	Quarter	Sales (£ hundred)	Index number of sales (Qtr 1 1990 = 100)
1990	1	200	100
	2	212	106
	3	256	128
	4	232	1161
1991	1	208	104
	2	217	108.5
	3	263	131.5
	4	186	93
1992	1	192	96
	2	225	112.5
	3	268	134
	4	235	117.5
1993	1	220	110
	2	237	118.5
	3	280	140
	4	237	118.5

quarter. In that column the actual quarterly sales have been expressed as a percentage of the sales in the first quarter of 1990, that is, the first quarter of 1990 is the base period. It is usual to write this fact more concisely as Qtr 1 1990 = 100. We would obtain a different series of index numbers if we used another quarter as the base period, so it is absolutely essential to state what the base period is whenever an index number is quoted.

It is a convention *not* to put in a % sign after the value of an index number. The statement that a number is the value of an index number automatically implies that it is in fact a percentage. The term 'index number' is often shortened to 'index'. The plural noun is either 'index numbers' or 'indices'. A series of index numbers evaluated at regular intervals of time, as in the final column of Table 10.1, is termed an *index series*.

From the final column of Table 10.1 it is easy to see that in two quarters, the last quarter of 1991 and the first quarter of 1992, sales fell below the level in the base period by 7% and by 4% respectively. We also note that the highest sales were achieved in the third quarter of 1993 when they were 40% above the level in the base period.

The index number we have calculated in Table 10.1 concerns the sales of just one product over a period of time. If the particular company selling this product also sells other items, we might wish to calculate, for example, an 'all products' index of profits. This task will not usually be so straightforward since the profits on each different product are not necessarily identical. Clearly the volume of sales of the products and the profit per unit will affect the index. The volume of sales as well as the profit will change with time and the construction of a suitable index number requires deeper consideration and we shall concern ourselves with such problems in the following sections of this Unit.

The variables that are most commonly put into index form are prices, output, sales, wages, profits, value of the £, and share prices. The commonly quoted index series, for example, the General Index of Retail Prices, the Index of Output of the Production Industries, the *Financial Times* Industrial Ordinary Share Index, the Index of Average Earnings, have many different components affecting their values and it is essential that you should appreciate the principles underlying the construction of index numbers and the limitations in their use.

10.2 Simple index numbers

When studying the principles of index number construction, it is convenient to restrict ourselves to one variable of interest and we shall follow the usual practice of taking that variable as price. Our aim is thus to make comparisons of prices over a period of time. Suppose we existed on a bill of fare that consisted only of bread, cheese and beer. Let us imagine that the average prices of these three commodities in two consecutive years are as shown in Table 10.2.

We want to be able to combine the price per given unit of each commodity so that we have a single index number comparing prices in Year 1 with those in Year 0. One approach is to express the total of the prices per given unit in

Table 10.2 Prices of basic commodities

	Price in Year 0	Price in Year 1
Bread	60p per loaf	70p per loaf
Cheese	160p per 500g	190p per 500g
Beer	135p per $\frac{1}{2}$ litre	110p per $\frac{1}{2}$ litre

Year 1 as a percentage of the total of prices per given unit in Year 0. The result is termed a *simple aggregate* (or *aggregative*) *index*:

$$\text{Simple aggregate price index for Year 1 (Year 0 = 100)} = \frac{(70 + 190 + 110)}{(60 + 160 + 135)} \times 100$$

$$= \frac{370}{355} \times 100$$

$$= 104.2$$

On the basis of this index number we conclude that prices of these goods taken as a whole rose 4.2% in the year. (We shall work out the values of index numbers correct to one decimal place. This is precise enough for most comparisons, particularly in years when price changes are rapid.)

Note that, although there are no units to an index number since we are expressing one price as a percentage of another price, the actual prices to be summed should be in the same units, in this case pence, when calculating an aggregate index.

This simple approach to index number construction can be criticized. Firstly, the value of the index depends on the units for which the prices are quoted. Suppose that the price for beer had been given 'per litre' instead of 'per half litre', that is as 270p per litre and 220p per litre respectively in the two years. We would then have:

$$\text{Simple aggregate price index for Year 1 (Year 0 = 100)} = \frac{(70 + 190 + 220)}{(60 + 160 + 270)} \times 100$$

$$= \frac{480}{490} \times 100$$

$$= 98.0$$

This implies that prices have fallen by 2% rather than risen by 4.2% as previously calculated. We thus conclude that this simple approach needs to be modified so that the units of measurement do not affect the value of the index number.

If we use the *ratio* of prices per unit quoted for a given item rather than the actual prices, we can avoid this problem with the units. The price of a loaf of bread in Year 1 as a percentage of its price in Year 0 is $70/60 \times 100 = 116.7$.

Table 10.3 Price relatives of the three basic commodities

	Price in Year 0	Price in Year 1	Price relative in Year 1 (Year 0 = 100)
Bread	60p per loaf	70p per loaf	116.7
Cheese	160p per 500g	190p per 500g	118.8
Beer	135p per $\frac{1}{2}$ litre	110p per $\frac{1}{2}$ litre	81.5

This ratio, 116.7 (again written conventionally without the % sign), is termed the *price relative* for bread in Year 1 (Year 0 = 100). The price relatives for all three items in the basic bill of fare are shown in Table 10.3. Note that the price relative for beer is 81.5 whether the price of beer is quoted per litre or per half litre, since the ratios 110/135 and 220/270 have the same value. We can work out the average of these price relatives to give us an index number called the *arithmetic mean of relatives index.*

$$\text{Arithmetic mean of price relatives index for Year 1 (Year 0 = 100)} = \frac{116.7 + 118.8 + 81.5}{3}$$

$$= \frac{317}{3}$$

$$= 105.7$$

On this basis, therefore, prices appear to have risen by 5.7%.

It is possible to use a geometric mean as the average in place of the arithmetic mean and this is in fact done when the *Financial Times Industrial Ordinary Share Index* is calculated. Using a geometric mean reduces the effect on the index of a large increase in the price of one share. Unless you are specifically requested to use a geometric mean, always assume if an examiner asks for a mean of relatives index to be calculated that an arithmetic mean is required.

A price relative type index has a second advantage over an aggregate index in that the prices for different commodities do not have to be in the same units. This is a particularly useful property when prices from several countries each quoted in its own currency are being compared.

10.3 Weighted index numbers

The simple index numbers calculated in Unit 10.2 do not take into account the relative importance of the three commodities in the consumer's diet, as no allowance has been made for the quantities consumed. For example, a typical family in the UK would be more concerned about a rise of 2p in the price of a pint of milk than a 10p rise in the price of a jar of salad dressing, since far more milk is consumed than is salad dressing. We shall now consider how to *weight* the prices of items in our basic bill of fare so as to attach greater importance to the prices of those commodities which are of more consequence in our

Table 10.4 Amounts of basic commodities consumed

	Quantity consumed in Year 0	Quantity consumed in Year 1
Bread	10 loaves	12 loaves
Cheese	1 kg	1 kg
Beer	2 litres	$2\frac{1}{2}$ litres

diet. This is not quite as straightforward as might be supposed since consumption patterns, as well as prices, change with time and we have to decide which particular consumption pattern is to be used when constructing a price index. In Table 10.4 are shown the quantities of the three basic commodities consumed in typical weeks in Year 0 and in Year 1. Using the information given in the table both the aggregate and price relative types of index numbers can be put into weighted form.

(a) Weighted aggregate index numbers

Let us first take the consumption pattern of Year 0, the base period, as typical. To calculate a *base weighted aggregate price index*, often called a *Laspeyres price index*, we first calculate the total expenditure on our three items in a typical week in the base period, Year 0. Next we work out what the expenditure in the current period, Year 1, would be if we were consuming the same quantities as in the base period, Year 0. We finally express this expenditure as a percentage of the expenditure in the base period, Year 0. The calculation is conveniently set out in a table. (See Table 10.5.)

$$\text{Laspeyres price index for Year 1 (Year 0 = 100)} = \frac{1520}{1460} \times 100 = 104.1$$

Table 10.5 Calculation of Laspeyres price index

	Year 0 Price	Year 0 Quantity consumed	Year 1 Price	Expenditure in Year 0 (pence)	Expenditure in Year 1 (pence) assuming Year 0 consumption
Bread	60p per loaf	10 loaves	70p per loaf	600	700
Cheese	160p per 500g	1 kg	190p per 500g	320	380
Beer	135p per $\frac{1}{2}$ l	2 l	110p per $\frac{1}{2}$ l	540	440
TOTAL				1460	1520

You might be asking yourself why we could not simply take the ratio of our actual expenditure in Year 1 to the actual expenditure in Year 0. If we did do that, we would have calculated an index of *expenditure*. We are trying to construct a *price* index so we must not let the altered pattern of *consumption* distort the price changes.

We can, however, take the consumption pattern in Year 1 as typical and calculate a *current weighted aggregate price index*, often called a *Paasche price index*. Current weighting uses the consumption pattern in the period for which we are calculating the index number. To calculate a Paasche price index, we work out the actual expenditure in the current period, Year 1, and express this as a percentage of what the expenditure would have been in the base period, Year 0, if the consumption then had been the same as it is in the current period, Year 1. Table 10.6 shows the layout of the calculation.

Table 10.6 Calculation of Paasche price index

	Year 0 Price	Year 1 Price	Year 1 Quantity consumed	Expenditure in Year 0 (pence) assuming Year 1 consumption	Expenditure in Year 1 (pence)
Bread	60p per loaf	70p per loaf	12 loaves	720	840
Cheese	160p per 500g	190p per 500g	1 kg	320	380
Beer	135p per $\frac{1}{2}$ l	110p per $\frac{1}{2}$ l	$2\frac{1}{2}$ l	675	550
TOTAL				1715	1770

$$\text{Paasche price index for Year 1 (Year 0 = 100)} = \frac{1770}{1715} \times 100 = 103.2$$

In this particular example the values of the Laspeyres and Paasche indices differ very little but this is not always the case and we shall discuss the relative merits of the two types of weighting in (*c*) below.

(b) Weighted mean of relatives index numbers

A *weighted arithmetic mean of relatives index number* is calculated as follows:

$$\text{Weighted price relative index} = \frac{\Sigma(\text{Price relative of a commodity} \times \text{Weighting for that commodity})}{\Sigma\ \text{Weighting}}$$

where the sums are taken over all the commodities. The simple mean of relatives index calculated in Unit 10.2 is the special case when the weightings are all equal to one, that is, each item was taken to be of the same importance. This is not usually the case—those items on which we spend most are the most important ones in our budget. Occasionally in an examination, weightings are given in the question for you to use without needing to know how their values have been arrived at. Away from the examination desk, however, you need to be able to calculate weightings from prices and consumption patterns such as those given in Tables 10.2 and 10.4. The weightings are taken as proportional to the amounts we spend on the commodities. Again we have the choice of taking either the base period or the current period as the typical consumption pattern. Table 10.7 shows the calculations necessary if base weighting is used.

$$\text{Base weighted price relative index for Year 1 (Year 0 = 100)} = \frac{152046}{1460} = 104.1$$

Table 10.7 Calculation of base weighted price relative index

	Year 0 Price	Year 0 Quantity consumed	Year 1 Price	Price relative in Year 1 (Year 0 = 100)	Expenditure (pence) in Year 0 (weighting)	Price relative × weighting (pence)
Bread	60p per loaf	10 loaves	70p per loaf	116.7	600	70020
Cheese	160p per 500g	1 kg	190p per 500g	118.8	320	38016
Beer	135p per ½ l	2 l	110p per ½ l	81.5	540	44010
TOTAL					1460	152046

Note that this index has the same value as the Laspeyres base weighted aggregate index calculated in (*a*). The values of these two indices always agree.

Table 10.8 shows the calculation using current weighting.

Table 10.8 Calculation of current weighted price relative index

	Year 0 Price	Year 1 Price	Year 1 Quantity consumed	Price relative in Year 1 (Year 0 = 100)	Expenditure (pence) in Year 1 (weighting)	Price relative × weighting (pence)
Bread	60p per loaf	70p per loaf	12 loaves	116.7	840	98028
Cheese	160p per 500g	190p per 500g	1 kg	118.8	380	45144
Beer	135p per ½	110p per ½ l	2½ l	81.5	550	44825
TOTAL					1770	187997

$$\text{Current weighted price relative index for Year 1 (Year 0 = 100)} = \frac{187\,997}{1770} = 106.2$$

This result is not the same as the Paasche current weighted aggregate index and in general these two methods using current weighting do not give identical results.

(c) Comparison of weighting systems

There is endless discussion among economists as to the relative merits of base period weighting and current period weighting. It is during periods of rapidly rising prices that the differences between the methods become especially important.

With a Laspeyres or base weighted index, the weights become out-of-date as the pattern of demand changes. If an index series is being constructed, far less work is involved if base weighting is used, since only current prices need to be determined. The consumption in the base period is recorded once and for all and no further consumption observations are required. This is satisfactory in the short term, but over a period of several years consumption does change and allowance for this needs to be made. When we look at a series of base weighted index numbers, the values in every period are directly comparable since the consumption pattern used in their calculation is the same.

With a Paasche or current weighted index, the consumption pattern is up-to-date but extra work is involved when calculating such an index since the current consumption as well as the current prices need to be determined. Also, in a current weighted index series each index is not strictly comparable with the rest since different consumption patterns have been used in each calculation.

There is of course a link between prices charged and the quantities consumed. Consumption tends to go down when price increases. Thus a base weighted index tends to overstate the increase in prices, since no account is taken of the fall in consumption. On the other hand a current weighted index understates the rise in prices since people would have continued at a higher rate of consumption but for the increase in prices. It is interesting to note that the UK Government in 1975 decided to use as weightings in the Retail Prices Index the expenditure pattern in the most recent year for which figures were available, rather than an average of the three most recent years as had been the custom previously. This had the effect of tending to understate the rise in prices and thus reduced inflation at a stroke.

As neither current nor base period weighting is entirely satisfactory, other methods of index number construction have been devised, some of which have desirable theoretical properties. These methods are beyond the scope of this book.

10.4 Quantity or volume index numbers

Although the index numbers we have calculated so far have referred to prices, index number construction, as we noted in Unit 10.1, is not confined to the comparison of prices. Often *volume or quantity* indices are of interest as, for example, in the Index of Output of the Production Industries which seeks to measure the changes in volume of output in a whole range of industries over a period of time.

Quantity index numbers can be calculated in a similar way to price indices. The price and consumption data for our bread, cheese and beer are given again in Table 10.9.

Our aim now is to compare the quantities consumed in the two years. The *quantity relative* of a commodity in Year 1 (Year 0 = 100) is the quantity consumed in the current period, Year 1, as a percentage of the quantity consumed in the base period, Year 0. Thus:

$$\text{Quality relative for bread in Year 1 (Year 0 = 100)} = \frac{12}{10} \times 100 = 120$$

$$\text{Quality relative for cheese in Year 1 (Year 0 = 100)} = \frac{1}{1} \times 100 = 100$$

$$\text{Quality relative for beer in Year 1 (Year 0 = 100)} = \frac{2.5}{2} \times 100 = 125$$

These values show that the consumption of beer has risen by the greatest proportion and that the consumption of cheese has remained constant. A simple mean of quantity relatives index for Year 1 (Year 0=100) is $\frac{120+100+125}{3}$, that is, 115. Using this method we see that consumption has increased by 15%.

The methods used in Unit 10.3 for constructing weighted price index numbers can be transposed to compare quantities. A *Laspeyres* or *base weighted aggregate quantity index* expresses the expenditure at current quantities but at base period prices as a percentage of the actual expenditure in the base year. The *Paasche* or *current weighted aggregate quantity index* expresses the actual

Table 10.9 Prices and consumption of basic commodities

	Year 0 Price	Year 0 Quantity consumed	Year 1 Price	Year 1 Quantity consumed
Bread	60p per loaf	10 loaves	70p per loaf	12 loaves
Cheese	160p per 500g	1kg	190p per 500g	1kg
Beer	135p per $\frac{1}{2}$ l	2 l	110p per $\frac{1}{2}$ l	$2\frac{1}{2}$ l

Table 10.10 Calculation of Paasche quantity index

	Year 0 Quantity consumed	Year 1 Quantity consumed	Year 1 Price	Expenditure (pence) in Year 0 assuming Year 1 prices	Expenditure (pence) in Year 1
Bread	10 loaves	12 loaves	70 p per loaf	700	840
Cheese	1 kg	1 kg	190 p per 500 g	380	380
Beer	2 l	$2\frac{1}{2}$ l	110 p per $\frac{1}{2}$ l	440	550
				1 520	1 770

expenditure in the current period as a percentage of what the expenditure in the base period would have been at current prices. The calculation of the Paasche quantity index for our basic commodities is laid out in Table 10.10

$$\text{Paasche quantity index for Year 1 (Year 0 = 100)} = \frac{1770}{1520} \times 100 = 116.4$$

A ***weighted quantity relative index*** is calculated using:

$$\text{Weighted quantity relative index} = \frac{\Sigma(\text{Quantity relative of a commodity} \times \text{Weighting for that commodity})}{\Sigma\,\text{Weighting}}$$

where the sums are taken over all the commodities. Again, the expenditure on each commodity is the appropriate weighting to use and either base period or current period expenditure can be employed. The calculation of the base year weighted quantity relative index for the basic commodities is shown in Table 10.11

$$\text{Base weighted quantity relative index for Year 1 (Year 0 = 100)} = \frac{171\,500}{1460} = 117.5$$

10.5 Formulae for index numbers

It is convenient at this stage to summarize the various methods of index number construction using formulae. Some of these formulae will usually appear on the formula lists provided in examinations.

We shall use p_0 to stand for the price of one unit of a commodity in the base period and p_1 for its price in the current period. (Some formulae booklets use p_n

Table 10.11 Calculation of base weighted quantity relative index

	Year 0 Quantity consumed	Year 0 Price	Year 1 Quantity consumed	Quantity relative in Year 1 (Year 0 = 100)	Expenditure (pence) in Year 0 (weighting)	Quantity relative × weighting (pence)
Bread	10 loaves	60p per loaf	12 loaves	120	600	72000
Cheese	1kg	160p per 500g	1kg	100	320	32000
Beer	2l	135p per $\frac{1}{2}$l	$2\frac{1}{2}$l	125	540	67500
TOTAL					1460	171500

in place of p_1.) Also q_0 denotes the number of these units consumed in the base period and q_1 the corresponding consumption in the current period. For any one commodity at any one time:

$$\text{Expenditure} = \text{Price per unit} \times \text{Quantity consumed}$$
$$\text{Expenditure in the base period} = p_0 q_0$$
$$\text{Expenditure in the current period} = p_1 q_1$$

Note that the above terminology looks at the situation from a consumer's point of view. From a producer's viewpoint, p is the price per unit of output and q is the number of units produced, that is, the *volume* of production. Thus for any one product at any one time we have:

$$\text{Value of output} = \text{Price per unit} \times \text{Volume produced}$$
$$\text{Value in the base period} = p_0 q_0$$
$$\text{Value in the current period} = p_1 q_1$$

Examiners do not necessarily provide information about the unit prices and the quantities consumed (or produced). Sometimes, for example, they give details of unit prices and the total expenditure on each item. Using the relationship shown above between unit price, quantity consumed and expenditure, however, you can work out the quantities consumed if this is necessary.

The formulae for the index numbers calculated in the previous sections are:

$$\text{Simple aggregate price index} = \frac{\Sigma p_1}{\Sigma p_0} \times 100$$

$$\text{Price relative of a commodity in the current period relative to the base period} = \frac{p_1}{p_0} \times 100$$

$$\text{Simple mean of relatives index} = \frac{\sum\left(\frac{p_1}{p_0} \times 100\right)}{n}$$

where n is the number of commodities and Σ indicates summation over all the commodities.

$$\text{Laspeyres base weighted aggregate price index} = \frac{\Sigma p_1 q_0}{\Sigma p_0 q_0} \times 100$$

$$\text{Paasche current weighted aggregate price index} = \frac{\Sigma p_1 q_1}{\Sigma p_0 q_1} \times 100$$

$$\text{Weighted price relative index} = \frac{\Sigma \left(\frac{p_1}{p_0} \times 100 \right) \times \textit{Weighting}}{\Sigma \textit{ Weighting}}$$

$$\text{Base weighted price relative index} = \frac{\Sigma \left(\frac{p_1}{p_0} \times 100 \right) (p_0 q_0)}{\Sigma p_0 q_0}$$

$$\text{Current weighted price relative index} = \frac{\Sigma \left(\frac{p_1}{p_0} \times 100 \right) (p_1 q_1)}{\Sigma p_1 q_1}$$

Quantity index numbers are obtained from these formulae by interchanging p_0 with q_0 and p_1 with q_1.

10.6 The practical problems of index number construction

In the previous sections we have illustrated the methods of index number construction using as data a simple diet consisting of only three items. Of course in practice, if you are given the task of designing an index or if you are interpreting the value of an official index number, the situation will usually be much more complicated than this.

Before embarking on the construction of an index, you must specify precisely what you are attempting to measure. Terminology such as the 'cost of living in the UK is widely used in the media but it has no one precise value in the way that the cost of a television licence in the UK has. As we have seen from Units 10.2 and 10.3 an index number is nothing more than a specialized type of average and its value in any particular circumstance depends on how it is constructed. We shall assume in this section that we are attempting to construct an index of prices.

A decision has to be made on which prices to include in the index and on how many items are to be priced. In constructing an index of retail prices it would be impossible to collect the retail price of every different commodity at every possible sales point at exactly the same time. We have to decide which items we are going to take as 'typical' or 'representative' of a whole group of items and we must also decide when and where we are going to collect the prices of these items. If we want to calculate a weighted index number, as is usually the case, in addition to information on prices we need details of consumption or expenditure and we must decide how to obtain this information.

The actual type of index to be calculated must be specified. Is it to be an aggregate or price relative index and is base period or current period weighting appropriate? The choice of a suitable base period has to be made. A base period should be one in which conditions are as 'normal' as possible since all future comparisons are going to be made relative to that period. A time of instability in the general economic situation would not be suitable. The requisite information about prices and consumption in the base period must be available. With most official UK index numbers such as the General Index of Retail Prices and the Index for Output of the Production Industries, the base period is brought up-to-date at fairly frequent intervals (see Unit 10.7). The *Financial Times* Stock Exchange (FT-SE) 100 share index, or the 'Footsie Index' as it is known (see Unit 10.10(e)), is based on 31 December 1983, with a base of 1000. This leads to values that have in recent times varied between 2800 and 3500. In the US it is a legal requirement that some index numbers do not change their base—this has resulted in extremely high values of the index numbers and defeats the prime object of an index number which is to enable short-term comparisons of relative changes to be made easily.

An alternative to changing the base of an index is to use a *chain base index* in which the index for the current period is based on the immediately preceding period (see Unit 10.8). This type of index is suitable only for short-term period-by-period comparisons.

10.7 Changing the base of an index

Suppose we have the following index series based on Year 2:

Year	0	1	2	3	4	5	6	7	8	9	10
Index	89	96	100	105	111	114	118	121	125	130	132

It has been decided to change the base year to Year 8. Each index in the series must be recalculated by expressing it as a percentage of the value in Year 8. The arithmetic is:

$$\text{Year 0:} \quad \frac{89}{125} \times 100 = 71.2$$

$$\text{Year 1:} \quad \frac{96}{125} \times 100 = 76.8$$

$$\text{Year 2:} \quad \frac{100}{125} \times 100 = 80.0$$

$$\text{Year 3:} \quad \frac{105}{125} \times 100 = 84.0$$

and so on to:

$$\text{Year 10:} \quad \frac{132}{125} \times 100 = 105.6$$

Thus the new series is:

Year	0	1	2	3	4	5	6	7	8	9	10
Index	71.2	76.8	80.0	84.0	88.8	91.2	94.4	96.8	100	104.0	105.6

You should check the intermediate values for yourself.

10.8 A chain base index

In a chain base index each index number is given as a percentage of that for the previous period. Consider the following fixed base index series:

Year	0	1	2	3	4
Index	100	105	111	114	118

The first term in the chain base index series is the value for Year 1 (based on Year 0) and as the fixed base series is based on Year 0, no calculation is needed and the chain base index for Year 1 is 105. The chain base index for Year 2 based on Year 1 is:

$$\frac{111}{105} \times 100 = 105.7$$

The chain base index for Year 3 on Year 2 is:

$$\frac{114}{111} \times 100 = 102.7$$

The chain base index for Year 4 on Year 3 is:

$$\frac{118}{114} \times 100 = 103.5$$

The chain base index series, therefore, is:

Year	1	2	3	4
Index	105	105.7	102.7	103.5

Note that we cannot give the value of the index for Year 0 since we do not know the value of the fixed base index in the year preceding Year 0.

It is extremely important to state clearly when you are using a chain base method otherwise mistakes in interpretation can occur.

10.9 The Retail Prices Index

(a) History of the index

The UK Retail Prices Index (RPI) is one of the most frequently quoted and widely studied of all index numbers. It is generally regarded as the barometer of the cost of living since from its value the current rate of inflation can be calculated. It is now common practice to link increases in savings, pensions and insurance premiums, for example, with the value of the RPI, and most wage claims refer to it.

The RPI is compiled by the Department of Employment and is published, in varying detail, in several government journals, for example, *Employment Gazette, Monthly Digest of Statistics, Annual Abstract of Statistics* and *Economic Trends*. Its stated purpose is to measure the monthly degree of change in the relative prices of goods and services over the whole field over which households distribute their expenditure.

The RPI cannot be said to be a 'cost of living index' since its coverage is not retricted to the cost of the essentials of life; some types of expenditure, most notably income tax and National Insurance contributions, are excluded. Its origins, however, lie in an interim index introduced in 1948 to replace a cost of living index. The interim index measured the average monthly changes in the prices of goods and services bought by *working class* households. Estimates of consumption were used in its calculation until the results of a Household Expenditure Survey of 20000 households of *all types* conducted in 1953 were available (13000 households co-operated).

In February 1956 the Cost of Living Advisory Committee (now called the Retail Prices Index Advisory Committee) submitted a report to the then Ministry of Labour suggesting the setting-up of the RPI based on January 1956 = 100. The results of the 1953 survey were used to construct a weighting system. The expenditures of the highest income groups and very low income groups, mainly pensioner households, were excluded when the weightings were calculated. Certain types of expenditure were excluded from the coverage of the index: income tax payments, National Insurance contributions, life insurance premiums, payments to pension funds, premiums for household insurance (other than insurance of the building), subscriptions to trades unions, friendly societies, hospital funds, church collections, etc., cash gifts, pools and other betting stakes, private medical fees and capital sums or mortgage payments for house purchase or major structural alterations.

The RPI was made up of ten groups: food, alcoholic drink, tobacco, housing, fuel and light, durable household goods, clothing and footwear, transport and vehicles, miscellaneous goods, and services. Each group was divided into sub-groups—for example, clothing and footwear was subdivided into: men's outer clothing, men's underclothing, women's outer clothing, women's underclothing, children's clothing, other clothing (including hose, haberdashery, hats and materials), footwear.

The RPI was revised in 1962 with a new base 1962 = 100 which was intended to be used for ten years. The spending pattern of households, which formed the basis of the weighting system was to be kept up-to-date by the Family Expenditure Survey of 5000 households annually (see (*c*) below) since spending patterns were changing rapidly at this time. The RPI was reweighted every February using data from the three years ending the previous June, adjusted to correspond with the level of prices ruling in the current year.

The RPI now consists of fourteen groups, as shown in Table 10.13.

The base year was changed to January 1974 = 100 and again to January 1987 = 100, and from February 1975 onwards the weights have been based on information collected from the Family Expenditure Survey for the *one* year ending the previous June. This was made possible by extending the Family Expenditure Survey to a greater number of households. Another alteration is that expenditure of owner-occupiers on mortgage interest payments (but not repayment of mortgage capital) is now included.

The booklet *A Short Guide to the Retail Prices Index* (HMSO) is informative. Recent changes are referred to in the monthly *Employment Gazette*.

(b) Collection of prices

The prices used are those in force on the Tuesday nearest the 15th of every month. For expenditure coming within the scope of the RPI, a representative list of items has been selected. For example, under the category of bread, a standard large white loaf, a standard small white loaf, a standard small brown loaf and a wrapped large white sliced loaf are the only items priced. Prices are obtained mainly by officers of the Department of Employment from retail shops typical of those from which the majority of households commonly make their purchases in some 200 areas covering urban centres and rural districts, throughout the UK. Some prices such as local authority rents, nationalized industry prices, postage and telephone costs can be obtained direct from the organizations concerned. As far as is possible the prices relate to goods and services of unchanged quality at successive dates. When the quality changes, the RPI is adjusted accordingly. Another problem is the emergence of new goods, for example, fish fingers, which have become so commonly used that they have been added to the list of representative items to be priced or have actually replaced the original representatives. From the prices collected at each outlet, a price relative for each item is calculated relative to the base month, at present January 1987, and all the price relatives for that item are averaged, without weighting, to find the UK price relative for that item.

(c) The Family Expenditure Survey

The RPI is a weighted price relative type index so, as well as information on prices, details of the relative importance of the various categories of household expenditure have to be found. The weighting system for the RPI is based on the results of the *Family Expenditure Survey* which is a continuing enquiry conducted by the Department of Employment covering all types of private households in the UK. A sample of some 11 000 addresses is selected annually from the Postcode Address File. An effective sample of approximately 10 000 households is obtained since those selected are not always contactable and others may not co-operate. About 70% of the sample households agree to co-operate. The households are visited in rotation throughout the year and they are asked in confidence to keep complete records of all their income and expenditure in a two week period and to give information covering a longer period for such payments as fuel bills, season tickets etc. Definitions of such terms as 'household', 'income' and 'expenditure' have to be precisely stated and you will find that these are given in every copy of the *Family Expenditure Survey.*

(d) Calculation of the RPI

Using the results of the Family Expenditure Survey the weighting system for the RPI is calculated. The total of weightings is chosen arbitrarily to be 1000. The highest income groups, approximately the top 4%, are excluded from consideration when the weightings are worked out as it is felt that their spending patterns are not typical of the vast majority of households. For each of the remaining households the proportion of expenditure coming under each subsection heading in the RPI is calculated; for example, all expenditure on any type of bread is allocated to the bread sub-group although information regarding prices (see (*b*) above) is collected for four representative types of bread. The proportions are then averaged for all the households and expressed as parts per thousand. These parts per thousand are the weights. The weightings for the subgroups in the clothing and footwear group are shown in column (2) of Table 10.12. From these weightings we can conclude that, on average, 58/1000, that is, 5.8%, of total household expenditure is on clothing and footwear and that 19/58, that is 32.8%, of clothing and footwear expenditure is on women's outerwear.

Using the UK price relatives for the representative items within each section, an index for each sub-group (see Table 10.12, column (1)) is calculated. We are thus able to say that the price of children's outerwear rose by 17.1% in the period from January 1987 to June 1993. The weighted average of the subgroup indices is the index for the clothing and footwear group as a whole.

The complete list of weightings in 1993 for the fourteen main categories of the RPI is given in column (2) of Table 10. 13. The 'all items' Retail Prices Index is the weighted average of the fourteen individual group indices. It is this value that is the most commonly quoted form of the RPI. The details of the calculation for June 1993 are shown in column (3) of Table 10.13. There are methods of shortening the arithmetic but we shall not use them here. We have:

$$\text{All items Retail Prices Index for June 1993 (Jan. 1987 = 100)} = \frac{\text{Group index} \times \text{Weighting}}{\Sigma\ \text{Weighting}}$$

$$= \frac{141\,940.4}{1000}$$

$$= 141.9$$

Our calculated value does not agree exactly with the official value of 141, because in our calculation the rounded values of the group indices in the *Monthly Digest of Statistics* have been used.

The statisticians at the Department of Employment have the unrounded values available to them and can thus reach a more accurate result.

Table 10.12 Weightings and indices in the clothing and footwear group of the RPI, 1993

Subgroup	(1) Index for June 1993 (Jan. 1987 = 100)	(2) Weighting
Men's outerwear	120.4	12
Women's outerwear	108.8	19
Children's outerwear	117.1	6
Other clothing	137.9	10
Footwear	126.1	11
		58

Table 10.13 Weightings and indices for the fourteen main categories of the RPI, 1988

Group	(1) Index for June 1993 (Jan. 1987 = 100)	(2) Weighting	(3) Index × weighting
Food	131.4	144	18921.6
Catering	155.8	45	7011.0
Alcoholic drink	155.1	78	12097.8
Tobacco	156.7	35	5484.5
Housing	150.4	164	24665.6
Fuel and light	125.7	46	5782.2
Household goods	128.1	79	10119.9
Household services	140.7	47	6612.9
Clothing and footwear	120.2	58	6971.6
Personal goods and services	147.3	39	5744.7
Motoring expenditure	146.9	136	19978.4
Fares and other travel costs	152.6	21	3204.6
Leisure goods	122.8	46	5648.8
Leisure services	156.4	62	9696.8
All items	141.0	1000	141940.4

(e) Criticisms of the RPI

No index number is a perfect measure. When the Retail Prices Index commands so much attention and affects the political and economic situation, as well as reflecting it, there is bound to be much criticism of the way in which it is constructed. Often however it is the interpretation of the Index rather than its construction that is at fault.

The *year on year rate of inflation* is the percentage increase in the RPI between the same months in consecutive years. In January 1991 the all items Index was 130.2 (Jan. 87 = 100) and in January 1992 it was 135.6 (Jan. 87 = 100) so the annual rate of inflation at January 1992 was

$$\left(\frac{135.6}{130.2} \times 100\right) - 100 = 4.1\%$$

When interpreting percentages however, one should always ask 'Percentage of what, when?' Here we are saying that prices as measured by the Index of Retail Prices in January 1992 had risen by 4.1% from those prevailing in January 1991. If we look at the value of the Index in January 1993, 137.9, we find the annual rate of inflation is

$$\left(\frac{137.9}{135.6} \times 100\right) - 100 = 1.7\%$$

In other words, prices from January 1992 to January 1993 rose by 1.7%. Note that this is a percentage of *1992* prices which were higher than 1991 prices. Even if inflation stays at the same low rate of 1.7% the *actual* increase in prices from 1993 to 1994 will be higher than from 1992 to 1993 since 1993 prices were higher than 1992 prices.

The choice of representative items whose prices form the basis of the RPI can be criticized and, as we noted in (*c*) above, the choice is modified as new products emerge. The fact that only a limited number of items is priced allows the valuc of the RPI to be manipulated by the Government. For example, under the UK Government's price control policy in 1975, the prices of certain food items were controlled. The types of bread whose prices were controlled were the representative items for the bread category in the RPI. The prices of other types of bread were not controlled but any price rises in this 'non-standard' bread were not recorded by the RPI.

A second major criticism is that the RPI is an average for the whole of the UK and over all types of income groups. Regional values for the RPI are not published although it is acknowledged that regional variations exist. As lower income groups have less choice on how to spend their incomes since they must buy the basic commodities, it is argued that the RPI understates the rise in prices for these groups. A separate pensioner index is calculated and this has greater weightings for the food and fuel and light categories so price increases in these groups are given more importance. However, no separate index is calculated for lower income households with children. A much larger Family Expenditure Survey would be needed to measure regional and income group patterns of expenditure more precisely.

The expenditure that is considered when calculating the weighting system of the RPI excludes many of the major outgoings of households. Life insurance contributions and repayments of mortgage capital are considered as investment expenditure. Income tax and National Insurance contributions cannot be considered in this light and in August 1979 a new Tax and Price Index (TPI) was introduced by the UK Government 'to provide a truer guide than the RPI to the changes in costs facing taxpayers'. Full details of the methodology of the TPI can be found in *Economic Trends*, August 1979. The TPI combines in a single index a measure of changes both in direct taxes and in retail prices. It does not apply to those on very low levels of income who do not pay these taxes. Its value initially gave a lower measured rate of annual inflation, then higher, but from January 1992 to December 1993 the rate of inflation as measured by the TPI was less than that measured by the RPI.

10.10 Economic indicators other than the RPI

This section contains brief descriptions of some of the index numbers which are regarded as important pointers to the economic well-being or otherwise of the UK economy. The list is by no means exhaustive and you should seek out current information on index numbers which are particularly relevant to your own studies using the *Guide to Official Statistics* and the *Annual Supplement to the Monthly Digest of Statistics*.

(a) Index numbers of producer prices

These indices are published monthly in the *Monthly Digest of Statistics*. They provide useful guides for industry and business to future costs and to increases in costs as compared with the base year. This base year is at present 1990. To establish weightings for the indices, prices are collected for materials and products from firms and trade associations.

Tables are published under various heading, such as

(i) Materials and fuel purchased by broad sectors of industry;
(ii) Materials and fuel purchased;
(iii) Outputs: home sale.

(b) Index of output of the production industries

This index is compiled by the Central Statistical Office. It replaced the Index of Industrial Production in September 1983. It is the main indicator of industrial activity in the UK. The base is now 1990 = 100.

The monthly index is intended to provide a general measure of monthly changes in the volume of output of the production industries. Agriculture, construction, distribution, transport, communications, finance and all other public and private services are excluded.

The index is a weighted arithmetic mean of separate indicators, each of which describes the activity of a small sector of industry. To combine the individual production series, each industry has been given a weight. These weights are shown in Table 10.14.

The index is published in *Employment Gazette, Annual Abstract of Statistics, Monthly Digest of Statistics and Economic Trends*.

Table 10.14 Industry weightings for the Index of Output of the Production industries, January 1994

Industry	Weighting
Oil and gas	59
Coal	16
Other mining and quarrying	2
Food, drink and tobacco	110
Textiles and textile products	45
Leather and leather products	2
Coke, ref. petrol and nuclear fuels	23
Chemicals and man-made fibres	86
Basic metals and metal products	74
Machinery and equipment	97
Electrical and optical equipment	107
Transport equipment	98
Wood and wood products	26
Pulp, paper, printing and publishing	90
Rubber and plastic products	37
Other non-metallic mineral products	39
Other manufacturing NES	10
Electricity, gas and water	79
TOTAL	1000

Source: *Monthly Digest of Statistics*

(c) Index of average earnings of all employees (monthly inquiry)

This index is published in *Employment Gazette and Monthly Digest of Statistics*. The tables give comparisons of monthly earnings for twenty-six industry groups, all manufacturing industries, production industries and the whole economy. The last three indices are given before and after seasonal adjustment. The present base is January 1985 = 100.

Prior to 1984, Indices of Wage Rates of Manual Workers and Normal Hours Worked were also published but this information is now tabulated as actual average weekly earnings and hours worked (in manufacturing and certain other industries), that is, it is not in the form of an index number. One reason for the discontinuation of the Index of Wage Rates was that it took no account of changes in hours worked nor of overtime payments.

(d) Import and export volume index numbers

These indices are compiled by the Department of Trade and Industry and together with Indices of Import and Export Unit Values are published monthly in the *Monthly Digest of Statistics*. The ratio of the Export Unit Value Index to the Import Unit Value Index indicates to the Government the *terms of trade*. Students of economics will be familiar with this concept. When this ratio is greater than 100 the terms of trade are favourable; when it is less than 100, they are adverse. The index series are based on 1985 = 100.

(e) The *Financial Times*–Stock Exchange (FT-SE) 100 share index

This index, popularly known as the 'Footsie Index', is based on the market value of the top 100 UK-registered companies. It is updated minute by minute by the International Stock Exchange's computers.

The FT-SE 100 Share Index was introduced by the Stock Exchange at the start of January 1984 as being more representative than the Financial Times Ordinary Share Index, which is based on only 30 shares. The Financial Times All Share Index, based on more than 700 shares, is more representative still, but is too large to update as frequently as the FT-SE Share Index.

The index is a weighted arithmetic index, the weights being the issued share capitals of the companies included in the index. The basic intention is that the index should consist of the largest 100 UK-registered companies, but there are some exclusions rules which prevent some of these being included. Membership of the index is reviewed quarterly.

The FT-SE Share Index is an important economic indicator. Its value is watched constantly and quoted in most news bulletins. A constant rise in its value is seen as being a sign of economic well-being, and a fall in its value has been known to throw the City of London and even the government into a panic.

10.11 Exercises

1. From the following details of the prices of an average family budget, calculate a price relative index for Year 1 (Year 0 = 100) and comment on the impact of the index on the average family's cost of living over this period.

Commodity	Weighting	Price per unit (£) Year 0	Price per unit (£) Year 1
A	2	0.20	0.16
B	1	0.75	1.50
C	2	0.80	1.20
D	3	1.00	2.50

2. What is the purpose of 'weighting' in the construction of index numbers? Describe the basis and use of weightings in the UK Retail Prices Index. How might the RPI be used by (*a*) trade unions and (*b*) businesses? How much reliance do you think can be placed on this index?

3. A wine and spirits merchant has the following sales records:

Item	1991 Price per bottle (£)	1991 Quantity sold	1992 Price per bottle (£)	1992 Quality sold	1993 Price per bottle (£)	1993 Quantity sold
Whisky	9.00	10000	10.00	11000	11.00	12000
Brandy	12.00	8000	13.00	8000	14.00	8000
Wine	3.00	12000	3.50	22000	4.00	25000
Beer	0.90	50000	1.00	45000	1.20	42000

Calculate, based on the year 1991: (*a*) Laspeyres price index numbers for 1992 and 1993, (*b*) Paasche price index number for 1992 and 1993. Say why the Laspeyres price index is the more commonly used index.

4. Suppose you were asked by the management of a firm for which you worked to construct a cost of living index for all employees within the company, a large manufacturing firm. How would you set about undertaking this exercise? What problems would you expect to meet?

5. Construct (*a*) Laspeyres and Paasche price indices, (*b*) base weighted and current weighted price relative indices, for 1994 based on 1990.

	1990 Price (£) per lb	1990 Quantity (lb)	1994 Price (£) per lb	1994 Quantity per (lb)
Butter	1.20	25	1.80	15
Beef	2.40	30	3.70	24
Salt	0.27	6	0.65	6

6. Outline the main differences between the Laspeyres and Paasche price indices and discuss the circumstances when one should be used rather than the other.

7. Explain the meaning of the following types of index numbers: (*a*) base weighted; (*b*) current weighted; (*c*) chain base; (*d*) fixed base.

8. Outline the methods used in the construction and compilation of the Family Expenditure Survey. Why is this survey important to the UK economy?

9. A company employs four grades of workers in its production plant—skilled, semi-skilled, unskilled and apprentices. The following table shows the number of workers employed and the rates paid over a three-year period:

Grade	1991 Rate per hour (£)	1991 Number of workers	1992 Rate per hour (£)	1992 Number of workers	1993 Rate per hour (£)	1993 Number of workers
Skilled	7.00	32	7.50	33	8.00	41
Semi-skilled	5.50	20	5.80	18	6.20	13
Unskilled	4.00	10	4.30	9	4.70	8
Apprentices	3.00	5	3.20	7	3.50	7

Using 1991 as the base year calculate index numbers for the average wage for 1992 and 1993 using: (*a*) the Laspeyres formula and (*b*) the Paasche formula.

10. The following statistics relate to the production of a standard component by a company.

Year	Saleable output (thousands)	Mean number of assembly operators	Mean daily output per operator	Hourly earnings per operator (pence)
1988	362	62	26	55
1989	358	60	26	58
1990	366	60	27	62
1991	365	56	28	67
1992	370	55	29	74
1993	367	52	31	90

Convert all information given to a comparable basis using index numbers. Construct a ratio scale graph and plot your results. Report trends shown in your graph in a brief memorandum to the management of the company.

(Answers at the end of the book.)

UNIT ELEVEN

The analysis of time series

11.1 Introduction

A time series is the recording of a series of measurements of a variable over a period of time. A line graph of a time series is called a *historigram*, not to be confused with a histogram which is used to depict a frequency distribution. The time element is shown on the horizontal axis of the historigram; the measurements of the variable being recorded are shown on the vertical axis. Fig. 11.1 is the graph of the time series of the sales of a product over four years given in Table 11.1.

Table 11.1 Sales of a product over four years

Year	Quarter	Sales (£ hundred)
1990	1	200
	2	212
	3	256
	4	232
1991	1	208
	2	217
	3	263
	4	186
1992	1	192
	2	225
	3	268
	4	235
1993	1	220
	2	237
	3	280
	4	237

We can see from the graph that there are ups and downs in the sales figures, but that there is an overall tendency for the sales to move slightly upwards. There are other variations in the values, however, and when we analyse time

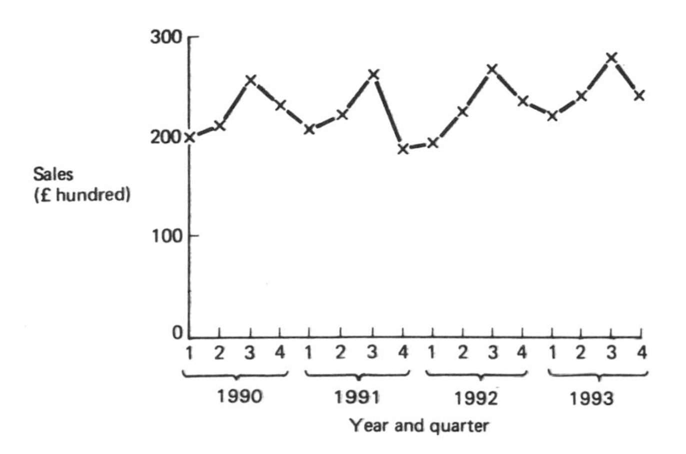

Fig. 11.1 Historigram showing sales of a product over four years

series we are interested in the components or movements in the time series which give rise to these variations. If we are able to analyse a time series into its components, we are then in a better position to predict the likely future movements of the series. The components of a time series are:

(i) The long-term or the secular trend This is the basic long-term movement of the time series. We can see from the graph of sales (Fig. 11.1) that there is a slight upward trend in the series since sales in later years, taken overall, are higher than in the early years. This could indicate a time of business expansion.

(ii) Seasonal variations These are short-term variations from the trend that occur regularly with the passage of time. Sales of many products—for example, warm coats, ice cream, are subject to such variations. In Fig. 11.1 it appears that the peak of sales occurs in quarter 3 of every year with a low in quarter 1. Seasonal variations do not necessarily coincide with seasons of the year. If the daily takings at a supermarket are analysed over many weeks, it might be found that peak sales for the week occur every Friday, with a low on Mondays. Such short-term regular variations are still termed seasonal, although they correspond to days of the week rather than to quarters or months of the year.

(iii) Cyclical variations These long-term variations about the trend are caused by general cyclical movements in trade and business: booms and slumps in the economy affecting business activity. These booms and slumps have been found to occur at intervals of between seven and twelve years and have been explained by economists over the years as being caused by a wide variety of factors. In order to detect cyclical variation in a time series, readings over many years are needed and in our example (see Table 11.1) we have insufficient data available to estimate the cyclical effect. Because of the large amount of data needed, you are unlikely to be asked to calculate cyclical variations under examination conditions.

(iv) Catastrophic variations These give an abnormally high or low value of the time series and arise from an unusual event such as a fire at the factory or a strike at the firm's suppliers. Such variations are readily discernible from the historigram. In Fig. 11.1, although the reading in 1991 quarter 4 is somewhat low, it does not appear to be sufficiently so as to be classified as catastrophic. If you find a catastrophic movement in a time series, you should remove it before proceeding with the analysis. If there is a seasonal pattern present, then the mean of the observations in the two adjacent corresponding seasons can be used in place of the abnormal reading. If there is no seasonal variation, the mean of the two adjacent observations can be used as a replacement.

(v) Residual variations These are variations from any other cause. We are unable to analyse them or predict their future effect but we expect them on average over a long period of time to have no overall effect on the time series, being as likely to lead to an increase in sales as to a decrease in sales.

11.2 Estimation of the trend

(a) Three-point method (method of semi-averages)

For many purposes the trend of a time series may be adequately estimated by drawing a freehand curve or line through the body of points on the historigram so that the fluctuations of the time series are smoothed out. This is a subjective method, however, and different results could be obtained from the same data by different analysts.

The three-point method (see Unit 9.3(*a*)) is an objective method of obtaining a straight line trend. We split the data into two equal time periods, ignoring the middle reading if there is an odd number of observations. We find the mean value of the time series in each of these two periods and plot the means at the midpoints of the time periods. We also plot the overall mean of the time series at the midpoint of the time period spanned by the series. The calculation for our sales data is set out in Table 11.2.

$$\text{Mean quarterly sales in first half} = \frac{1774}{8} = 221.75$$

$$\text{Mean quarterly sales in second half} = \frac{1894}{8} = 236.75$$

$$\text{Overall mean quarterly sales} = \frac{3668}{16} = 229.25$$

Table 11.2 The trend by the three-point method

Year	Quarter	Sales (£ hundred)	
1990	1	200	
	2	212	
	3	256	
	4	232	
1991	1	208	
	2	217	
	3	263	
	4	186	Total = 1774
1992	1	192	
	2	225	
	3	268	
	4	235	
1993	1	220	
	2	237	
	3	280	
	4	237	Total = 1894
TOTAL		3668	

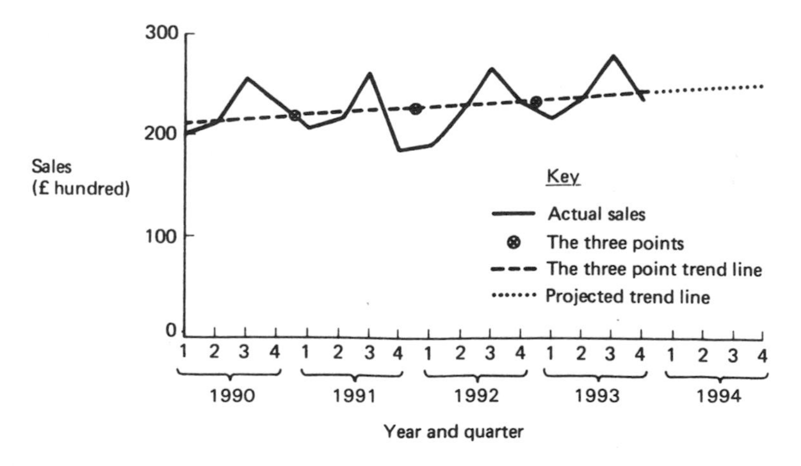

Fig. 11.2 Sales of a product over four years, showing the trend by the three-point method

All the sales values are in units of £ hundreds. The three mean values are plotted halfway between 1990 quarter 4 and 1991 quarter 1, halfway between 1992 quarter 4 and 1993 quarter 1, and halfway between 1991 quarter 4 and 1992 quarter 1, respectively. These points are joined with a straight line (Fig. 11.2). To forecast the trend of sales in 1994 we extend the trend line to the right to obtain:

1994	Quarter 1	Quarter 2	Quarter 3	Quarter 4
Forecast for trend of sales (£ hundred)	247	249	251	253

The three-point method is easy to execute but, as noted in Unit 9, it is rather crude as a lot of information is lost in the averaging process. It always results in a *linear* trend, which may not be appropriate.

(b) The least squares method

To obtain mathematically the line of best fit to the data we can use the least squares method (see Unit 9.3(*b*)). The equation of the least squares line is:

$$y = a + bx$$

where:

$$b = \frac{n\Sigma xy - \Sigma x \Sigma y}{n\Sigma x^2 - (\Sigma x)^2}$$

$$a = \bar{y} - b\bar{x}$$

x is the time variable measured from a suitable origin. It is usually simplest, when using a calculator, to measure time from the period of the first reading and to work in units of the time between consecutive readings. In our sales example (see Table 11.3) it is thus convenient to measure time in quarters from quarter 1. Column (3) shows the values of x corresponding to each time period. y is the dependent variable—in this example the sales in £ hundreds. n is the number of observations ($n = 16$). The calculation is set out in Table 11.3.

Substituting in the formula for b gives:

$$b = \frac{(16 \times 28\,239) - (120 \times 3\,668)}{(16 \times 1\,240) - (120)^2}$$

$$= \frac{451\,824 - 440\,160}{19\,840 - 14\,400}$$

$$= \frac{11\,664}{5\,440}$$

$$= 2.14 \text{ to 3 significant figures}$$

Substituting in the formula for a gives:

Table 11.3 The trend by the least squares method

(1) Year	(2) Quarter	(3) x	(4) y	(5) xy	(6) x^2
1990	1	0	200	0	0
	2	1	212	212	1
	3	2	256	512	4
	4	3	232	696	9
1991	1	4	208	832	16
	2	5	217	1085	25
	3	6	263	1578	36
	4	7	186	1302	49
1992	1	8	192	1536	64
	2	9	225	2025	81
	3	10	268	2680	100
	4	11	235	2585	121
1993	1	12	220	2640	144
	2	13	237	3081	169
	3	14	280	3920	196
	4	15	237	3555	225
TOTALS		120	3668	28239	1240

$$a = \frac{3668}{16} - \left(\frac{11664}{5440}\right)\left(\frac{120}{16}\right)$$

$$= 229.25 - 16.08$$

$$= 213.17$$

$$= 213 \text{ to 3 significant figures}$$

The least squares trend line, therefore, is given by the equation:

$$y = 213 + 2.14x$$

where x is measured in quarters from 1990 quarter 1, y is the sales in £ hundreds and the coefficients are given correct to three significant figures. The least squares trend line is shown in Fig. 11.3. The forecasts for the trend of sales in 1994 may be obtained from the graph by extending the trend line to the right or by substituting in the equation for y. The results are:

1994	Quarter 1 ($x = 16$)	Quarter 2 ($x = 17$)	Quarter 3 ($x = 18$)	Quarter 4 ($x = 19$)
Forecast for trend of sales (£ hundred)	247	249	251	254

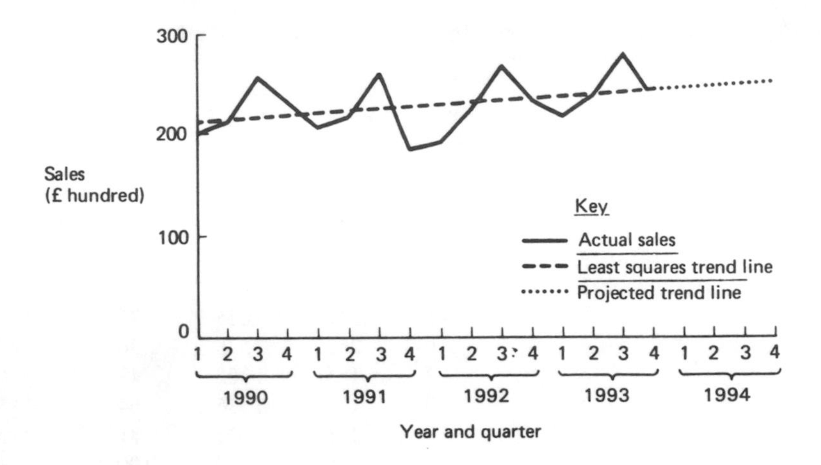

Fig. 11.3 Sales of a product over four years, showing the least squares trend line

In this particular example, the forecasts provided by the three-point method and the least squares method are in close agreement.

The least squares method of fitting is to be preferred from a mathematical viewpoint but it does involve more arithmetic. We can easily predict the trend in any future period by substituting the appropriate values of x in the equation for y. The trend by this method will always be linear, although the method can be extended so that a *least squares curve* may be fitted. One has to decide beforehand what is a suitable curve to fit—for example, should its equation contain an x^2 or x^3 term? Although the use of computer programs enables this procedure to be executed routinely, we shall consider only least square *lines* in this book.

(c) The method of moving averages

The method of moving averages may be used to find the trend when there appear to be regular intervals between corresponding movements of the given time series. To pick out the interval between the fluctuations we look again at the historigram (Fig. 11.1). In our sales example, the fluctuations are repeated on an annual basis, that is, quarter 3 always shows the highest sales and quarter 1 almost always the lowest. If we take the average of the sales in the four quarters of 1990 (see Table 11.4), we average out these seasonal variations. Similarly, if we take the year from 1990 quarter 2 to 1991 quarter 1 and find the average sales in the four quarterly period, we have again averaged out the seasonal variations since we have incorporated every type of quarter when working out the average. We can continue this process for any four consecutive quarters. The resulting averages (see Table 11.4, column (5)) are termed *moving averages*. Note that it is convenient to include the four quarterly totals in the table (column (4)). The four quarterly totals and the four quarterly averages are positioned at the centre of the time period to which they refer so

Table 11.4 The trend by the method of moving averages

(1) Year	(2) Quarter	(3) Sales (£ hundred)	(4) 4 quarterly total	(5) 4 quarterly average	(6) Centred average i.e. trend
1990	1	200			
	2	212			
			900	225	
	3	256			226
			908	227	
	4	232			227.625
			913	228.25	
1991	1	208			229.125
			920	230	
	2	217			224.25
			874	218.5	
	3	263			216.5
			858	214.5	
	4	186			215.5
			866	216.5	
1992	1	192			217.125
			871	217.75	
	2	225			223.875
			920	230	
	3	268			233.5
			948	237	
	4	235			238.5
			960	240	
1993	1	220			241.5
			972	243	
	2	237			243.25
			974	243.5	
	3	280			
	4	237			

that for example the first moving average, 225, lies between 1990 quarter 2 and 1990 quarter 3. We could plot these moving averages on the historigram to give a picture of the trend and this would be satisfactory if we were concerned only with the projection of the trend. Usually, however, we wish to continue the analysis and to examine the fluctuations about the trend. It is more convenient to have the values of the trend at exactly the same times as the original movements. In order to do this we average each adjacent pair of moving averages to obtain the *centred average* which gives the trend value at the same time as the original observation. This centring of the moving averages is only necessary when the moving average is taken over an even number of readings such as four quarters or twelve months. If it is appropriate to take a moving average over an odd number of periods, such as the three working shifts of a day or the five working days in a week, then there is no need to centre—each moving average will automatically be positioned at a time corresponding to one of the original measurements. The full calculation of the trend is shown in Table 11.4 and its graph is shown in Fig. 11.4.

You may have noticed from Table 11.4 that to obtain the centred averages, all that is necessary is to add adjacent four quarterly totals and to divide these by 8. The arithmetic using this approach is set out in Table 11.5. It is usually easier to do the calculation in this way and rounding errors are less likely to creep in.

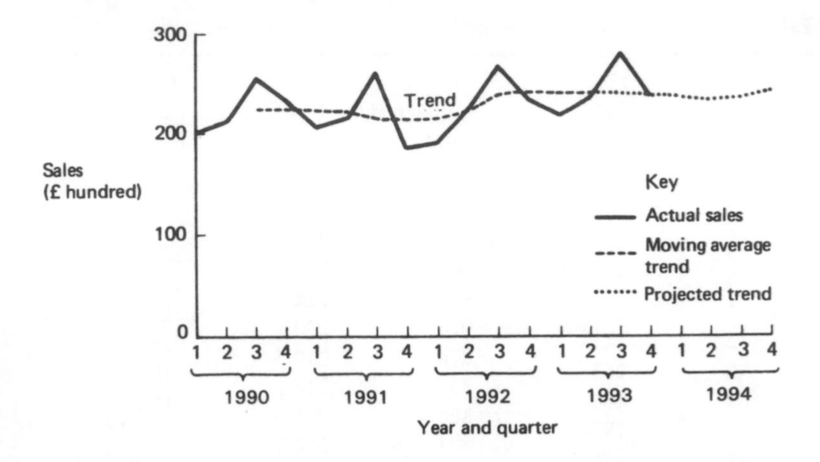

Fig. 11.4 Sales of a product over four years, showing the moving average trend

From Fig. 11.4 we see that using the method of moving averages the trend is not restricted to a straight line and its movement reflects the continually changing sales pattern. The method suffers the disadvantage that trend values are not available at the beginning and end of the series because of the averaging

Table 11.5 The trend by the method of moving averages – alternative method

(1) Year	(2) Quarter	(3) Sales (£ hundred)	(4) 4 quarterly total	(5) Centred total	(6) Trend (5) ÷ 8
1990	1	200			
	2	212	900		
	3	256	908	1808	226
	4	232	913	1821	227.625
1991	1	208	920	1833	229.125
	2	217	874	1794	224.25
	3	263	858	1732	216.5
	4	186	866	1724	215.5
1992	1	192	871	1737	217.125
	2	225	920	1791	223.875
	3	268	948	1868	233.5
	4	235	960	1908	238.5
1993	1	220	972	1932	241.5
	2	237	974	1946	243.25
	3	280			
	4	237			

process. The loss at the beginning of the series is usually not important as earlier sales figures, if required, could be obtained by reference to older sales records. The loss at the end of the series is serious because we do not have trend values corresponding to the most recent sales figures. This makes forecasting more prone to error. In order to make a forecast of the trend using the method of moving averages, you extend the trend curve on the graph as you think fit – there is no mathematical equation to give the answer. Obviously, background knowledge of the likely state of the market, competitors' future plans and so on, can be incorporated but only short-term projections, that is one or at the most two years ahead, are reliable. The suggested projection of the trend for the sales figures is shown by the dotted curve in Fig. 11.4 giving the following results:

1994	Quarter 1	Quarter 2	Quarter 3	Quarter 4
Forecast for trend of sales (£ hundred)	237	235	237	245

These forecasts are less than those using the earlier methods.

In our sales example it was easy to see that a *four quarterly* moving average was appropriate. When we are presented with a time series of annual readings, that is one observation for each year, it is not always obvious how many years to include in the moving average to find the trend. It is preferable to choose an odd number of years since then there is no problem of centring the data. A period of three, five, or more commonly seven or nine years is usually appropriate but to some extent it is a matter of trial and error to find a suitable period. In annual data it is cyclical rather than seasonal fluctuations that are present. Too short a period will not smooth out the data sufficiently whereas too long a period will lead to greater loss of information about the trend at the end of the series.

The technique of *exponential smoothing* is a development of the method of moving averages. It takes all the past observations into account but gives more weight to the most recent values. It enables forecasts to be made routinely. You will find details of the technique in the books by Kazmier and Pohl, Mansfield, and Mills listed in the Further Reading section.

11.3 Models of time series

There are two basic models for time series analysis: the *additive model* and the *multiplicative model.*

In the additive model the various components are considered to be *added* together to give the observation, thus:

$$\text{Sales} = \text{Trend} + \underset{\text{variation}}{\text{Seasonal}} + \underset{\text{variation}}{\text{Cyclical}} + \underset{\text{variation}}{\text{Catastrophic}} + \underset{\text{variation}}{\text{Residual}}$$

In a multiplicative model, the components are *multiplied* together:

$$\text{Sales} = \text{Trend} \times \underset{\text{variation}}{\text{Seasonal}} \times \underset{\text{variation}}{\text{Cyclical}} \times \underset{\text{variation}}{\text{Catastrophic}} \times \underset{\text{variation}}{\text{Residual}}$$

The additive model is appropriate for short-term forecasting and when there is little change in the trend. Using this model the seasonal component is the same *actual* amount in the corresponding season of each year. In the multiplicative model the seasonal component affects the trend by the same *proportion* in the corresponding season of each year. This is more reasonable as a long-term model and in cases where there are marked changes in trend. The analysis using the additive model is generally more straightforward than with the multiplicative model.

11.4 Seasonal analysis

(a) Estimation of seasonal variations using the additive model

In our analysis we shall assume that there is no cyclical or catastrophic component present. We assume also that the trend has been found using the method of moving averages—this assumption is not essential but it is usually made and examiners expect this method to be used prior to seasonal analysis, unless they specifically state otherwise.

Using the additive model we have:

$$\text{Sales} = \text{Trend} + \underset{\text{variation}}{\text{Seasonal}} + \underset{\text{variation}}{\text{Residual}}$$

Thus, the difference between the actual sales and the trend value in each quarter (see Table 11.6) will give an estimate of the seasonal and residual components:

$$\text{Sales} - \text{Trend} = \underset{\text{variation}}{\text{Seasonal}} + \underset{\text{variation}}{\text{Residual}}$$

Some of the fluctuations in the final column of Table 11.6 are positive, indicating quarters where the sales were higher than the trend and some are negative when the actual sales are lower than the trend.

We now assume that these fluctuations from the trend arise solely from seasonal factors and we take each type of quarter in turn and find the average fluctuation from the trend for that quarter of the year. This calculation is set out in Table 11.7.

The average seasonal variations should add up to zero but they usually do not, because the assumption that the fluctuations are entirely due to seasonal factors is not true. It is usual therefore to adjust the average seasonal variations to make the total zero as follows:

(i) If the excess is *positive*, divide it by the number of columns and *subtract* the result from each average seasonal variation;
(ii) If the excess is *negative*, divide it by the number of columns and *add* the result to each average seasonal variation.

Table 11.6 The differences between sales and trend values

Year	Quarter	Sales (£ hundred)	Trend (£ hundred)	Sales–Trend (£ hundred)
1990	1	200		
	2	212		
	3	256	226	30
	4	232	227.625	4.375
1991	1	208	229.125	−21.125
	2	217	224.25	−7.25
	3	263	216.5	46.5
	4	186	215.5	−29.5
1992	1	192	217.125	−25.125
	2	225	223.875	1.125
	3	268	233.5	34.5
	4	235	238.5	−3.5
1993	1	220	241.5	−21.5
	2	237	243.25	−6.25
	3	280		
	4	237		

Table 11.7 Calculation of seasonal variations (additive model)

Year	Quarter 1	2	3	4
1990	–	–	30	4.375
1991	−21.125	−7.25	46.5	−29.5
1992	−25.125	1.125	34.5	−3.5
1993	−21.5	−6.25	–	–
TOTALS	−67.75	−12.375	111.0	−28.625
Average seasonal variations	−22.583	−4.125	37.0	−9.542

The excess in Table 11.7 is +0.75 so we must subtract 0.1875 from each seasonal variation to give:

	Quarter 1	Quarter 2	Quarter 3	Quarter 4
Adjusted seasonal variation	−22.7705	−4.3125	36.8125	−9.7295

Strictly speaking, the adjustment to each average seasonal variation should be in proportion to its size, so there is some leeway here for making arbitrary adjustments. Up to this stage we have done virtually no rounding in our calculations, so as to prevent the build up of rounding errors. Since the original sales values, however, were given correct to the nearest hundred pounds it is unrealistic to quote estimates of trend and seasonal variation to any greater accuracy. Rounded to the nearest hundred pounds the seasonal variations are −23, −4, 37 and −10 respectively. We would thus predict, if the sales pattern continues as previously, that every first quarter of the year sales will be £2300 below the trend value. We are now able to forecast the sales for 1994 by adding the seasonal variations to the moving average trend values calculated in Unit 11.2(*c*):

1994	Quarter 1	Quarter 2	Quarter 3	Quarter 4
Sales forecast (£ hundred)	214	231	274	235

(b) Seasonally adjusted series

You will often have seen in press articles, particularly with reference to unemployment figures and food prices, phrases such as 'allowing for seasonal factors'. To do this, using the additive model, the seasonal variation for each

Table 11.8 Seasonally adjusted sales figures

Year	Quarter	Sales (£ hundred)	Seasonal variation	Seasonally adjusted sales
1990	1	200	−23	223
	2	212	−4	216
	3	256	37	219
	4	232	−10	242
1991	1	208	−23	231
	2	217	−4	221
	3	263	37	226
	4	186	−10	196
1992	1	192	−23	215
	2	225	−4	229
	3	268	37	231
	4	235	−10	245
1993	1	220	−23	243
	2	237	−4	241
	3	280	37	243
	4	237	−10	247

period is subtracted from the original time series. This process is known as *seasonally adjusting or deseasonalizing* the data. We have:

$$\text{Sales} - \underset{\text{variation}}{\text{Seasonal}} = \text{Trend} + \underset{\text{variation}}{\text{Cyclical}} + \underset{\text{variation}}{\text{Catastrophic}} + \text{Residual variation}$$

The seasonally adjusted data can therefore often give a truer picture of the underlying movement of the time series that may have been masked with the seasonal variations present. Table 11.8 shows the seasonally adjusted sales data which are plotted on the historigram in Fig. 11.5. From the graph we can see that the fluctuations are less great in the seasonally adjusted series. The marked dip in sales in the fourth quarter of 1991 is still apparent, however, and would merit further investigation. In practice, of course, you would usually have some background knowledge of unusual factors influencing the sales and this would help you to decide whether to treat the reading in this particular quarter as being influenced by a catastrophic movement.

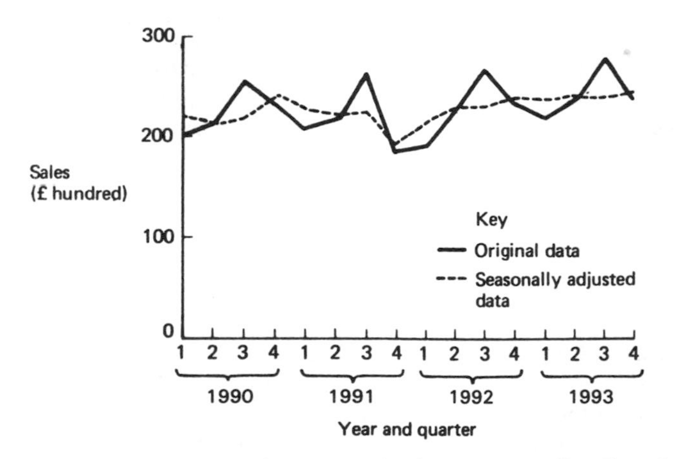

Fig. 11.5 Sales of a product over four years, showing the seasonally adjusted graph

(c) Estimation of seasonal variations using the multiplicative model

The multiplicative model postulates that:

$$\text{Sales} = \text{Trend} \times \underset{\text{variation}}{\text{Seasonal}} \times \underset{\text{variation}}{\text{Cyclical}} \times \underset{\text{variation}}{\text{Castrophic}} \times \underset{\text{variation}}{\text{Residual}}$$

We are assuming there is no cyclical or catastrophic component present so we may put these two factors equal to one, giving:

$$\text{Sales} = \text{Trend} \times \underset{\text{variation}}{\text{Seasonal}} \times \underset{\text{variation}}{\text{Residual}}$$

The quotient Sales/Trend (see Table 11.9) will give an estimate of the seasonal and residual factors. Quarters where the sales are greater than the trend have a Sales/Trend ratio greater than one.

Table 11.9 The quotients of sales and trend values

Year	Quarter	Sales	Trend	$\frac{\text{Sales}}{\text{Trend}}$ (Correct to 4 decimal places)
1990	1	200		
	2	212		
	3	256	226	1.1327
	4	232	227.625	1.0192
1991	1	208	229.125	0.9078
	2	217	224.25	0.9677
	3	263	216.5	1.2148
	4	186	215.5	0.8631
1992	1	192	217.125	0.8843
	2	225	223.875	1.0050
	3	268	233.5	1.1478
	4	235	238.5	0.9853
1993	1	220	241.5	0.9110
	2	237	243.25	0.9743
	3	280		
	4	237		

We now need to average these ratios to remove the residual variations and to obtain the seasonal components. This is done in Table 11.10 (*cf.* Table 11.7).

The total of the average seasonal variations should be 4. Here they add up to 4.0043 so we multiply each seasonal component by the factor 4/4.0043 to bring their sum to 4:

	Quarter 1	Quarter 2	Quarter 3	Quarter 4
Adjusted seasonal variation (to 3 decimal places)	0.900	0.981	1.164	0.955

In the multiplicative model the seasonal variations are usually referred to as *seasonal indices* and you will sometimes find them expressed in percentage form, that is, 90.0, 98.1, 116.4 and 95.5 respectively. We would thus predict that if the same pattern of sales continues, the sales in the first quarter of the year will be 90.0% of the trend value. Our forecast of sales for 1994 using the multiplicative model and the moving average trend values calculated in Unit 11.2(*c*) is:

Table 11.10 Calculation of seasonal variations (multiplicative model)

Year	Quarter 1	2	3	4
1990	–	–	1.1327	1.0192
1991	0.9078	0.9677	1.2148	0.8631
1992	0.8843	1.0050	1.1478	0.9853
1993	0.9110	0.9743	–	–
TOTALS	2.7031	2.9470	3.4953	2.8676
Average seasonal variations	0.9010	0.9823	1.1651	0.9559

1994	Quarter 1	Quarter 2	Quarter 3	Quarter 4
Sales forecast (£ hundred)	213	231	276	234

These values are close to those obtained using the additive model because this is a short-term forecast and the series does not show any marked change in trend.

In practice, one would use the model which has given the most reliable results in the past. Sophisticated mathematical techniques do not guarantee accurate forecasts since unforeseen developments may affect any predictions.

We could continue the analysis of the sales data using the multiplicative model by removing the seasonal components. This procedure is analogous to that in (*b*) above, but wherever previously we added or subtracted, we must now multiply or divide:

$$\text{Seasonally adjusted sales} = \frac{\text{Sales}}{\text{Seasonal variation}} = \text{Trend} \times \text{Cyclical variation} \times \text{Catastrophic variation} \times \text{Residual variation}$$

You should attempt this analysis for yourself as an exercise and plot the seasonally adjusted data on a historigram.

11.5 Residual variations

Using the additive model and assuming no cyclical or catastrophic variation, we postulated that:

$$\text{Sales} = \text{Trend} + \text{Seasonal variation} + \text{Residual variation}$$

Table 11.11 Calculation of the residual variation

Year	Quarter	Sales (£ hundred)	Trend	Seasonal variation	Residual variation
1990	1	200			
	2	212			
	3	256	226	37	−7
	4	232	228	−10	14
1991	1	208	229	−23	2
	2	217	224	−4	−3
	3	263	216	37	10
	4	186	−10	−20	
1992	1	192	217	−23	−2
	2	225	224	−4	5
	3	268	234	37	−3
	4	235	238	−10	7
1993	1	220	242	−23	1
	2	237	243	−4	−2
	3	280			
	4	237			

We have now estimated the trend and the seasonal variation and by subtracting these from the original sales data we can find the residual variation (see Table 11.11). We have used the *rounded* values of the trend and the seasonal variations as we are interested in the overall pattern of the residuals rather than their precise values. From the final column of Table 11.11, you can see that there is no particular pattern to the residuals and they are relatively small. This is as it should be, since a periodic pattern might indicate that we had not removed the seasonal variations adequately. Any large residual variation would indicate that there was a catastrophic component that we had ignored. The residual value of −20 for the fourth quarter of 1991 has the highest magnitude but this is compensated for by the two positive residual values, 14 and 10, in the fourth quarter of 1990 and the third quarter of 1991 respectively, so it would appear that we were justified in not treating the sales in the fourth quarter of 1991 as an abnormal value. A study of residuals over the long term would give an indication of any cyclical component that should be considered. If the residuals had all been of the same sign, then we could not have estimated the trend accurately.

The residual variation may be estimated using the multiplicative model by dividing the sales by the trend and by the seasonal variation (see Unit 11.4(*c*)). These residual variations should have values close to one but again there should be no noticeable pattern to them. You might like to calculate these residual variations as an exercise.

11.6 Deflation of time series

In recent years, as a result of the increasing concern with inflation, it has often become important to assess the 'real' value or cost of sales, imports, wages, assets and so on, allowing for inflation. To *deflate* a time series we must divide each observation by the corresponding value of a suitable price index number. The Retail Prices Index is often used for this but in some cases a price index specific to the industry or product concerned, if details of such an index are available, may be more appropriate.

In Table 11.12 are shown the average weekly earnings of female employees from 1984 to 1992. The data are taken from the *Monthly Digest of Statistics*. The values of the Retail Price Index are listed in column (3) of Table 11.12. Two sets of figures are shown for 1987 because in that year the base year for the RPI was changed from 1974 = 100 to 1987 = 100, and the way the earnings are shown in the *Monthly Digest of Statistics* also changed. Column (4) of Table 11.12 shows the average earnings for 1984–87 adjusted to 1974 values by dividing the actual earnings (column (2)) by the corresponding RPI (1974 = 100) *expressed as a proportion*. The results are in column (4)—the 'real' earnings in 1984 based on 1974 values are $\frac{£96.30}{3.518}$ or £27.37, and similarly for later years up to 1987.

Table 11.12 Deflation of a time series

(1) Year	(2) Average earnings (3)	(3) RPI 1977 = 100 (annual average)	(4) 'Real' earnings (3) based on 1974 values	(5) RPI 1980 = 100 (annual average)	(6) 'Real' earnings (£) based on 1987 values
1984	96.30	351.8	27.37	89.0	108.20
1985	103.21	373.2	27.65	94.6	109.10
1986	110.48	385.9	28.63	97.8	112.97
1987	118.79 (Jan)	394.5	30.11	100.0	118.79
1987	145.5			101.9	142.79
1988	161.6			106.9	151.17
1989	179.3			115.2	155.64
1990	198.6			126.1	157.49
1991	220.0			133.5	164.79
1992	238.8			138.5	172.42

In column (6) of Table 11.12 the RPI based on 1987 = 100 is shown. From 1987 onwards it is the actual RPI, but prior to 1987 the values of the RPI based on 1974 = 100 have been recalculated based on 1987 = 100. To do this we divide the data in column (3) by 394.5, the value of the RPI in 1987 (1974 = 100). To obtain the earnings at 1987 values we divide the actual earnings (column (2)) by the corresponding values of the RPI (1987 = 100) in column (5). The results are shown in column (6).

From the calculations on the whole series we can see that although the average earnings rose from £96.30 to £238.80, an increase of $\left(\frac{238.8}{96.30} \times 100 - 100\right)$%, or 148% from 1984 to 1992, in real terms the increase based on 1987 values was only from £108.20 to £172.42, an increase of $\left(\frac{172.42}{108.2} \times 100 - 100\right)$%, or 59%.

11.7 Z charts

A *Z chart* is a type of historigram that records a time series in three ways on one diagram. It can be used for displaying sales, purchases, production and other important business series. Usually the chart covers data for one year, although several years may be included on one graph for comparative purposes. The three ways in which the series is recorded on the graph are:

(*a*) the recording of the actual values for each time period in the year, which may be weeks, months or even days;
(*b*) the *cumulative total* of the actual values over the year;
(*c*) the *moving annual total* for the year starting with the annual total for the previous year. This gives us a trend line on the graph.

Table 11.13 Monthly sales figures over two years

	Sales, 1992 (£ thousand)	Sales, 1993 (£ thousand)	Cumulative total 1993	Moving annual total, 1993
January	102	101	101	1357
February	105	101	202	1353
March	101	103	305	1355
April	107	106	411	1354
May	108	107	518	1353
June	112	110	628	1351
July	113	111	739	1349
August	120	117	856	1346
September	118	120	976	1348
October	122	118	1094	1344
November	125	120	1214	1339
December	125	122	1336	1336
TOTALS	1358	1336		

Sometimes a Z chart is drawn with more than one scale on the vertical axis, since the sets of values being recorded are of different magnitude. The actual values for each time period in the year may only creep just above the horizontal axis unless they are recorded on a larger scale than that used for the moving annual and cumulative totals.

Table 11.13 gives the monthly sales figures for a company over two years.

The moving annual total is the total of sales for the 12 months up to and including the month we are interested in. Before we can calculate a moving annual total we need to know the annual sales for the preceding year. For example to calculate the moving annual total for January 1993, subtract from the total sales for 1992 the sales for January 1992 and add on the sales for January 1993:

$$1358 - 102 + 101 = 1357$$

We then have sales for the period from the beginning of February 1992 to the

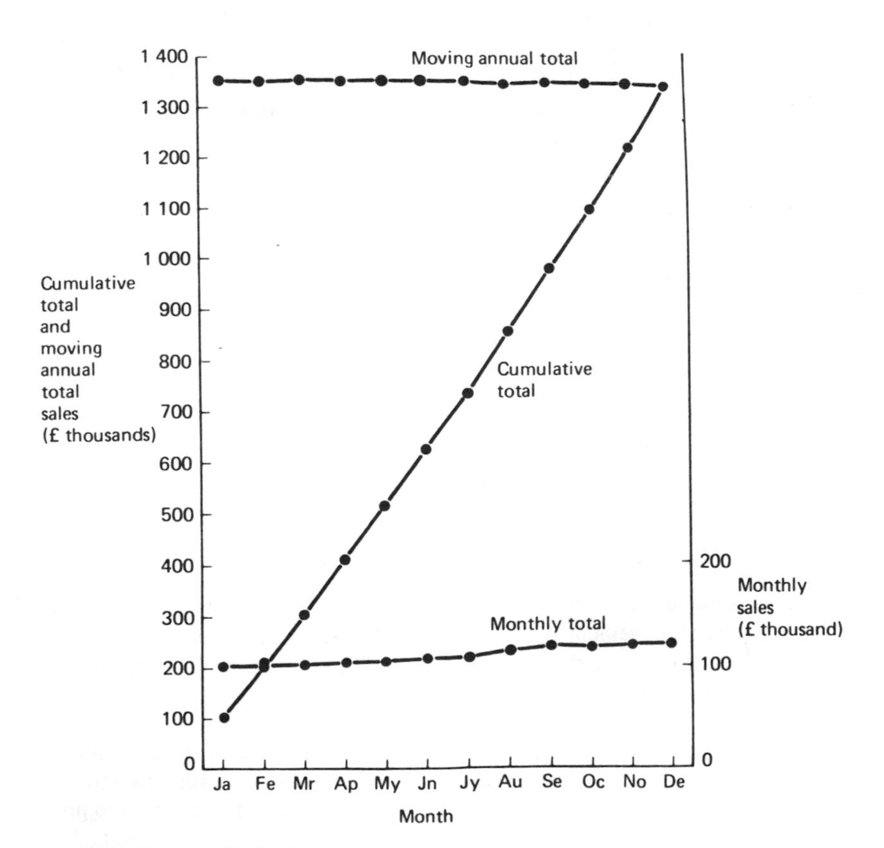

Fig. 11.6 Z chart of sales in 1993

end of January 1993. The moving annual total for February 1993 is the above total, less sales for February 1992 and plus sales for February 1993:

$$1357 - 105 + 101 = 1353$$

Repeating the process for each month of 1993 gives the moving annual totals shown in Table 11.13. Note that the moving annual totals are positioned opposite the last month making up the total. This is a different procedure from that employed in positioning the four quarterly totals in Unit 11.2(*c*). This procedure is adopted so that the graph of the moving annual totals does join up with the cumulative monthly sales to form the top part of a Z shape (Fig. 11.6).

The moving annual total removes seasonal fluctuations and shows that although sales made an upward trend month by month during 1993, the overall trend, taking 1992 sales into account, is downwards since the sales in 1993 are generally down on the corresponding month in 1993. The Z chart for these values is shown in Fig. 11.6. The monthly sales figures are recorded on the right-hand axis on a scale larger than that used for the moving annual total and the cumulative sales figures. This is not essential but it enables the month by month variations to be seen more clearly.

The lines on the graph form a distinct 'Zed' shape, hence the title Z chart. (The Z chart is sometimes referred to as a *Zee* chart, presumably as a concession to American pronunciation.)

11.8 Exercises

1. The values in the following table are for notifiable offences recorded by the police in England and Wales, 1990–92 (in thousands):

	1990	1991	1992
Quarter 1	1080.9	1251.2	139.2
Quarter 2	1113.3	1326.2	1376.7
Quarter 3	1110.3	1317.2	1363.4
Quarter 4	1241.0	1381.6	1458.8

Source: *Monthly Digest of Statistics*

(*a*) Assuming that cyclical variation is zero, calculate both the trend and seasonal movement by the method of moving averages.

(*b*) Give reasons for deriving: (i) seasonal variations and (ii) seasonally adjusted figures.

2. The following data give the mean seasonal output in a manufacturing department. Plot the data on a graph and by means of a suitable moving average remove the seasonal variation from the data and plot the trend line.

	1989	1990	1991	1992	1993
Quarter 1	34	36	33	31	38
Quarter 2	44	41	49	51	48
Quarter 3	61	62	67	65	61
Quarter 4	43	42	41	41	40

3. The following figures show average weekly earnings of male manual workers on adult rates in the UK for 1977–83, and for full-time male employees on adult rates in the manufacturing industries for 1984–92. All figures are taken from the *Monthly Digest of Statistics*. Draw a graph, calculate the trend in the data and show that trend on the graph. Briefly describe the method you used.

Year	Wage (£)	Year	Wage (£)
1977	73.56	1985	187.2
1978	84.77	1986	202.3
1979	98.28	1987	217.0
1980	113.06	1988	236.3
1981	125.58	1989	257.3
1982	137.06	1990	282.2
1983	149.13	1991	299.5
1984	171.2	1992	319.8

4. The sales in tonnes of a product manufactured by a company were:

	Quarter 1	Quarter 2	Quarter 3	Quarter 4
1990	–	–	–	13
1991	14	16	9	14
1992	16	17	12	17
1993	18	20	13	–

(*a*) Find the seasonal variation and adjust the series for seasonal variations.
(*b*) Draw a trend line on a graph of the series and from it estimate sales for the next four quarters.

5. From the following data on the population (in thousands) of the UK, 1966–90, from the *Annual Abstract of Statistics*, draw a graph and show on that graph trend lines using: (*a*) the moving average method and (*b*) the least squares method.

1966	54643	1971	55928	1976	56216	1981	56352	1986	56763
1967	54959	1972	56097	1977	56190	1982	56306	1987	56930
1968	55214	1973	56223	1978	56178	1983	56347	1988	57065
1969	55461	1974	56236	1979	56240	1984	56460	1989	57236
1970	55632	1975	56226	1980	56330	1985	56618	1990	57411

6. The following table shows rainfall (in cm) in a country during the period 1989–93:

Year	Quarter 1	Quarter 2	Quarter 3	Quarter 4
1989	152	204	134	213
1990	187	198	311	299
1991	210	257	231	323
1992	210	225	236	318
1993	180	–	–	–

(*a*) Plot a trend line for these data by means of a centred four quarterly moving average.
(*b*) Calculate the average seasonal variation.

7. A company manufacturing consumer goods may find considerable variations in sales and production over a long period of time. Explain why these variations may occur and why it is necessary to isolate these variations from the statistics.

8. Here are the monthly sales figures for a company for 1992 and 1993:

	1992	1993
January	6000	5700
February	6200	6000
March	6300	6500
April	6100	5800
May	6500	6000
June	6700	6200
July	6600	6500
August	6800	6300
September	7500	6700
October	7200	6900
November	7500	6900
December	7600	7000

Show the 1993 figures on a Z chart.

(Answers at the end of the book.)

UNIT TWELVE

An introduction to probability

12.1 Measuring probability

You will have met the concept of *probability* in everyday life. If an event is likely to occur, we say that it is probable; if it is not likely to happen, it is improbable. Statisticians take this vague idea and try to introduce some precision into the definition by measuring probability on a scale from zero to one. If an event can never happen or has never been known to occur, it is assigned a probability of zero; the probability that I shall never die is zero. At the opposite end of the scale, if an event is certain to occur or has always happened previously, it is assigned a probability of one.

For events which may or may not happen, the probability lies somewhere between zero and one. When we toss a coin into the air, it must come down heads or tails (we discard the possibility of its landing on its edge). Assuming that the coin is not weighted or worn in any way, we expect there to be an equal chance of the coin falling heads or falling tails. Thus, the probability of a coin falling heads is the same as the probability of the coin falling tails and as no other outcome is possible, this probability must be halfway along the probability scale and take the value $\frac{1}{2}$. Consider tossing an unbiased die. (Die is the singular noun; dice the plural noun.) Each time it is thrown there are six possible equally likely outcomes so the probability for any one face is $\frac{1}{6}$. If you have a bag containing six white sweets and two black sweets and you are asked to pick a sweet from the bag without looking, the probability of picking a white sweet is $\frac{6}{8}$, or $\frac{3}{4}$ in lowest terms.

Probabilities of this kind which we assess from our prior knowledge of the situation are called *a priori* probabilities. When we toss a coin or roll a die, we say we are conducting a *trial* of a *statistical experiment*. An *outcome* is one specific result of this trial. The list of all possible outcomes is referred to as the *sample space*. Thus, when tossing a coin, the two possible outcomes are 'heads' and 'tails'. When rolling a die there are six possible outcomes: 'score of 1', 'score of 2', 'score of 3', 'score of 4', 'score of 5' and 'score of 6'. When picking a sweet from the bag containing six white and two black sweets, there are eight possible outcomes—six when a white sweet is chosen and two when a black sweet is chosen.

An *event* is the particular set of outcomes that we are interested in and can comprise more than one outcome. For example, obtaining a head, obtaining a score of less than three when rolling a die and choosing a white sweet are all

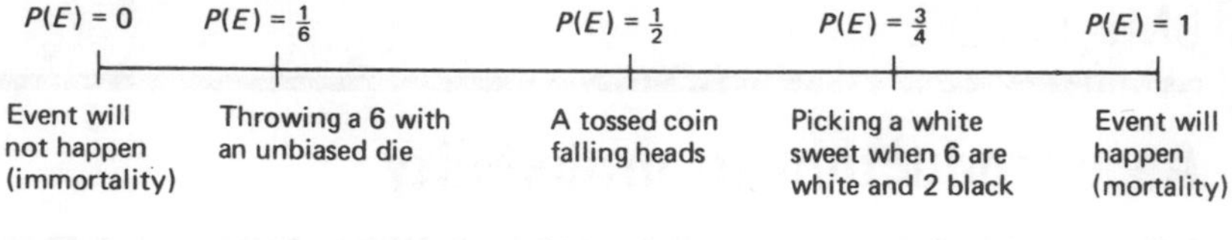

$P(E) = 0$ — Event will not happen (immortality)

$P(E) = \frac{1}{6}$ — Throwing a 6 with an unbiased die

$P(E) = \frac{1}{2}$ — A tossed coin falling heads

$P(E) = \frac{3}{4}$ — Picking a white sweet when 6 are white and 2 black

$P(E) = 1$ — Event will happen (mortality)

Fig. 12.1 The scale of probability

examples of events in the contexts of our three illustrative experiments. It is usual to write the probability of an event occurring as *P*(event occurring) or *P*(*E*), as in Fig. 12.1, where *E* (or any other capital letter) stands for the desired event. We can thus write for example:

$$P(\text{heads when one coin is tossed}) = P(H) = \tfrac{1}{2}$$

Probabilities do not have to be written as vulgar fractions. They can be given as decimal fractions:

$$P(\text{heads when one coin is tossed}) = 0.5$$

Occasionally you will find probabilities given in percentage terms such as in the statement that there is a 50% chance of an unbiased coin landing heads.

When the sample space consists of equally likely outcomes, the probability of an event *E* occurring is given by:

$$P(E) = \frac{\text{Number of outcomes in } E}{\text{Total number of outcomes in the sample space}}$$

Thus, in the coin tossing experiment:

$$P\left(\begin{array}{l}\text{heads when one}\\ \text{coin is tossed}\end{array}\right) = \frac{\text{Number of different ways coin can land to give heads}}{\text{Total number of different ways coin can land}}$$

$$= \frac{1}{2}$$

When rolling the die:

$$P(\text{score of } 6) = \frac{\text{Number of ways die can land showing 6}}{\text{Total number of different scores the die can show}}$$

$$= \frac{1}{6}$$

$$P\left(\begin{array}{r}\text{score less}\\ \text{than 3}\end{array}\right) = \frac{\text{Number of ways die can land showing a score of 1 or 2}}{\text{Total number of different scores the die can show}}$$

$$= \frac{2}{6}$$

$$= \frac{1}{3}$$

When selecting a sweet from the bag:

$$P(\text{white sweet}) = \frac{\text{Number of ways of selecting a white sweet}}{\text{Total number of sweets in bag}}$$

$$= \frac{6}{8}$$

$$= \frac{3}{4}$$

In real life of course it is not always possible to assess probabilities beforehand. What is the probability of passing a statistics examination at the first attempt? What is the probability of a company exporting more than ten million pounds worth of goods next year? What is the probability that the next invoice a clerk checks contains an error? In such cases we have to resort to observation and we count the *relative frequency* with which such an event has occurred in the past and we call this the probability of the event occurring. In the examination example, we would consult records to find the total number of students taking the statistics examination and the number of these who passed at their first attempt. We would then have:

$$P(\text{passing a statistics examination at the first attempt}) = \frac{\text{Number of students who have passed at the first attempt}}{\text{Total number of students who took the examination}}$$

This method can provide us only with an estimate of the probability. As the years go by and more students sit the examination, we would expect to obtain a better estimate. Probabilities assessed in this way are called *empirical probabilities*.

12.2 The addition rule

Let us return to tossing coins. The total probability of all the possible outcomes of an event must be 1 because it is certain that one of the possible outcomes *must* occur. If we toss a coin into the air, it must come down showing either heads or tails. The probability of it falling heads is $\frac{1}{2}$ and the probability of it falling tails is $\frac{1}{2}$, assuming the coin is unbiased. It must come down either heads, with the probability of $\frac{1}{2}$, or tails, also with the probability of $\frac{1}{2}$. The probability that it will fall either heads or tails is therefore $\frac{1}{2} + \frac{1}{2} = 1$. This extremely obvious statement illustrates one of the rules of probability, the *addition rule of probability*. This rule states that the total probability of *either* of two mutually exclusive events occurring is the sum of the probabilities of each of the events occurring:

$$P(A \text{ or } B) = P(A) + P(B)$$

The inclusion of the term *mutually exclusive* means that this rule applies in cases where the occurrence of one event excludes the possibility of the second event. A coin falls heads *or* it falls tails: the one excludes the other. For this reason the rule is often referred to as the *or* rule.

The probability of throwing either a one or a six with an unbiased die is:

$$\frac{1}{6} + \frac{1}{6} = \frac{2}{6} = \frac{1}{3}$$

Work out for yourself the probability of picking a white sweet or a black sweet from a bag containing six white sweets, two black sweets and five brown sweets. (The answer is $\frac{8}{13}$.)

The addition rule holds for any number of mutually exclusive events:

$$P(A_1 \text{ or } A_2 \text{ or } A_3 \ldots \text{ or } A_n) = P(A_1) + P(A_2) + P(A_3) + \ldots + P(A_n)$$

In the coin tossing illustration, the two events, heads or tails, are *exhaustive*, that is one of the two is certain to happen. In the die rolling experiment, the events, scores of one or six, although mutually exclusive, are not exhaustive since other scores, two, three, four or five, are possible.

A particular event either occurs or does not occur and we are *certain* that one or other of these situations exists. Thus, the probability of an event occurring plus the probability of the event not occurring must add up to one:

$$P(A) + P(\text{not } A) = 1$$

P(not *A*) stands for the probability of event *A* not occurring. A' or $\bar{A}$ can be used as a symbol for 'not *A*'. *A* and 'not *A*' are termed *complementary events*. It is often easier to find the probability of an event not occurring than to find the probability that it does occur. We shall illustrate the procedure by showing alternative methods for working out the probability of obtaining a score greater than two when a die is rolled.

Method (i)

$$\begin{aligned} P(\text{score greater than } 2) &= P(\text{score 3 or 4 or 5 or 6}) \\ &= P(\text{score 3}) + P(\text{score 4}) + P(\text{score 5}) + P(\text{score 6}) \\ &= \frac{1}{6} + \frac{1}{6} + \frac{1}{6} + \frac{1}{6} \\ &= \frac{4}{6} \\ &= \frac{2}{3} \end{aligned}$$

Method (ii)

$$
\begin{aligned}
P(\text{score greater than } 2) &= 1 - P(\text{score not greater than } 2) \\
&= 1 - P(\text{score 1 or 2}) \\
&= 1 - [P(\text{score } 1) + P(\text{score } 2)] \\
&= 1 - \left[\frac{1}{6} + \frac{1}{6}\right] \\
&= 1 - \frac{1}{3} \\
&= \frac{2}{3}
\end{aligned}
$$

12.3 The multiplication rule

We now look at compound situations where the outcomes of interest are made up of two or more simpler outcomes. When we toss a coin twice, for example, how can we work out the probability of obtaining two heads? If the coin is unbiased, we know that the chance of the coin falling heads each time is $\frac{1}{2}$ or 0.5. The total probability for all possible outcomes of an experiment must be 1. When we toss a coin into the air twice it must come down each time and it can fall in any one of four, equally likely ways:

heads and heads
heads and tails
tails and heads
tails and tails

Thus the probability of obtaining two heads is $\frac{1}{4}$ or 0.25. This illustrates the second rule of probability, known as the *multiplication rule*, or sometimes as the *and* rule. This states that the probability of two or more independent events occurring is the product of their individual probabilities:

$$P(A \text{ or } B) = P(A) \times P(B)$$

Thus:

$$
\begin{aligned}
&P(\text{heads on first throw}) = \tfrac{1}{2} \\
&P(\text{heads on second throw}) = \tfrac{1}{2} \\
&P(\text{two heads}) = \tfrac{1}{2} \times \tfrac{1}{2} = \tfrac{1}{4}
\end{aligned}
$$

Events are *independent* when the probability of either of them occurring is not affected by the occurrence or non-occurrence of the other.

Let us consider a second example of the application of this rule. Out of a crate of 40 bottles, 2 are underfilled. An inspector selects one bottle at random from the crate. What is the probability that it is not underfilled? In a second crate of 40 bottles, 4 are underfilled. The inspector randomly selects a bottle from this crate. What is the probability that both bottles selected are not underfilled?

1st crate:

$$P(\text{underfilled}) = \frac{2}{20} = \frac{1}{20}$$

$$P(\text{not underfilled}) = 1 - \frac{1}{20} = \frac{19}{20}$$

2nd crate:

$$P(\text{underfilled}) = \frac{4}{40} = \frac{1}{10}$$

$$P(\text{not underfilled}) = \frac{9}{10}$$

so:

$$P(\text{both bottles not underfilled}) = \frac{19}{20} \times \frac{9}{10} \text{ since the events are independent}$$

$$= \frac{171}{200}$$

$$= 0.855$$

12.4 The general addition rule

The next problem is that of finding the probability of either event A or event B occurring when A and B are not mutually exclusive. The rule of probability that applies here is the *general rule for the addition of probabilities*. It applies when we want to find the probability of event A or event B occurring, including the probability that both occur simultaneously.

Suppose, for example, the head of a department has to select a new trainee from an intake of 20 trainees joining the company. There are 8 female trainees; 3 of the females and 3 of the males have A levels. The head of department would like a female trainee, preferably with A levels. However, it is decided to put the names of all the trainees into a hat and to draw out a name at random so that there is an equal chance of selecting any of the trainees. What is the probability that either a female trainee or one with A levels will be selected?

At first glance, it might seem that the simple addition rule would apply. As 8 of the trainees are female, the probability of selecting a female is $\frac{8}{20}$; 6 trainees have A levels so the probability of picking a trainee with A levels is $\frac{6}{20}$; thus it might appear that the probability of selecting either a female trainee or one with A levels is $\frac{8}{20} + \frac{6}{20} = \frac{14}{20}$. This is incorrect, however, because 3 females also have A levels and we must not count them twice, that is, as females and again as having A levels. The probability of the selected trainee being both female and having A levels is $\frac{11}{20}$. We have:

$$P\left(\begin{array}{l}\text{selected trainee is either}\\ \text{female or with A levels}\end{array}\right) = \frac{6}{20} + \frac{8}{20} - \frac{3}{20}$$

$$= \frac{11}{20}$$

This example illustrates the general rule for the addition of probabilities:

$$P(A \text{ or } B) = P(A) + P(B) - P(A \text{ and } B)$$

where $P(A$ and $B)$ is the probability that events A and B both occur at the same time.

It is often easier to work out probabilities if we draw a *Venn diagram*. We represent the sample space by a rectangle, S, and events in the sample space are drawn as enclosed areas, usually circles but the shape and size are not important (Fig. 12.2).

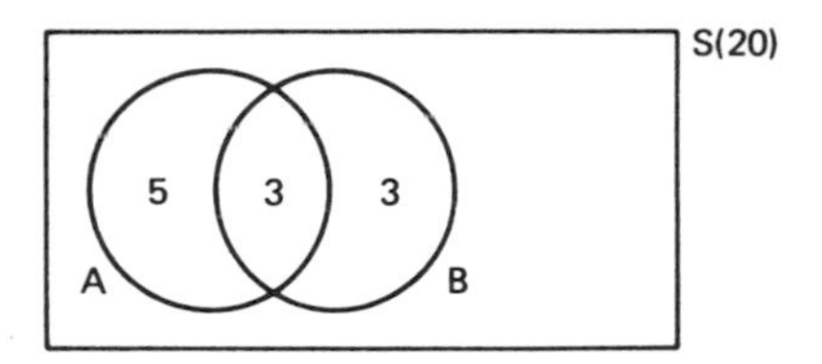

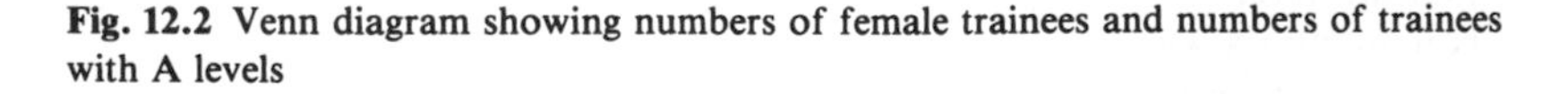

Fig. 12.2 Venn diagram showing numbers of female trainees and numbers of trainees with A levels

In the trainee example, S represents all 20 trainees. Circle A represents the female trainees. Circle B represents those trainees with A levels. A and B overlap because some of the trainees are female and have A levels. To enable us to find the probability of selecting either a female trainee or one with A levels, we first work out the number of trainees enclosed by the circles, being careful not to count those in the overlap twice. We are told that 3 female trainees have A levels so there are 3 trainees in the overlap. There are 3 male trainees with A levels so there are 3 trainees in that part of circle B not overlapping with circle A. There are 8 females altogether so there must be 5 who do not have A levels. These trainees are in circle A but not in the overlap. Altogether we have $5 + 3 + 3 = 11$ trainees in A or B or both, so again we have:

$$P(\text{selected trainee is either female or with A levels}) = \frac{11}{20}$$

Analysing the general rule for the addition of probabilities using a Venn diagram:

$$P(A \text{ or } B) = P(A) + P(B) - P(A \text{ and } B)$$

(*A* or *B*) is represented by that part of the diagram within circles *A* and *B*. Using the language of set theory, it is the *union* of *A* and *B* and can be written as $A \cup B$. Event *A* is represented by that part of the diagram inside circle *A* and similarly for *B*. (*A* and *B*) is indicated by the portion common to both *A* and *B* and is referred to as the *intersection* of *A* and *B*, denoted by $A \cap B$. For mutually exclusive events, the Venn diagram appears as in Fig. 12.3—there is no overlap between circles *A* and *B* so *P*(*A* and *B*) is zero. The general rule for addition of probabilities then reduces to the simple rule. (See Unit 12.2.)

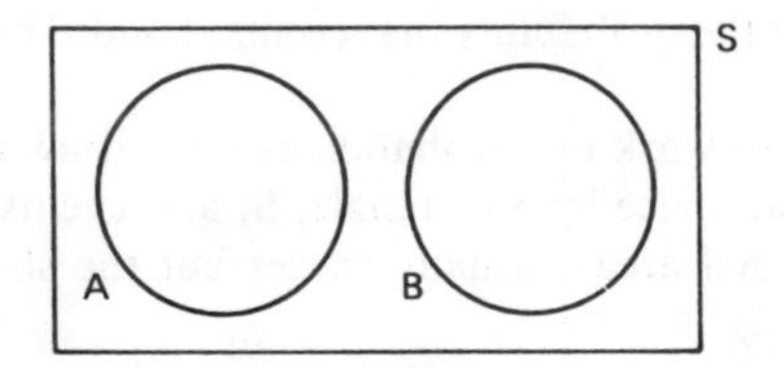

Fig. 12.3 Venn diagram for mutually exclusive events

12.5 The general multiplication rule

When we are considering the probabilities of compound events, it can happen that the outcome at one stage does affect the outcome at a later stage. In a bag containing six white and two black sweets, if one person has already taken a white sweet from the bag and has eaten it, there is less chance of a second person obtaining a white sweet. The general multiplication rule applies when events are not independent.

Suppose, for example, the head of department (see example in Unit 12.4) selects the names of two trainees at random from the hat. What is the probability that two females are selected, assuming that the first name is not replaced before selecting the second? There are 8 females out of 20 trainees, so the probability of picking a female on the first selection is $\frac{8}{20}$. When it comes to the second selection, however, if the first selection was a female, there are only 7 females left out of the remaining 19 trainees. The probability of selecting a second female is then $\frac{7}{19}$. Thus the probability of obtaining a female on the first selection and on the second selection is:

$$\frac{8}{20} \times \frac{7}{19} = \frac{14}{95}$$

In general terms, the probability of two events *A* and *B* occurring is:

$$P(A \text{ and } B) = P(A) \times P(B/A)$$

where $P(A)$ is the probability of A occurring and $P(B/A)$ (read as the probability of B given A) is the *conditional probability* of the event B occurring, that is the probability that B occurs given that A has already occurred.

Try working out for yourself the probability that the head of department selects two male trainees.

A compound experiment, that is, one with more than one component part, is regarded as a sequence of simpler trials. In the above example, the choice of two trainees is considered as the choice of one trainee followed by the choice of a second. A *tree diagram* enables us to construct an exhaustive list of the mutually exclusive outcomes of a compound experiment – every possible outcome is considered and if one of the outcomes of the compound experiment occurs, then the others cannot.

Fig. 12.4 shows the diagram for the selection of two trainees. We work from left to right in the diagram. The first selection may be male or female so we draw two branches corresponding to these two possibilities. On each branch we write the probability of this particular outcome. There were 8 females out of 20 trainees and we have:

$$P(\text{female trainee}) = P(F) = \frac{8}{20}$$

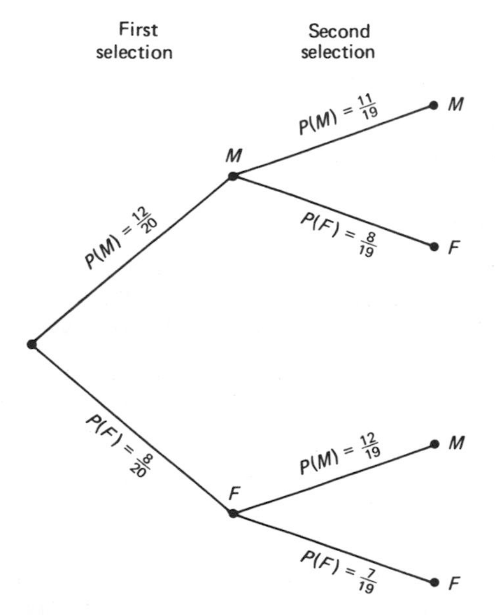

Fig. 12.4 Tree diagram for selection of two trainees

$$P(\text{male trainee}) = P(M) = \frac{12}{20}$$

Next we consider the selection of the second trainee. Whether the first selection was male or female, the second selection can result in either a male or a female, so we mark in these two possibilities at the end of each of the two branches on the existing tree diagram. We enter on this second group of branches the *conditional* probabilities associated with them. Thus for the uppermost branch, we write in the probability that the second trainee is male given that the first trainee was male. This probability is $\frac{11}{19}$ as there are 19 trainees left to choose 19 from of whom 11 are male. We write the probabilities on the other branches similarly. Each complete branch from start to tip represents one possible outcome of the compound experiment and each branch is mutually exclusive. To obtain the probability of a particular outcome of the compound experiment we multiply the probabilities along all sections of the branch using the general multiplication rule for probabilities. We obtain:

$$P(\text{two males}) = \frac{12}{20} \times \frac{11}{19} = \frac{33}{95}$$

$$\left.\begin{array}{l} P(\text{a male then a female}) = \dfrac{12}{20} \times \dfrac{8}{19} = \dfrac{24}{95} \\[2ex] P(\text{a female then a male}) = \dfrac{8}{20} \times \dfrac{12}{19} = \dfrac{24}{95} \end{array}\right\} P(\text{a male and a female}) = \frac{48}{95}$$

$$P(\text{two females}) = \frac{8}{20} \times \frac{7}{19} = \frac{14}{95}$$

As a check on our working, we note that $\frac{33}{95} + \frac{48}{95} + \frac{14}{95} = 1$ so we have an exhaustive list of the possible outcomes.

12.6 Exercises

1. There were 12 wet days in July last year. If a name is selected at random from a directory what is the probability that the person's birthday was on a wet July day last year (assuming that all birthdays are equally likely)?

2. Two machines operate in a factory. The probability that machine A will last for another four years is $\frac{1}{5}$, while the probability that machine B will last for the same period of time is $\frac{1}{4}$. What is the probability that:

   (*a*) both machines will be operating in 4 years;
   (*b*) neither will be operating;
   (*c*) at least one will be operating?

3. A machine has three working parts all of which must work for the machine to function. The probabilities of each part working are $\frac{1}{7}$, $\frac{1}{5}$ and $\frac{1}{3}$. What is the probability that the machine works?

4. A box contains 40 components, six of which are defective. Two components

are selected separately at random from the box, the first not being replaced before the second is selected. What is the probability that:

(*a*) the first selected is defective;
(*b*) both are defective;
(*c*) one is defective and one is not;
(*d*) the first selected is defective and the second is not?

5. A manufacturing company receives a component from two firms A and B. It receives three times as many components from A as from B. On average 3 per cent of the components from A are rejects while from firm B the average reject rate is 5 per cent. If two components are selected at random at the premises of the manufacturing company what is the probability:

(*a*) that they both come from firm A;
(*b*) that one comes from firm A and one from B;
(*c*) that they are both rejects;
(*d*) that one is a reject from firm A and the other a reject from B?

6. There are 25 students in a class. What is the probability that at least two of them have the same birthday?

7. Two companies are attempting to obtain building contracts independently of each other. For each contract available, company A has a probability of $\frac{1}{3}$ of winning it while company B has a probability of $\frac{2}{3}$ of winning it. Calculate the probabilities of:

(*a*) firm A winning the next two available contracts;
(*b*) firm B winning the next two available contracts;
(*c*) firm A winning one and firm B winning the other.

8. A restaurant owner has four times as many male customers as female. 40% of men and 70% of women take the set lunch, the remainder choosing from the optional menu items. Of men choosing the set lunch 60% have an alcoholic drink, the remainder having a soft drink. The corresponding proportion for those choosing the optional items is 90%. Among the female customers 40% of those having the set lunch and 60% of those having the optional lunch choose an alcoholic drink. Using a tree diagram (or by some other method) find:

(i) the proportion of people taking the set lunch;
(ii) the proportion of set lunch takers having an alcoholic drink;
(iii) the proportion of women customers who choose a soft drink;
(iv) the probability that a randomly chosen alcoholic drinker will be male.

9. (i) Four rose bushes are planted in a row. They have all lost their labels, but it is known that three are pink and one is yellow. What is the probability that the three pink ones come up next to each other in the row?
(ii) If five rose bushes are planted, three pink and two yellow, what is the probability of the three pink being next to each other?

(Answers as the end of the book.)

UNIT THIRTEEN

Probability distributions

13.1 Introduction

In Unit 4.1 we gave examples of attribute variables, discrete variables and continuous variables. Where the variation in the value a variable can take is random variation, that is, solely due to chance, we describe the variable as a *random variable* or *variate*. Examples of random variables are the score obtained when rolling a die and the number of heads occurring when three coins are tossed. A random variable can take only numerical values. It is conventional to denote a random variable, such as the score on a die or the weight of a randomly selected pig, by a capital letter – X or Y. If an experiment is conducted and the outcome, and hence the value of a random variable, observed, we denote this observed value by x. When we are concerned with a discrete random variable (see Unit 4.1(ii)), the table containing the values the random variable can assume together with the associated probabilities is called a *probability distribution*. The symbol $P(X = x)$ stands for the probability that the random variable, X, assumes the value x.

Examples of probability distributions are given in Tables 13.1, 13.2 and 13.3. We have met the probabilities in the first two tables before. In Table 13.3 we have to calculate the probabilities. Listing all the different possible outcomes we find there are 36 equally likely ways in which the two dice can fall. From the table you can see, for example, that there are six ways in which the score of seven can occur so the probability of a score of seven, $P(X = 7)$, is $\frac{6}{36}$ or $\frac{1}{6}$. You will notice that in each of the tables, the sum of the probabilities

Table 13.1 Probability distribution of number of heads when an unbiased coin is tossed

Number of heads $X = x$	$P(X = x)$
0	$\frac{1}{2}$
1	$\frac{1}{2}$
TOTAL	1

Table 13.2 Probability distribution of score on fair die

Score $X = x$	$P(X = x)$
1	$\frac{1}{6}$
2	$\frac{1}{6}$
3	$\frac{1}{6}$
4	$\frac{1}{6}$
5	$\frac{1}{6}$
6	$\frac{1}{6}$
TOTAL	1

Table 13.3 Probability distribution of total score when two dice are thrown

Total score $X = x$	Arrangements resulting in x	Number of arrangements resulting in x	$P(X = x)$
2	1 and 1	1	$\frac{1}{36}$
3	1 and 2, 2 and 1	2	$\frac{2}{36}$
4	1 and 3, 2 and 2, 3 and 1	3	$\frac{3}{36}$
5	1 and 4, 2 and 3, 3 and 2, 4 and 1	4	$\frac{4}{36}$
6	1 and 5, 2 and 4, 3 and 3, 4 and 2, 5 and 1	5	$\frac{5}{36}$
7	1 and 6, 2 and 5, 3 and 4, 4 and 3, 5 and 2, 6 and 1	6	$\frac{6}{36}$
8	2 and 6, 3 and 5, 4 and 4, 5 and 3, 6 and 2	5	$\frac{5}{36}$
9	3 and 6, 4 and 5, 5 and 4, 6 and 3	4	$\frac{4}{36}$
10	4 and 6, 5 and 5, 6 and 4	3	$\frac{3}{36}$
11	5 and 6, 6 and 5	2	$\frac{2}{36}$
12	6 and 6	1	$\frac{1}{36}$
TOTALS		36	1

equals one, that is, $\Sigma P(X = x) = 1$. This is as we would expect since we have listed all the possible outcomes of the three experiments.

There are some standard types of probability distributions that are particularly useful and we shall discuss these in Units 13.3 and 13.4.

13.2 Observed and expected frequencies

If you took a pair of dice and threw them a large number of times, say 360 (the number 360 has been chosen purely for convenience because it gives us whole numbers to deal with; it does not matter how many times you throw the dice as long as it is a sufficiently large number), you would record the *number* of times you obtained each different score, i.e. the frequency of each score. Table 13.4 shows the frequencies that a student observed when doing this experiment. We could divide each observed frequency by 360 to find the relative frequency of each score and these would be *empirical probabilities* (see Unit 12.1) to compare with the expected *a priori* probabilities in Table 13.3. It is more convenient, however, because of the statistical tests that are available, to compare the *observed frequencies* with the frequencies we would expect to obtain if our dice were unbiased. To obtain these *expected frequencies* we multiply each probability in Table 13.3 by 360 since this particular experiment involved 360 throws. The expected frequencies are shown in the final column of Table 13.4.

Table 13.4 Observed and expected frequencies of total scores when two dice are thrown 360 times

Total score $X = x$	Observed frequency	Expected frequency
2	4	10
3	16	20
4	38	30
5	48	40
6	46	50
7	55	60
8	55	50
9	36	40
10	25	30
11	28	20
12	9	10
TOTALS	360	360

The expected and observed frequencies in Table 13.4 provide the basis for some statistical analysis. We know what we expected to happen and we can see what actually did happen, and there are differences between the expected and the observed. Obviously it would be remarkable if the observed frequencies had exactly matched those expected: there are bound to be some differences due simply to random factors and because we can only make a finite number of observations. We will be interested, however, when the observed frequencies are markedly different from the expected, and especially in whether any differences are what is termed *statistically significant*. In Unit 15 we shall look at a method of testing to see if the differences between expected and observed frequencies are significant.

When the differences prove to be significant, we want to know why there are these differences. There can be three reasons why the expected frequencies are not the same as those observed:

(i) The assumption or *hypothesis* that we are working on is not correct. In this case the hypothesis is that the dice are unbiased.
(ii) The observations were not made correctly so a random sample has not been obtained. In this case it would mean that the dice had not been well shaken. Should you doubt that a pair of dice can be thrown incorrectly, try playing a dice game with some young children!
(iii) A rare sample has been obtained by chance.

We shall see how we could perform a *significance test* on the results in Table 13.4 in Unit 15.5(*b*) and Unit 15.7 (Exercise 11).

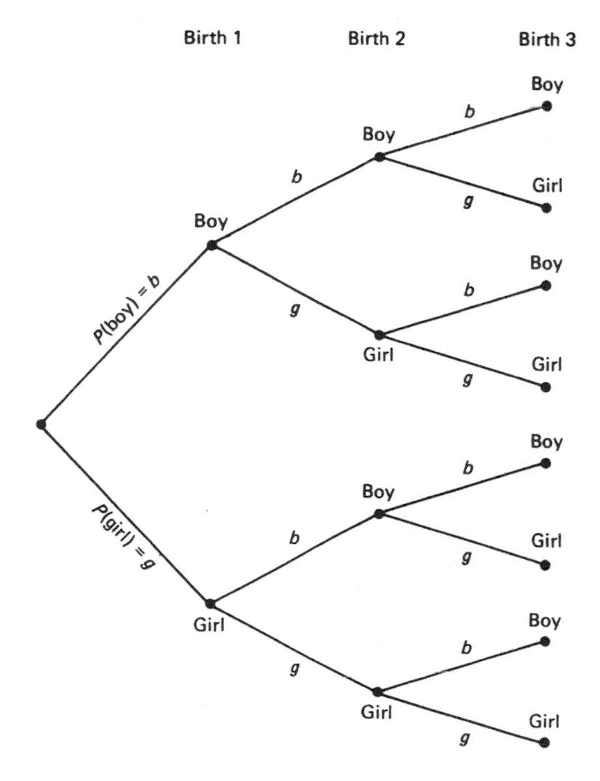

Fig. 13.1 Tree diagram of births into a three-child family

13.3 The binomial distribution

We often meet situations where each trial of an experiment has just two possible mutually exclusive outcomes. For example, an outcome of an experiment is a success or a failure; a coin lands either heads or tails; a birth into a family is either a boy or a girl; a light-bulb is either defective or not defective. If we repeat the trial several times and the probability of success remains constant from trial to trial, then the binomial distribution tells us the probability of any specified number of successes in all these trials. We can work out, for example, the probability of obtaining six heads when a coin is tossed ten times, the probability of having three daughters and a son in a family of four children or the probability of four defective light-bulbs in a batch of one hundred.

Let us look at a tree diagram (see Unit 12.5) for the possible births in a three-child family (Fig. 13.1). Each birth can be either a boy or a girl and we shall exclude the possibility of twins. Let us call the probability of a boy b and the probability of a girl g where $b + g = 1$ since the baby *must* be a boy or a girl. The first birth is shown by the two branches on the left of the diagram. The second birth results in either a boy or a girl so we have four possibilities for two births: boy and boy, boy and girl, girl and boy and girl and girl. Using the multiplication rule of probability (see Unit 12.5) the respective probabilities of these outcomes are:

$$b^2 \quad bg \quad gb \quad g^2$$

For three children there are eight ways in which the births can occur as shown in Table 13.5.

Table 13.5 Births into a three-child family

Order of birth	Probability
Boy Boy Boy	b^3
Boy Boy Girl	b^2g
Boy Girl Boy	$bgb = b^2g$
Boy Girl Girl	bg^2
Girl Boy Boy	$gb^2 = b^2g$
Girl Boy Girl	$gbg = bg^2$
Girl Girl Boy	$g^2b = bg^2$
Girl Girl Girl	g^3

Let us group together the results we have so far, to see if we can discover a pattern.

(i) One birth

Event	*Probability*
0 girls	b
1 girl	g

(ii) Two births

Event	*Probability*
0 girls	b^2
1 girl	$2bg$
2 girls	g^2

(iii) Three births

Event	*Probability*
0 girls	b^3
1 girl	$3b^2g$
2 girls	$3bg^2$
3 girls	g^3

You may recognize the probability terms as occurring in the binomial expansions $(b + g)$, $(b + g)^2$, $(b + g)^3$. If you have not met binomial expansions previously, you should consult the appendix at the end of this Unit. In fact as the family grows the probabilities of the possible combinations of children in the family are given by the successive powers of the binomial term $(b + g)$, hence the title *binomial distribution*.

Knowing this, we can quickly work out the probability of a four-child family having exactly three girls. The expansion of $(b + g)^4$ is:

$$(b + g)^4 = b^4 + 4b^3g + 6b^2g^2 + 4bg^3 + g^4$$

The terms give us, in order, the probabilities of 0 girls, 1 girl, 2 girls, 3 girls and 4 girls. Thus the probability of exactly three girls is $4bg^3$. Statistics of births show that the probability of a male birth is 0.51 and of a female birth, 0.49, so $4bg^3 = 0.24$ to two decimal places. Thus almost a quarter of all four-child families have three girls.

If we go on increasing the family size, the same method would enable us to find, for instance, the probability of having three girls in a family of seven children. We would expand $(b + g)^7$ and pick out the term containing b^4g^3.

A useful device for helping us to find the number of ways a particular family or other combination can occur is *Pascal's triangle*:

$$\begin{array}{c}
1 \\
1 \quad 1 \\
1 \quad 2 \quad 1 \\
1 \quad 3 \quad 3 \quad 1 \\
1 \quad 4 \quad 6 \quad 4 \quad 1 \\
1 \quad 5 \quad 10 \quad 10 \quad 5 \quad 1 \\
1 \quad 6 \quad 15 \quad 20 \quad 15 \quad 6 \quad 1 \\
1 \quad 7 \quad 21 \quad 35 \quad 35 \quad 21 \quad 7 \quad 1 \\
\cdots\cdots\cdots\cdots\cdots\cdots
\end{array}$$

The triangle shows the number, called a *binomial coefficient*, multiplying each successive power of b and g in the expansion of $(b + g)^n$ where n can be any non-negative integer. We have already seen, for example, that:

$$(b + g)^3 = b^3 + 3b^2g + 3bg^2 + g^3$$

and we can now write down the expansion of $(b + g)^7$:

$$(b+g)^7 = b^7 + 7b^6g + 21b^5g^2 + 35b^4g^3 + 35b^3g^4 + 21b^2g^5 + 7bg^6 + g^7$$

We conclude that the probability of having three girls in a family of seven children is:

$$35b^4g^3 = 35(0.51)^4\,(0.49)^3 \text{ assuming } b = 0.51 \text{ and } g = 0.49$$
$$= 0.28 \text{ to 2 decimal places}$$

To obtain binomial coefficients for larger values of n, Pascal's triangle may be extended. Each line always starts and ends with a one and each of the coefficients in the body of the triangle is the sum of the two coefficients above and to either side of it. Thus the next line, corresponding to $n = 8$ is:

$$1\ (1 + 7)\ (7 + 21)\ (21 + 35)\ (35 + 35)\ (35 + 21)\ (21 + 7)\ (7 + 1)\ 1$$

or:

$$1\quad 8\quad 28\quad 56\quad 70\quad 56\quad 28\quad 8\quad 1$$

so we can write:

$$(b+g)^8 = b^8 + 8b^7g + 28b^6g^2 + 56b^5g^3 + 70b^4g^4 + 56b^3g^5 + 28b^2g^6 + 8bg^7 + g^8$$

For values of n greater than about ten, Pascal's triangle becomes impracticable because of the numbers of terms involved. Also we are often interested in only one particular term rather than in the complete expansion. Fortunately there is a formula giving the binomial coefficient. The number of ways in which x girls can occur in a family of n children is:

$$\frac{n!}{x!\,(n-x)!}$$

$n!$ is *factorial n*. It is a shorthand way of writing the product of n and all the integers below n down to 1. Factorial 5, 5!, therefore, is:

$$5 \times 4 \times 3 \times 2 \times 1 = 120$$

and factorial 10, 10!, is:

$$10 \times 9 \times 8 \times 7 \times 6 \times 5 \times 4 \times 3 \times 2 \times 1 = 3628800$$

Factorial 1, 1!, is 1 and factorial zero, 0!, is defined to be 1.

To find the number of ways three girls can occur in a family of seven children, we evaluate

$$\frac{7!}{3!\,(7-3)!} = \frac{7!}{3!4!}$$
$$= \frac{7 \times 6 \times 5 \times 4 \times 3 \times 2 \times 1}{3 \times 2 \times 1 \times 4 \times 3 \times 2 \times 1}$$

$$= \frac{7 \times 6 \times 5}{3 \times 2 \times 1}$$

$$= 7 \times 5$$

$$= 35$$

This agrees with the result obtained using Pascal's triangle. Always cancel as many factors as possible before doing any multiplication. This helps to keep the numbers small.

The symbol $\binom{n}{x}$ is used to denote the binomial coefficient $\frac{n!}{x!\,(n-x)!}$. You will also find the notation ^nC_x or $_nC_x$ commonly used.

We must now generalize our results so they are in a form that is easy to apply to any binomial experiment. In a general binomial experiment each trial has just two mutually exclusive outcomes which, for convenience, can be called 'success' and 'failure'. We shall denote the probability of a success in one trial as p and the probability of failure as q, so $p + q = 1$ since one or other of these outcomes is bound to happen. If the trial is performed n times, then the probabilities of 0, 1, 2, 3, . . . up to n successes are given by successive terms in the binomial expansion of $(q + p)^n$. The probability of exactly x successes in n trials is:

$$P(X = x) = \binom{n}{x} q^{n-x}p^x \text{for } x = 0, 1, 2, \ldots, n$$

p and q must remain constant from trial to trial. Tables giving values of $\binom{n}{x}$ for various values of n and x are published and these are particularly useful when n is large. These tables, however, are not usually provided in examinations.

We shall work out one further binomial distribution application to illustrate its usefulness in the business world. A retail sales manager will accept delivery of a large consignment of goods if a random sample of twelve items contains not more than one defective. If, in fact, 2% of the producer's total output is defective, what is the probability that delivery of a consignment will be accepted?

Each item selected is either defective or not defective, thus:

$$P(\text{defective}) = 2\% = 0.02 = P(\text{success}) = p$$

$$P(\text{not defective}) = 1 - 0.02$$

$$= 0.98 = P(\text{failure}) = q$$

One of the conditions for using the binomial distribution is that p must be constant throughout the series of trials. This means that if we are taking items from a batch and not replacing each item before the next item is taken, then the binomial distribution does not strictly apply because the batch is fractionally smaller by one item each time. In practice, however, when the batch from which a sample is being taken is very large compared with the sample, the binomial distribution is a satisfactory approximation. As a rough guide you can consider the batch to be very large if it is more than about ten times the sample size. In the example considered here we are told this is a 'large' consignment of goods and a sample of only twelve is taken, so it is in order to use the binomial distribution.

We require the probability of not more than one defective in a sample of twelve items – which is the same as the probability of zero or one defectives in a sample of twelve:

$$P(\text{zero or one defectives}) = P(X = 0 \text{ or } 1)$$
$$= P(X = 0) + P(X = 1) \text{ using the simple addition rule}$$

$$P(X = x) = \binom{n}{x} q^{n-x} p^x \text{ using binomial distribution formula}$$
$$= \binom{12}{x} (0.98)^{12-x} (0.02)^x$$

Therefore:

$$P(X = 0) = \binom{12}{0} (0.98)^{12} (0.02)^0$$
$$= \frac{12!}{0!\,(12 - 0)!} (0.98)^{12} \times 1$$
$$= \frac{12!}{1 \times 12!} (0.98)^{12}$$
$$= 1 \times (0.98)^{12}$$
$$= 0.7847$$

and:

$$P(X = 1) = \binom{12}{1} (0.98)^{11} (0.02)^1$$
$$= \frac{12!}{1!\,(12 - 1)!} (0.98)^{11} (0.02)$$
$$= \frac{12!}{1 \times 11!} (0.98)^{11} (0.02)$$

$$= 12(0.98)^{11}(0.02)$$
$$= 0.1922$$

Thus:

$$P(\text{zero or one defectives}) = 0.7847 + 0.1922$$
$$= 0.977 \text{ to 3 decimal places}$$

We conclude that the probability that delivery of a consignment will be accepted is 0.977, that is, 97.7% of all consignments pass the manager's quality control check.

13.4 The Poisson distribution

The binomial distribution is useful in situations where we consider a fixed number of trials and count the number of successes. Sometimes we encounter cases where we do not have a definite number of trials and the binomial distribution cannot be used. For example, if we are studying the number of people entering a shop in a five-minute period, we can count the number of people who enter the shop but we have no way of knowing how many people did not enter the shop. In such cases, provided the only variation is of a random nature, the *Poisson distribution* enables us to find the probability of a specified number of events occurring when we know only the *mean* (average) number of times the events occur in a given time period.

The Poisson distribution probability for x occurrences of the desired event in a given time period is:

$$P(X = x) = e^{-\mu}\frac{\mu^x}{x!} \quad \text{where } x = 0, 1, 2, \ldots$$

where μ is the mean number of times the event occurs in the time period and e is a constant, called the *exponential number*. The number e has the value 2.7183 correct to four decimal places.

Suppose we have found that, on average, four people enter a shop in a five-minute period. The manager would like to know the probability of not more than five customers entering the shop in a five-minute period because the shop assistants cannot cope with more than five customers in this time. Using the information given we have $\mu = 4$, since this is the mean number of customers in a five-minute period. The Poisson probability of x customers entering in a five-minute period is:

$$P(X = x) = \frac{e^{-4} \times 4^x}{x!}$$

There are several ways of evaluating e^{-4}:

(i) Some calculators have the facility to work out e^{-x} directly. Others have a y^x facility so you key in $y = e$, that is, 2.7183 and $x = -4$, to evaluate e^{-4}.

(ii) Some books of tables give you the value of e^{-x} directly (see Table E). In fact tables of Poisson probabilities are published although these are not usually available for use in examinations. From Table E, $e^{-4} = 0.01832$.

(iii) e^{-4} can be rewritten as $\frac{1}{e^4}$. The value of e is 2.7183 so:

$$e^{-4} = \frac{1}{(2.7183)^4}$$

$$= \frac{1}{54.5996} \text{ using a calculator}$$

$$= 0.0183 \text{ to 3 significant figures}$$

If no calculator is available, then logarithm tables can be used to work out the power of e. This procedure has the advantage that it can be used even when μ is not a whole number. To find e^{-4} using logarithms:

$$\log e^{-4} = -4 \log e = \begin{array}{r} 0.4343 \times \\ -4 \\ \hline -1.7372 \end{array}$$

$$= -2 + 0.2628$$

$$= \bar{2}.2628$$

(Note that only the whole number part of a logarithm may be negative.)

$$\text{Antilog } \bar{2}.2628 = 0.01831$$

Thus e^{-4} equals 0.0183 to three significant figures as we found previously.

We can now evaluate the Poisson probabilities:

$$P(X=0) = \frac{e^{-4} \times 4}{0!} = \frac{e^{-4} \times 1}{1} = 0.0183$$

$$P(X=1) = \frac{e^{-4} \times 4^1}{1!} = \frac{e^{-4} \times 4}{1} = 0.0732$$

$$P(X=2) = \frac{e^{-4} \times 4^2}{2!} = \frac{e^{-4} \times 16}{2} = 0.1464$$

$$P(X=3) = \frac{e^{-4} \times 4^3}{3!} = \frac{e^{-4} \times 64}{6} = 0.1952$$

$$P(X=4) = \frac{e^{-4} \times 4^4}{4!} = \frac{e^{-4} \times 256}{24} = 0.1952$$

$$P(X = 5) = \frac{e^{-4} \times 4^5}{5!} = \frac{e^{-4} \times 1024}{120} = 0.1562$$

Thus, the probability of not more than five customers entering the shop in a five-minute period is the sum of these probabilities, which is 0.78 to two decimal places. We are not justified in writing down any further places of decimals as we have used a rounded value for e^{-4}. The manager has some cause for concern since the assistants are only able to cope for about 78% of the time. It might be beneficial to employ a further assistant, although the cost of this would have to be weighed against the cost of possible orders lost when too many customers enter the shop at any one time and decide to leave before making a purchase.

The Poisson distribution occurs in many real life situations. It can be used to describe the distribution of rare events such as floods, accidents and strikes which can be considered to occur randomly in time. For the Poisson distribution to be applicable, the mean number of occurrences of these events in a given period of time must remain constant. Note that there is no upper limit to the number of events that could occur in any one period of time although as x becomes very large, the Poisson probabilities become negligible.

13.5 The normal distribution

We looked at the normal distribution in Unit 8.9 and you should revise that section now. The normal distribution differs from the Binomial and Poisson distributions in that it applies to a *continuous* random variable rather than a discrete variate. We cannot write down the probability distribution for each value of X as we did in the discrete case but provided we know the arithmetic mean and the standard deviation of the distribution, we can work out the probability of the random variable lying between any two values $X = a$ and $X = b$, by finding the area under the normal curve between a and b (Fig. 13.2). You will remember that the area under the normal curve gives us the proportion of times that X lies between a and b and we now know that this is what we consider to be the probability that X lies between a and b.

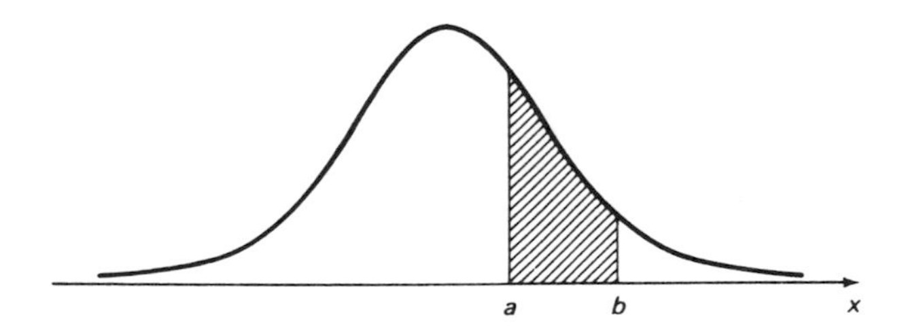

Fig. 13.2 Normal curve showing the probability that x lies between a and b

To remind you of the use of normal distribution tables let us look at an example where we need to work out the expected normal *frequencies* over a range of values of X. Suppose a manufacturer of men's sweaters wants to know how many to make on a production run of 10000 sweaters in each of the chest size ranges shown in Table 13.6.

Table 13.6 Size range of men's sweaters

Size range (cm)
80–85
85–90
90–95
95–100
100–105
105–110
110–115

It finds out that the arithmetic mean, μ, of men's chest sizes in the population where the sweaters will be sold is 98 cm, with a standard deviation, σ, of 5 cm, and it assumes that the distribution of sizes is normal. The distribution is shown in Fig. 13.3.

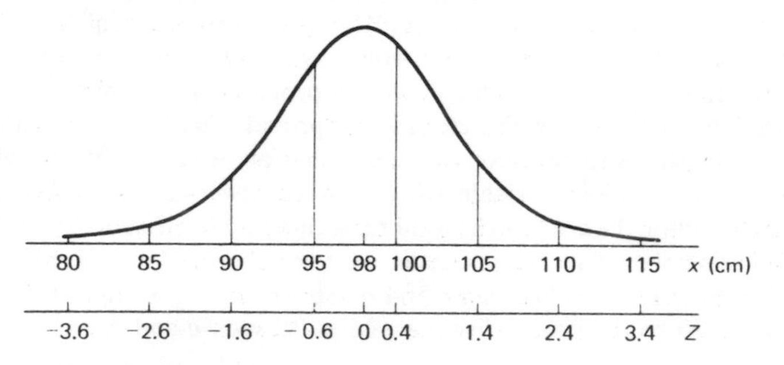

Fig. 13.3 Normal distribution of men's sweater sizes

We can use our knowledge of Z scores (see Unit 8.10) and the table of areas under the normal curve (Table D) to estimate the proportion of the population in each size range. To find the proportion between 80 cm and 85 cm we first calculate the Z score for 80 cm:

$$Z = \frac{x - \mu}{\sigma} = \frac{80 - 98}{5} = -3.6$$

From Table D we find that 0.4998 of the population lies between -3.6 and the mean, that is between 80 cm and 98 cm. Next we find the Z score for 85 cm:

$$Z = \frac{85 - 98}{5} = -2.6$$

From Table D, 0.4953 of the population is between −2.6 and the mean, that is, between 85 cm and 98 cm. We conclude that (0.4998–0.4953) of the population lies between 80 and 85 cm so the probability for chest size 80–85 cm is 0.0045.

Continuing with this method we find the probabilities for each size range (Table 13.7). You should check these probabilities for yourself.

Table 13.7 Probabilities and expected frequencies in each size range

Size range (cm)	Probability	Number to manufacture
80–85	0.0045	45
85–90	0.0501	501
90–95	0.2195	2195
95–100	0.3811	3811
100–105	0.2638	2638
105–110	0.0726	726
110–115	0.0079	79
TOTALS	0.9995	9995

The manufacturer has a production run of 10000 sweaters so the number to manufacture in each size range is found by multiplying the probability of each size range by 10000 (see final column of Table 13.7).

The total number to manufacture does not come exactly to 10000 because there are a few people at each end of the distribution not in these size ranges. Presumably to make up the production run of 10000 sweaters the manufacturer would make a few more in the middle ranges. Techniques such as these ensure that people who are very small or very large are not able to get clothes to fit them!

13.6 The relationship between the normal, binomial and Poisson distributions

(a) The normal approximation to the binomial distribution

In applications of the binomial distribution when n, the number of trials, becomes very large, the arithmetic becomes very involved. Fortunately, however, we can use the normal distribution as an approximation to the binomial distribution when n is sufficiently large.

Suppose we take a sample of 1000 items from a production line where the probability, p, of any one item being defective is 0.1 and we wish to know the probability of the sample containing more than 120 defectives. The binomial probabilities are given by the expansion of $(0.9 + 0.1)^{1000}$ and we require $P(X = 121) + P(X = 122) + P(X = 123) + \ldots + P(X = 1000)$. Even after realizing that this is the same as $1 - [P(X = 0) + P(X = 1) + \ldots + P(X = 120)]$, the arithmetic is very tedious.

It has been found that when the products np and nq for a binomial distribution are both greater than five, the normal distribution is a good approximation to the binomial distribution. It can be proved that the arithmetic mean, μ, of any binomial distribution is np while the standard deviation, σ, is $\sqrt{npq}$. You are not required to know the proofs of these results. In our example, we have:

$$p = 0.1$$

$$q = 1 - p = 0.9$$

$$n = 1000$$

$$\mu = np = 1000 \times 0.1 = 100$$

$$\sigma = \sqrt{npq} = \sqrt{1000 \times 0.1 \times 0.9} = \sqrt{90} = 9.49 \text{ to 2 decimal places}$$

There is one final problem before we can calculate a Z score. Using the binomial distribution we can only work out the probability of a *whole number* of defectives occurring. The normal distribution, being a continuous distribution, allows for *fractions* of defectives even though these cannot actually occur in practice. We thus allocate all the area under the normal curve between 120.5 and 121.5, for example, as the probability of obtaining exactly 121 defectives and the area between 121.5 and 122.5 as the probability of obtaining exactly 122 defectives and so on. To work out the probability of more than 120 defectives, we require the probability that X is greater than 120.5 so we need to find the Z score corresponding to 120.5 not 120. Using this procedure is called *applying a continuity correction*. The Z score is:

$$Z = \frac{x - \mu}{\sigma} = \frac{120.5 - 100}{9.49} = 2.16$$

Looking up a Z score of 2.16 in the table of the areas under the normal curve (Table D) tells us that 0.0154 of the population has more than 120 defectives, so the probability of more than 120 defectives is 0.0154.

(b) The normal approximation to the Poisson distribution

When μ, the mean of a Poisson distribution, becomes large, the Poisson distribution approaches a normal distribution and it is easier to work out the normal probabilities than the Poisson probabilities. The normal distribution is a satisfactory approximation to the Poisson for values of μ greater than about five. The mean of the Poisson distribution is, of course, μ and the standard deviation can be proved to be $\sqrt{\mu}$.

Suppose, for example, we found that on average 50 cars passed a certain checkpoint per hour and we wanted to know the probability of more than 60 cars passing in an hour, assuming a random traffic flow. Using the normal approximation and remembering the need for a continuity correction, we calculate a Z score in the usual way:

$$Z = \frac{x - \mu}{\sigma} = \frac{60.5 - 50}{\sqrt{50}} = 1.48$$

The normal curve table (Table D) tells us that 0.0694 of the population will be above this Z score so the probability of more than 60 cars passing is 0.0694.

(c) The Poisson approximation to the binomial distribution

Returning to the binomial distribution (see Unit 13.3), when n is large and p is small, the binomial distribution approaches the Poisson distribution; in fact when p is as small as 0.1 and n is larger than about 30, the two distributions are almost identical. Since the binomial distribution involves some awkward arithmetic when n is large, this fact can be very useful.

Suppose a company is producing components which it sells in boxes of 100, and it is known by the production manager that on average 2% of the output proves to be faulty. It is also known that the purchasers of the components will expect to be able to return and have replaced any boxes that have 6 or more faulty components in their contents. The management want to know how many boxes they must expect to replace out of the planned output of 200000 boxes in the next sales period.

For a box of 100 components, we could use the binomial distribution with $n = 100$ and $p = 0.02$ to find the probability of obtaining six or more faulty components in a box. As n is large, however, and p is small, it is easier to use the Poisson approximation to the binomial distribution. The mean, μ, of the distribution is the average number of faulty components in a box so:

$$\mu = np = 100 \times 0.02 = 2$$

For a Poisson distribution, the probability of obtaining exactly x defectives is:

$$P(X = x) = \frac{e^{-\mu}\mu^x}{x!}$$

We require:

$$\begin{aligned} P(X \geqslant 6) &= 1 - P(X < 6) \\ &= 1 - [P(X = 0) + P(X = 1) + P(X = 2) + P(X = 3) \\ &\quad + P(X = 4) + P(X = 5)] \end{aligned}$$

The value of e^{-2} is 0.1353 correct to four decimal places. We have:

$$P(X = 0) = \frac{0.1353 \times 2^0}{0!} = 0.1353 \qquad P(X = 3) = \frac{0.1353 \times 2^3}{3!} = 0.1804$$

$$P(X = 1) = \frac{0.1353 \times 2}{1!} = 0.2706 \qquad P(X = 4) = \frac{0.1353 \times 2^4}{4!} = 0.0902$$

$$P(X=2) = \frac{0.1353 \times 2^2}{2!} = 0.2706 \qquad P(X=5) = \frac{0.1353 \times 2^5}{5!} = 0.0361$$

$$\therefore P(X<6) = 0.9832$$

The probability of six or more faulty components in a box is approximately $1 - 0.9832 = 0.0168$. In a production run of 200000 boxes we would thus expect 200000×0.0168 to contain six or more faulty components so approximately 3400 boxes (to two significant figures) are likely to be returned in the next sales period.

13.7 Expected values

We have already commented that probability distributions are somewhat similar to frequency distributions. We can summarize various characteristics of a frequency distribution using measures of central tendency and dispersion. It is useful to be able to do the same for a probability distribution.

For the mean of a frequency distribution we have (see Unit 7.4):

$$\bar{x} = \frac{\Sigma fx}{\Sigma f}$$

The corresponding formula for the probability distribution of a discrete variable is:

$$\mu = \frac{\Sigma P(X=x) \times x}{\Sigma P(X=x)}$$ where the summation takes place over all possible values of x

$$= \Sigma P(X=x) \times x \text{ since } \Sigma P(X=x) = 1$$

μ is called the mean of the probability distribution or, alternatively, the *expected value* or *expectation* of the random variable X—it tells us the value of the variable that on average we would expect to obtain. Just as in the case of our observed frequency distributions, the value of μ does not necessarily correspond to a value of the variable that can actually occur.

Let us work out the expected value of X for each of the probability distributions discussed in Unit 13.1. It is best to set out the calculation in a table (see Tables 13.8, 13.9 and 13.10).

The expected number of heads when an unbiased coin in tossed is found to be $\frac{1}{2}$.

The expected score on a fair die is 3.5.

The expected total score when two dice are thrown is 7.

Note that in Table 13.8 the expected number of heads agrees with the formula we gave for the mean of a binomial distribution:

$$\mu = np = 1 \times \tfrac{1}{2} = \tfrac{1}{2}$$

The concept of expected value has many practical applications ranging from investment decisions to games of chance.

Table 13.8 Expected number of heads when an unbiased coin is tossed

Number of heads $X = x$	$P(X = x)$	$P(X = x).x$
0	$\frac{1}{2}$	0
1	$\frac{1}{2}$	$\frac{1}{2}$
TOTALS	1	$\frac{1}{2}$

Table 13.9 Expected score on a fair die

Score $X = x$	$P(X = x)$	$P(X = x)$
1	$\frac{1}{6}$	$\frac{1}{6}$
2	$\frac{1}{6}$	$\frac{2}{6}$
3	$\frac{1}{6}$	$\frac{3}{6}$
4	$\frac{1}{6}$	$\frac{4}{6}$
5	$\frac{1}{6}$	$\frac{5}{6}$
6	$\frac{1}{6}$	$\frac{6}{6}$
TOTALS	1	$\frac{21}{6} = 3.5$

Suppose you were asked to play a game of chance in which the banker rolls a die. You have to pay 50p for each roll of the die and you will be paid back £2 if a 6 is rolled, £1 if a 5 is rolled and nothing if a 1, 2, 3 or 4 is rolled. Do you consider that this is a fair game to play?

For a game to be fair your expected winnings and the banker's expected takings should balance in the long run and should both be zero. We thus need to work out the expected winnings. If they are zero, the game is fair. If they are positive, you stand to win more than you lose in the long run. If they are negative, you will lose money in the long run and the game is to the benefit of the bank. The calculation of the expected winnings is shown in Table 13.11.

The result is that your expected winnings are zero so it is a fair game to play.

13.8 Exercises

1. Eighty per cent of the employees taken on by an organization stay with the firm for at least five years. Of the next five employees to be engaged what

Table 13.10 Expected total score when two dice are thrown

Total Score $X = x$	$P(X = x)$	$P(X = x)$
2	$\frac{1}{36}$	$\frac{2}{36}$
3	$\frac{2}{36}$	$\frac{6}{36}$
4	$\frac{3}{36}$	$\frac{12}{36}$
5	$\frac{4}{36}$	$\frac{20}{36}$
6	$\frac{5}{36}$	$\frac{30}{36}$
7	$\frac{6}{36}$	$\frac{42}{36}$
8	$\frac{5}{36}$	$\frac{40}{36}$
9	$\frac{4}{36}$	$\frac{36}{36}$
10	$\frac{3}{36}$	$\frac{30}{36}$
11	$\frac{2}{36}$	$\frac{22}{36}$
12	$\frac{1}{36}$	$\frac{12}{36}$
TOTALS	1	$\frac{252}{36} = 7$

Table 13.11 Expected winnings in a game of chance

Score	Winnings (£) $X = x$	$P(X = x)$	$P(X = x).x$
6	1.50	$\frac{1}{6}$	$1.50 \times \frac{1}{6}$
5	0.50	$\frac{1}{6}$	$0.50 \times \frac{1}{6}$
1, 2, 3, 4	−0.50	$\frac{4}{6}$	$-0.50 \times \frac{4}{6}$
TOTALS		1	$\frac{2}{6} - \frac{2}{6} = 0$

is the probability that (*a*) one, (*b*) at least four and (*e*) all five stay with the organization for at least five years?

2. Ten per cent of the output of a production process is defective. Find the probability that out of a random sample of ten units taken from the production line (*a*) two will be defective; (*b*) at least eight will be good.

3. An average of 10 per cent of the employees of a company are absent on any day. The company employs six switchboard operators.

   (*a*) What is the probability of all switchboard operators being present on any day?

(*b*) What is the probability that at least four of them will be present?

4. It has been found that on average three serious accidents occur every month in an engineering factory. What is the probability that more than five serious accidents will occur in one month?

5. When can the Poisson distribution be used as an approximation to the binomial distribution? On audit checks in the past it has been found that 2.5% of all invoices contain errors. What is the chance that a random sample of 100 invoices contains 4 or more incorrect invoices?

6. The mean weight of sacks of a raw material is 150 kg and the standard deviation is 15 kg. Assuming the weights are normally distributed calculate what proportion of sacks weigh (*a*) between 125 and 160 kg, (*b*) more than 185 kg and (*c*) less than 130 kg.

7. A large organization runs an internal secretarial training course. To pass the course a trainee must attain a shorthand speed of 80 words per minute. Those trainees who attain a speed of 120 wpm receive a certificate of distinction. Past records show that the average shorthand speed attained by trainees is 95 wpm, with a standard deviation of 9 wpm, and the distribution of average speeds is normal. What is the probability of a trainee failing? What is the probability of a trainee receiving a certificate of distinction?

8. A large office employs 160 clerks. Absences average 40 per day and are normally distributed with a standard deviation of 8. The office manager maintains that if attendance falls to 110 the organization will collapse. What is the probability of that event occurring? If there are 200 working days in a year how many times a year does the organization collapse?

9. Ten per cent of men use a particular after-shave lotion. Use the normal approximation to the binomial distribution to find the probability that more than 60 men in a random sample of 500 use the after-shave.

10. A garage employs six mechanics each of whom can handle two breakdowns per day. On average eight breakdowns occur each day. Work out the probability of that garage not being able to help you if you break down (*a*) using the Poisson distribution and (*b*) using the normal approximation to the Poisson distribution.

11. You are asked to buy a lottery ticket. You can win £500 as first prize, £100 as second prize or £10 as third prize with probabilities 0.0001, 0.0005, and 0.001 respectively. What is a fair price to pay for the ticket?

12. A store opens at 0900 hours. The first ten-minute sessions commence at 0910 and from then there are 50 consecutive sessions. What is the expected number of sessions with 0, 1, 2, 3, 4 or more customers?

13.9 Appendix

A *binomial* is an algebraic expression containing *two* parts—for example, $(a + b)$, $(b + g)$, $(x + y)$, $(p + q)$. In our statistical work we are interested only in binomials raised to non-negative whole number powers—for example, $(a + b)^3$,$(b + g)^{10}$,$(x + y)^{20}$. In particular we wish to expand such expressions and collect together like terms since these give us the probabilities of particular events occurring. The lowest power is zero and any number raised to the power zero is one. Thus:

$$(a + b)^0 = 1$$

Any number raised to the power one is the number itself. Thus:

$$(a + b)^1 = a + b$$

We use the rules of algebra to work out higher powers:

$$\begin{aligned}
(a + b)^2 &= (a + b)\,(a + b) \\
&= a^2 + ab + ba + b^2 \\
&= a^2 + 2ab + b^2 \\
(a + b)^3 &= (a + b)\,(a + b)^2 \\
&= (a + b)\,(a^2 + 2ab + b^2) \\
&= a^3 + 2a^2b + ab^2 + ba^2 + 2ab^2 + b^3 \\
&= a^3 + 3a^2b + 3ab^2 + b^3
\end{aligned}$$

Higher powers can be obtained similarly (try working out $(a + b)^4$) but easier ways of obtaining the binomial coefficients (that is the numbers multiplying the powers of a and b) are given in Unit 13.3.

UNIT FOURTEEN

Sampling theory

14.1 Introduction

In Unit 2 we discussed reasons for taking samples from a population and the methods that can be used for sampling. In Units 7 and 8 ways of calculating statistical measures of location and dispersion from samples were described. We shall now turn to the problem of estimating a statistical measure for the *whole population* from the statistics calculated from a *sample* taken from that population, and the amount of confidence we can place in our estimate. In other words, we shall be finding out how to make realistic estimates about a whole population when all we have to go on are values obtained from a sample taken from that population. Throughout our discussion of sampling theory we shall assume we are dealing with *random* samples (see Unit 2.4(i)).

14.2 The sampling distribution of the arithmetic mean

(a) Formation of the distribution

Suppose, for example, that we wish to estimate the mean height of the adult male population of a large town. In theory it would be possible to calculate this mean height exactly if we knew the heights, x, of all the men. The task of finding all these heights, however, would be impracticable because of the expense and time involved and also because complete co-operation of the population is unlikely. It would be difficult for the observations to be carried out in a short period of time and changes in the population due to deaths, migration and immigration will occur. A sampling method involving fewer measurements and a consequent saving on expense and time has obvious attractions.

Let us denote the mean height of the population by μ and the standard deviation of the population by σ. Both of these quantities are usually unknown and our primary aim is to estimate μ. From a random sample of heights taken from the population, we can calculate the mean height of the sample, denoted by $\bar{x}$, and the standard deviation of the sample. We shall use n to stand for the *sample size*, that is, the number of observations in the sample. It is unlikely that the mean of the sample will be exactly the same as the mean of the population. The difference between μ and $\bar{x}$ is termed the *sampling error*. If we took a second random sample of size n from the population we would obtain another value of $\bar{x}$. In theory we could continue and take all possible random samples of size n and we could calculate $\bar{x}$ for each of these samples. If we listed all these values of $\bar{x}$, we would have another population – *the distribution of the sample means* or *the sampling distribution of the (arithmetic) mean.*

(b) Theorems concerning the distribution

Before we can use the sampling distribution of the mean we need to know how it is related to the population mean, μ, that we are trying to estimate. We shall now give several theoretical results. You do not need to know the proofs at this stage because they involve advanced mathematics but you *must* learn the results and how to use them.

(i) The mean of the sampling distribution of the mean is the same as the mean, μ, of the original, parent, population from which the samples were taken.

(ii) The standard deviation of the sampling distribution of the mean, referred to as the *standard error of the mean* and denoted by $\sigma_{\bar{x}}$, is equal to the standard deviation of the parent population divided by the square root of the sample size, $\sigma/\sqrt{n}$.

(iii) If the parent population has a normal distribution, the sampling distribution of the mean also follows a normal distribution regardless of the sample size. Even if the parent population is not normal, the sampling distribution of the mean approaches a normal distribution as the sample size becomes larger. For most practical purposes a sample size greater than 30 is large enough for the sampling distribution of the mean to be considered normal. This result is known as the *Central Limit Theorem.* It is one of the most important results in statistical method and demonstrates why the normal distribution is fundamental to advanced statistical work.

The relationship between two parent populations and the sampling distribution of the mean is shown in Fig. 14.1(*a*) and (*b*).

(*a*) Normal parent standard deviation σ

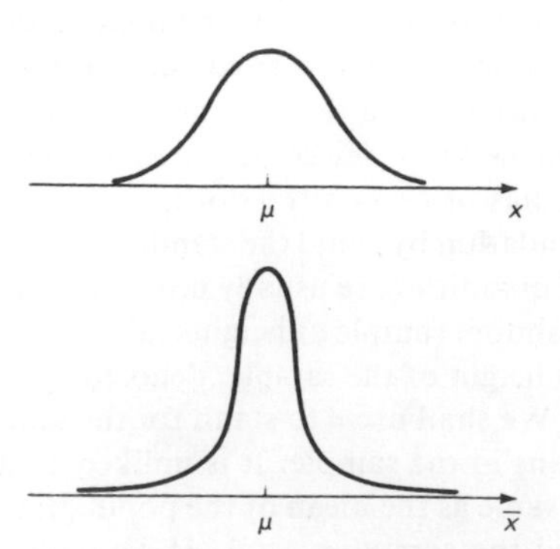

Normal sampling distribution, all n, standard error $\sigma/\sqrt{n}$

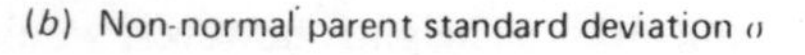

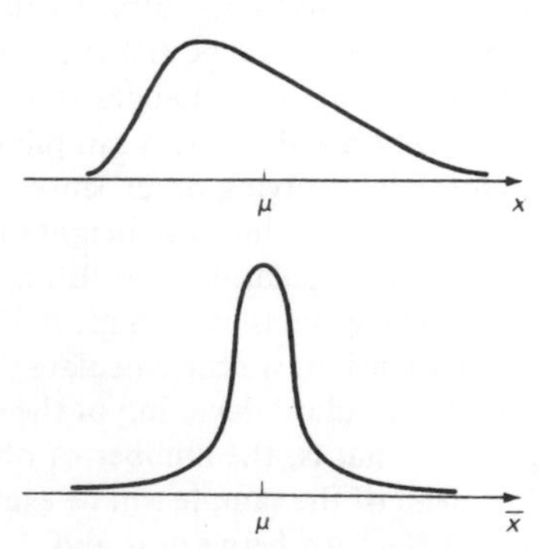

Normal sampling distribution, large n, standard error $\sigma/\sqrt{n}$

Fig. 14.1 Relationship between the sampling distribution of the mean and the parent population

(c) Estimation of a population mean

We are now going to use the theoretical results to enable us to estimate the mean of a population when all we know are the results from one sample of size n. We assume that either the parent population is normal or that n is large so that the sampling distribution of the mean can be considered normal (see Fig. 14.2).

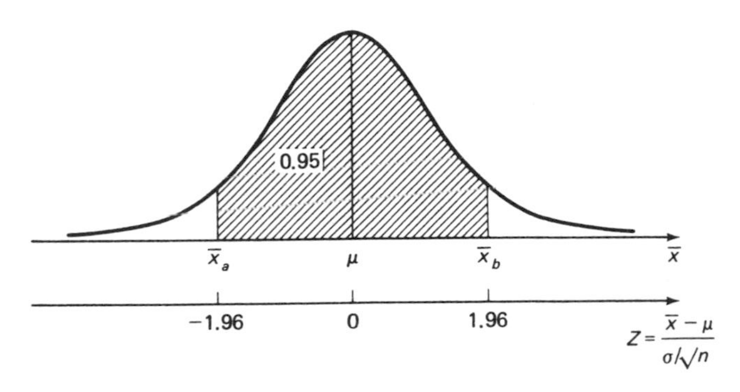

Fig. 14.2 The sampling distribution of the mean

The sample mean, $\bar{x}$, can take any value but it is most likely to be near μ. Looking at Fig. 14.2, where $\bar{x}_a$ and $\bar{x}_b$ are equidistant from μ on opposite sides of the mean, we can say that there is a probability of 0.95 that $\bar{x}$ will lie between $\bar{x}_a$ and $\bar{x}_b$ as 95% of the area under the normal curve, that is an area of 0.95 since the total area is 1, lies between $\bar{x}_a$ and $\bar{x}_b$. We can use Table D in reverse to find the Z scores of $\bar{x}_a$ and $\bar{x}_b$. The area between μ and $\bar{x}_b$ is $0.95/2 = 0.475$ and from Table D we find the Z value corresponding to this area and thus to $\bar{x}_b$ is 1.96. By symmetry, the Z value corresponding to $\bar{x}_a$ is -1.96.

We can now write down an expression for Z in terms of $\bar{x}$, μ, σ and n by remembering (see Unit 8.10) that a Z score is the difference between any value of the variable in a normal distribution and the arithmetic mean of that distribution divided by the standard deviation of the distribution. Here our normal variate is $\bar{x}$ and from the earlier theorems we know that the mean of the sampling distribution is μ and its standard deviation is the standard error $\sigma/\sqrt{n}$. We thus have:

$$Z = \frac{\bar{x} - \mu}{\sigma/\sqrt{n}}$$

The following three statements are, therefore, equivalent.

(i) There is a probability of 0.95 that $\bar{x}$ lies between $\bar{x}_a$ and $\bar{x}_b$.
(ii) There is a probability of 0.95 that Z lies between -1.96 and 1.96.
(iii) There is a probability of 0.95 that $\dfrac{\bar{x} - \mu}{\sigma/\sqrt{n}}$ lies between -1.96 and 1.96.

Statement (iii) can be rewritten using symbols—there is a probability of 0.95 that

$$-1.96 < \frac{\bar{x} - \mu}{\sigma/\sqrt{n}} < 1.96$$

By rearranging this inequality, the algebra of which you are not required to know, we obtain what is termed a 95% *confidence interval for* μ:

$$\bar{x} - 1.96\frac{\sigma}{\sqrt{n}} < \mu < \bar{x} + 1.96\frac{\sigma}{\sqrt{n}}$$

The quantities $\left(\bar{x} - 1.96\frac{\sigma}{\sqrt{n}}\right)$ and $\left(\bar{x} + 1.96\frac{\sigma}{\sqrt{n}}\right)$ are referred to as the 95% confidence limits for μ and can be evaluated provided we know the sample mean, $\bar{x}$, the sample size, n, and the population standard deviation, σ.

You must take great care in the interpretation of a confidence interval. μ is a fixed quantity whose value we do not know. When we take a sample of size n from the population and measure its mean, $\bar{x}$, there is a probability of 0.95 that the interval $\left(\bar{x} - 1.96\frac{\sigma}{\sqrt{n}}\right)$ to $\left(\bar{x} + 1.96\frac{\sigma}{\sqrt{n}}\right)$ contains μ. In other words, when we take a sample from the population, this procedure has a probability of 0.95 of providing us with an interval containing μ.

The result can also be stated as a *point estimate* for μ: we can assert with 95% confidence that $\mu = \bar{x} \pm 1.96\frac{\sigma}{\sqrt{n}}$.

Usually we do not know the value of σ and we use the standard deviation of the sample in place of σ. This approximation is possible only with large samples. For small samples taken from normal populations, the t distribution (see Unit 15) may be used. It can be proved that the formula:

$$s = \sqrt{\frac{\Sigma(x - \bar{x})^2}{n - 1}}$$

is a better estimator of σ than the formula:

$$\text{SD} = \sqrt{\frac{\Sigma(x - \bar{x})^2}{n}}$$

that we used in Unit 8.6 for measuring the dispersion of a frequency distribution. The formula with $(n - 1)$ in the denominator should always be used when estimating σ from sample measurements and in this book we shall use the symbol s to stand for a sample standard deviation calculated in this way. Obviously if n is large it will make little difference whether we have n or $n - 1$ in the denominator. You should assume if you are given the value of a standard deviation of a sample that it has been calculated using the formula with $n - 1$ in the denominator. The formula for s can be rewritten more conveniently as:

$$s = \sqrt{\frac{1}{n-1}\left[\Sigma x^2 - \frac{(\Sigma x)^2}{n}\right]}$$

or

$$s = \sqrt{\frac{1}{n-1}[\Sigma x^2 - n\bar{x}^2]}$$

The original formula with n in the denominator should be used when calculating the standard deviation of a frequency distribution merely to measure the dispersion and when evaluating the standard deviation of a population.

Let us now suppose that a random sample of 100 adult males from the population of a large town has a mean height of 179.5 cm with a standard deviation of 5 cm. We shall construct a 95% confidence interval for μ, the mean height of all adult males in the town. From the sample we know:

$$\bar{x} = 179.5 \text{ cm} \quad n = 100 \quad s = 5 \text{ cm}$$

We do not know σ, the standard deviation of the population, but as n is large we may use s in its place. The 95% confidence interval for μ is:

$$\left(179.5 - 1.96 \times \frac{5}{\sqrt{100}}\right) \text{cm} < \mu < \left(179.5 + 1.96 \times \frac{5}{\sqrt{100}}\right) \text{cm}$$

i.e.

$$(179.5 - 0.98) \text{ cm} < \mu < (179.5 + 0.98) \text{ cm}$$

i.e.

$$178.52 \text{ cm} < \mu < 180.48 \text{ cm}$$

Thus, the 95% confidence interval for μ is 178.52–180.48 cm. 1.96 is very nearly 2 and for many purposes it is satisfactory to take the 95% confidence limits for μ as:

$$\bar{x} \pm 2 \quad \text{standard errors}$$

In the above numerical example this procedure makes a difference of only 0.02 cm in each limit.

(d) Degrees of confidence

We cannot obtain the exact values of μ without measuring the whole of the population. From a sample we can work out the limits within which the population mean will lie for a specified degree of confidence. 95% confidence is most commonly used but any degree of confidence can be considered, for example 90%, 98%, 99% or 99.9% confidence limits are often found. In the formula for a 95% confidence interval, we have to replace 1.96 by the corresponding Z score for the required degree of confidence.

For example, at 98% confidence the area between the mean and Z is 0.49 giving a Z score of 2.33, and at 99% confidence the area between the mean and Z is 0.495, with a Z score of 2.58. Make sure that you can obtain these values from Table D. With a higher degree of confidence, the confidence interval is wider because we wish to be more certain that the interval contains μ. The 99% confidence interval for μ using the data of (*c*) above is:

$$\left(179.5 - 2.58 \times \frac{5}{\sqrt{100}}\right) \text{cm} < \mu < \left(179.5 + 2.58 \times \frac{5}{\sqrt{100}}\right) \text{cm}$$

i.e.

$$(179.5 - 1.29) \text{ cm} < \mu < (179.5 + 1.29) \text{ cm}$$

i.e.

$$178.21 \text{ cm} < \mu < 180.79 \text{ cm}$$

We conclude that there is a probability of 0.99 that the interval 178.21 cm–180.79 cm contains the unknown mean μ.

(e) The size of the sample

Before we take any sample a decision has to be made on its size. There would be little point taking a sample of 100 adult males in our example of (*c*) above, if we wished to estimate μ within a range of 1 cm with 95% confidence, because we found that the 95% confidence interval for μ based on a sample size of 100 spanned approximately 2 cm, from 178.52–180.48 cm.

If we wish to reduce the 95% confidence interval to span only 1 cm at most, then 1.96 standard errors must be 0.5 cm at most:

$$1.96 \frac{\sigma}{\sqrt{n}} < 0.5$$

We are again faced with the problem of not knowing σ, the standard deviation of the population. It may be that earlier work on the topic has been published and we can obtain a value for σ from secondary data. Alternatively we may have to carry out a preliminary survey in order to estimate σ. We shall assume that such a survey gave an estimate for σ of 5 cm. We thus have:

$$1.96 \times \frac{5}{\sqrt{n}} < 0.5$$

Remembering that inequalities may be manipulated like equations provided no negative numbers are involved, this becomes:

$$1.96 \times \frac{5}{0.5} < \sqrt{n}$$

i.e.

$$1.96 < \sqrt{n}$$

Squaring both sides we obtain:

$$384.16 < n$$

Thus the sample size must be at least 385 in order to estimate μ to within ± 0.5 cm with 95% confidence.

To summarize—in order to estimate the sample size n, we have to decide on the *precision* desired in our estimate of μ (± 0.5 cm) and the *degree of confidence* (95%), and we must have an *estimate* of σ.

Note that the sample size (385) for estimating μ to within ± 0.5 cm is almost four times as great as it was (100) when we estimated μ to within $\pm$ 0.98 cm. Because n occurs as a square root in the expression for the standard error, this means that to have a confidence interval we must quadruple the sample size and to reduce a confidence interval to a tenth of its former size we must take a sample one hundred times as large. You can see why sampling is said to be subject to the law of diminishing returns.

14.3 Sums and differences of random variables

In Unit 14.2 we considered the problem of estimating a single population mean. Often we are interested in comparing two or more populations and we want to be able to estimate differences between population means. Before we can do this we need to state some more results of statistical theory whose applications are not restricted to sampling problems.

It can be proved that if X and Y are two independent random variables with means μ_X, μ_Y and standard deviations σ_X, σ_Y respectively, then the random variable $T = X + Y$ has mean equal to $\mu_X + \mu_Y$ and standard deviation $\sqrt{\sigma_X^2 + \sigma_Y^2}$. Also the random variable $D = X - Y$ has mean equal to $\mu_X - \mu_Y$ and standard deviation $\sqrt{\sigma_X^2 + \sigma_Y^2}$. Note the plus sign in both cases for the standard deviation. A further result is that if X and Y have normal distributions then the variables $(X + Y)$ and $(X - Y)$ are also distributed normally. The results can be extended to the sums and differences of any number of independent random variables.

We now illustrate the uses of these results in two worked examples.

Example (i) A mass-produced electrical component consists of two separately manufactured and randomly selected parts A and B which are clipped together (Fig. 14.3(a)). Records of past production show that the lengths of part A have a normal distribution with mean 10 cm and standard deviation 0.3 cm and the lengths of part B have a normal distribution with mean 20 cm and standard deviation 0.8 cm. If the assembled component is more than 32 cm long, it has to be rejected because it will not fit into the final product. What proportion of components will be rejected for this reason?

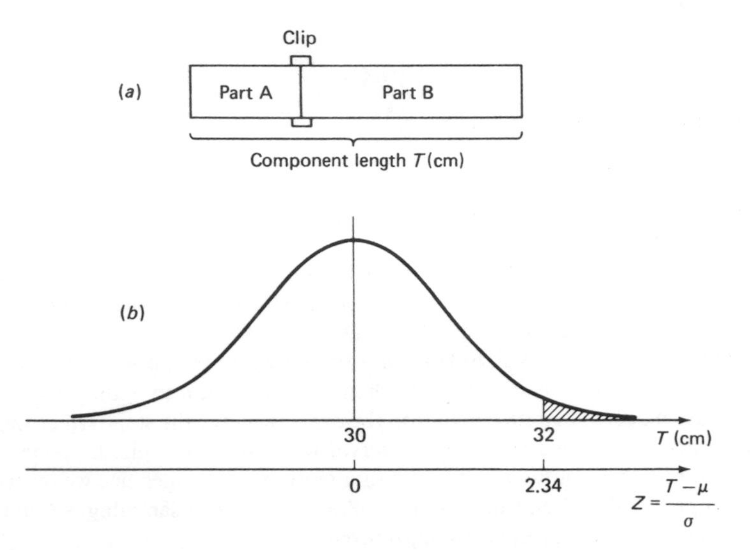

Fig. 14.3 The distribution of lengths of an electrical component

Let X be the length of part A and Y the length of part B. The length of part A is a normal variate and the length of part B is a normal variate so when we randomly select one of each type of part from the production line, the total length, T, of the two parts will also vary. The theoretical results tell us that T has a normal distribution with mean, μ, and standard deviation, σ, given by:

$$\mu = 10 + 20 = 30 \text{ cm}$$
$$\sigma = \sqrt{0.3^2 + 0.8^2}$$
$$= \sqrt{0.73}$$
$$= 0.854 \text{ cm to 3 significant figures}$$

In Fig. 14.3(*b*), we wish to find the proportion of components with a length greater than 32 cm so we need to find the Z score corresponding to $T = 32$:

$$Z = \frac{32 - 30}{0.854}$$
$$= 2.34$$

From Table D, we find the area beyond $Z = 2.34$ is 0.0096 so a proportion of 0.0096 (or 0.96%) of components will be rejected.

Example (ii) A dairy company sells yogurt in cartons. The weights of the cartons before filling are known to follow a normal distribution with mean 65 g and standard deviation 1.0 g. The gross weights of the filled cartons also follow a normal distribution with mean 165 g and standard deviation 2.5 g. The company has found that a customer will make a complaint if the weight of yogurt in a carton falls below 90 g. How many customer complaints can be expected for every batch of 100000 cartons sold?

The weight, D, of yogurt in a carton is a normal random variable since it is the difference between the gross weight of the filled carton and the weight of the unfilled carton, both of which are normal variates. The distribution of D has a mean, μ, and standard deviation, σ, given by:

$$\mu = 165 - 65 = 100\,\text{g}$$

$$\begin{aligned}\sigma &= \sqrt{2.5^2 + 1.0^2} \\ &= \sqrt{7.25} \\ &= 2.69\,\text{g to 3 significant figures}\end{aligned}$$

We must first find the proportion of yogurt weights less than 90 g (see Fig. 14.4) so we need to find the Z score corresponding to $D = 90$:

$$\begin{aligned}Z &= \frac{90 - 100}{\sqrt{7.25}} \\ &= -3.71\end{aligned}$$

From Table D we find the area beyond $Z = -3.71$ is 0.0001. Thus, in a batch of 100000 cartons, $100000 \times 0.0001 = 10$ cartons would be expected to have a yogurt weight of less than 90 g, so the dairy company can expect to receive 10 customer complaints.

14.4 Sampling distribution of the difference between two means

We are now going to apply the theoretical results of Unit 14.3 to the problem of finding a confidence interval for the difference between the means of two populations.

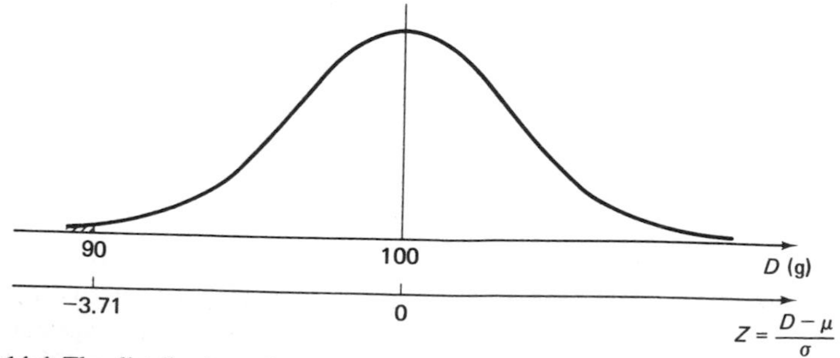

Fig. 14.4 The distribution of yogurt weights

Suppose we have a population with mean μ_1, and standard deviation σ_1. We take a sample of size n_1, from this population and work out its mean, $\bar{x}_1$ and its standard deviation s_1. From the central limit theorem we know that if we consider the distribution of the means of all samples of size n_1 this will be a normal distribution, provided n_1 is large or the parent population normal, with mean μ_1 and standard error $\sigma_1/\sqrt{n_1}$. Similarly if we consider samples of size n_2, mean $\bar{x}_2$, taken from a second population whose mean is μ_2 and standard deviation is σ_2, the distribution of $\bar{x}_2$ will also be normal, provided n_2 is large or the parent population normal, with mean μ_2 and standard error $\sigma_2/\sqrt{n_2}$.

Using the results of Unit 14.3, we may now consider the distribution of the difference between the sample means drawn from each population—the distribution of $\bar{x}_1 - \bar{x}_2$. We are dealing with the difference of two independent normal random variables so $(\bar{x}_1 - \bar{x}_2)$ will have a normal distribution, its mean, μ_D, will be $\mu_1 - \mu_2$ and its standard error will be:

$$\sqrt{\frac{\sigma_1^2}{n_1} + \frac{\sigma_2^2}{n_2}}$$

We can now use exactly the same type of reasoning as in Unit 14.2(*c*) to write down a 95% confidence interval for the difference between the two population means μ_1 and μ_2:

$$(\bar{x}_1 - \bar{x}_2) - 1.96\sqrt{\frac{\sigma_1^2}{n_1} + \frac{\sigma_2^2}{n_2}} < (\mu_1 - \mu_2) < (\bar{x}_1 - \bar{x}_2) + 1.96\sqrt{\frac{\sigma_1^2}{n_1} + \frac{\sigma_2^2}{n_2}}$$

The values of $\bar{x}_1$, $\bar{x}_2$, n_1 and n_2 are known once the sample has been taken, but the values of σ_1^2 and σ_2^2 are usually not known. Provided n_1 and n_2 are large, s_1^2 and s_2^2, the sample variances, may be used in place of σ_1^2 and σ_2^2 respectively. This procedure assumes that the population variances are unequal. If it is known or assumed that the population variances can be considered to be equal, the best estimate of the standard error of the difference between means is:

$$\sqrt{\frac{s^2}{n_1} + \frac{s^2}{n_2}} = s\sqrt{\frac{1}{n_1} + \frac{1}{n_2}}$$

where:

$$s^2 = \frac{(n_1 - 1)s_1^2 + (n_2 - 1)s_2^2}{n_1 + n_2 - 2}$$

This is referred to as a *pooled estimate*.

Suppose we know that a sample of 100 light bulbs from Company A has an average burning life of 1400 hours with a standard deviation of 80 hours. A sample of 144 light bulbs from Company B is found to have an average burning life of 1300 hours with a standard deviation of 100 hours. What are the 95%

confidence limits for the difference in mean burning times between light bulbs from these two companies?

We are given that:

$$n_1 = 100 \quad \bar{x}_1 = 1\,400\text{ hr} \quad s_1 = 80\text{ hr}$$

$$n_2 = 144 \quad \bar{x}_2 = 1\,300\text{ hr} \quad s_2 = 100\text{ hr}$$

The sample sizes are large so the distributions of $\bar{x}_1$ and $\bar{x}_2$ can be considered normal and hence the distribution of $(\bar{x}_1 - \bar{x}_2)$ will be normal.

We know that the 95% confidence limits for the difference between μ_1 and μ_2, the mean population burning times, are:

$$(\bar{x}_1 - \bar{x}_2) \pm 1.96\sqrt{\frac{\sigma_1^2}{n_1} + \frac{\sigma_2^2}{n_2}}$$

The population variances σ_1^2 and σ_2^2 are unknown but as the sample sizes are large, we may replace them by s_1^2 and s_2^2 respectively:

$$(\bar{x}_1 - \bar{x}_2) \pm 1.96\sqrt{\frac{s_1^2}{n_1} + \frac{s_2^2}{n_2}}$$

$$= (1\,400 - 1\,300) \pm 1.96\sqrt{\frac{80^2}{100} + \frac{100^2}{144}}$$

$$= 100 \pm 1.96\sqrt{64 + 69.44}$$

$$= 100 \pm 1.96\sqrt{133.44}$$

$$= 100 \pm 1.96 \times 11.55$$

$$= 100 \pm 22.64$$

We conclude that the 95% confidence limits for the difference between the two population means are 77.4 hours and 122.6 hours, correct to one decimal place.

14.5 Attribute sampling

(a) The distribution of sample proportions

The sampling distribution of the mean is only one of many sampling distributions we could construct. We may, for example, be interested in the distribution of the sample standard deviation or the sample median. Sometimes we are not dealing with a continuous variable such as height or burning hours but instead we are interested in whether or not an individual or an item has a certain *attribute* (see Unit 4.1(i)). We want to be able to estimate the proportion of a parliamentary constituency that will vote for a given political party at a general election or the proportion of consumers in a region who use a particular product or the proportion of total output that is defective coming off a production line. Again we wish to use sampling methods to obtain these estimates.

We take a sample of size n from the population and we count the number of items, x, in the sample that have the desired attribute. The remaining $(n - x)$ items will not have the attribute. We use p to stand for the proportion of the population that has the attribute. p is unknown to us and it is this value we wish to estimate. From the binomial distribution (see Unit 13.3) we know the probability that exactly x out of n items have the attribute, when the chance of any one item having the attribute is p, is given by:

$$P(X = x) = \binom{n}{x} p^x q^{n-x}$$

where $q = 1 - p, x = 0, 1, 2, 3, \ldots, n$ and the mean *number* of items with the attribute in the sample is np with standard deviation $\sqrt{npq}$.

To construct a confidence interval for the population proportion, p, it is more convenient to deal with the *proportion* of the sample with the attribute rather than with the *number* of items having this attribute. The proportion of the sample with the attribute is x/n and the probability of a sample of size n having this proportion with the attribute is $\binom{n}{x} p^x q^{n-x}$ as before. The mean proportion with the attribute is $\dfrac{np}{n} = p$ with standard deviation

$$\frac{\sqrt{npq}}{n} = \sqrt{\frac{pq}{n}} = \sqrt{\frac{p(1-p)}{n}}$$

We are saying that if we took all possible samples of size n from the population and measured the proportion of each sample with the attribute, the distribution of these sample proportions would have a mean equal to the true proportion of the population having the attribute and the standard deviation of this sampling distribution of proportions is $\sqrt{\dfrac{p(1-p)}{n}}$, referred to as the *standard error of the sample proportion*.

In Unit 13.6(*a*) we stated that when n is large enough for np and nq both to be greater than five, the normal distribution could be used as an approximation to the binomial distribution. We shall assume in all our estimation of proportions that this condition is satisfied. It is not too restrictive—a minimum sample size of thirty is usually adequate, provided p is not close to 0 or 1. Dealing with the proportion of the sample having the attribute rather than with the number in the sample is merely a change of scale and the distribution of sample proportions will also be normal. We can thus set up confidence intervals for the population proportion, p, in a way similar to that in Unit 14.2(*c*). It is convenient to denote the sample proportion x/n by the symbol $\hat{p}$ (pronounced p-hat).

From Fig. 14.5 we see that there is a probability of 0.95 that:

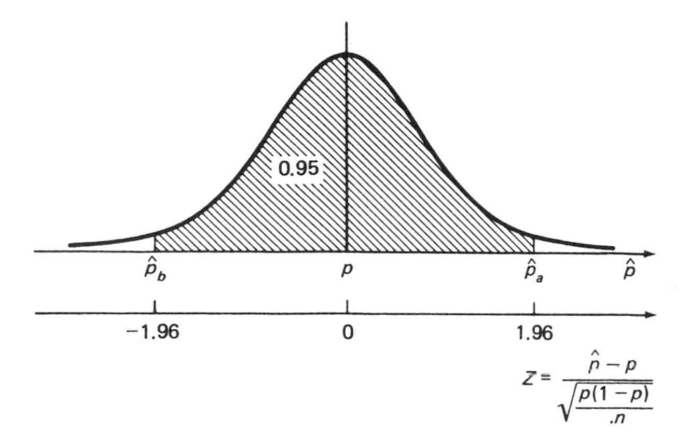

Fig. 14.5 The distribution of sample proportions

$$-1.96 < Z < 1.96$$

$$-1.96 < \frac{\hat{p} - p}{\sqrt{\dfrac{p(1-p)}{n}}} < 1.96$$

Rearranging this we have that there is a probability of 0.95 that:

$$\hat{p} - 1.96\sqrt{\frac{p(1-p)}{n}} < p < \hat{p} + 1.96\sqrt{\frac{p(1-p)}{n}}$$

We now have a 95% confidence interval for p but, in order to work out the confidence limits, we need to have an estimate for p to use in the square root terms. Provided the sample is large it is acceptable to use $\hat{p}$, the sample proportion, in place of p in the square root terms.

Suppose, for example, we wish to estimate 95% confidence limits for the proportion of the voting population in a parliamentary constituency who will vote Labour at the next general election. We send out an interviewer to question 100 randomly selected voters and she finds that 60 of the sample say they will vote Labour.

The size of the sample, n, is 100. $\hat{p}$, the proportion of the sample voting Labour is $\frac{60}{100} = 0.6$. $\hat{p}$ may be used in place of the population, p, in the standard error $\sqrt{\frac{p(1-p)}{n}}$ since n is large. The 95% confidence limits for p are:

$$\hat{p} \pm 1.96\sqrt{\frac{\hat{p}(1-\hat{p})}{n}} = 0.6 \pm 1.96\sqrt{\frac{0.6 \times 0.4}{100}}$$

$$= 0.6 \pm 1.96\sqrt{0.0024}$$
$$= 0.6 \pm 0.096$$

We conclude that the 95% confidence limits for the proportion of the population voting Labour are 0.504 and 0.696 or 50.4% and 69.6%. It would seem justifiable to predict a Labour victory in this constituency on the basis of this sample.

(b) The size of the sample

We must now consider the problem of how large a sample to take in order to estimate a population proportion, p, to a specified precision with the desired degree of confidence. This is of great importance to opinion pollsters and to market researchers. As they tend to deal with percentages rather than proportions, we shall obtain our results in this sub-section in terms of percentages.

We shall use P to denote the *percentage* of the population having the attribute so $P = 100p$ or $p = P/100$. We wish to know the sample size, n, that will enable us to be within $\pm L\%$ (that is a proportion of $L/100$) of the true value $P\%$ with 95% confidence. As noted previously other degrees of confidence such as 90% or 99% may be used. The precision, $\pm L\%$, desirable in the estimate of P depends to some extent on the value of P. If P is thought to be around 5%, then a precision of $\pm\frac{1}{2}\%$ may be appropriate whereas if P is about 80%, a precision of $\pm 4\%$ would probably be sufficient. In any event the precision must be specified at the outset.

Using similar reasoning to that used in Unit 14.2(*e*), 1.96 times the standard error of proportions must span at most $L\%$ (i.e. $L/100$). We thus have:

$$1.96 \times \text{standard error of proportions} < \frac{L}{100}$$

$$\therefore 1.96\sqrt{\frac{p(1-p)}{n}} < \frac{L}{100}$$

or

$$1.96\sqrt{\frac{\frac{P}{100}\left(1-\frac{P}{100}\right)}{n}} < \frac{L}{100}$$

Rearranging this inequality we find:

$$n > \frac{3.8416P(100-P)}{L^2}$$

Because 3.8416 is close to 4, you will often find the result quoted as n must be at least $4P(100-P)/L^2$. We need some idea of the value of P before n can be evaluated. A preliminary survey or previous work may give some guide. The greatest value of $P(100 - P)$ occurs when P is 50% so putting $P = 50$ will ensure a sample size that is large enough.

Suppose we wish to specify the percentage voting Labour in a parliamentary constituency to within ±2%. From a pilot survey we know that approximately 60% intend to vote Labour. What sample size do we need to take, working with 95% confidence?

We have:

$$L = 2 \quad P \simeq 60$$

Thus:

$$n > \frac{3.8416P(100 - P)}{L^2}$$

$$\text{i.e.} \quad n > \frac{3.8416 \times 60 \times 40}{2^2}$$

$$\text{i.e.} \quad n < 2304.96$$

A sample of at least 2305 is needed.

(c) The difference between two proportions

As in the case of sample means, we often wish to consider the difference in the proportions having a particular attribute in two distinct populations. For example, we might want to know by how much the proportion of consumers buying a particular brand of washing powder differs between two regions of the country.

We take a sample, size n_1, from a population which has an unknown proportion, p_1, with the desired attribute. We count the number, x_1, in the sample with the attribute and thus the proportion, $\hat{p}_1$, of the sample with the attribute is x_1/n_1. We know from sub-section (*a*) that $\hat{p}_1$ is a random variable with mean p_1 and standard error $\sqrt{\frac{p_1 q_1}{n_1}}$ where $q_1 = 1 - p_1$. Independently we take a sample, size n_2, from a second population with proportion p_2 having the attribute. x_2 is the number in the sample with the attribute and the sample proportion $\hat{p}_2$ is x_2/n_2 where we know that $\hat{p}_2$ is a random variable with mean p_2 and standard error $\sqrt{\frac{p_2 q_2}{n_2}}$ with $q_2 = 1 - p_2$. From the theoretical results of Unit 14.3 we can deduce that the variate $(\hat{p}_1 - \hat{p}_2)$ has a distribution with mean $(p_1 - p_2)$ and the standard error of this difference between proportions is:

$$\sqrt{\frac{p_1 q_1}{n_1} + \frac{p_1 q_2}{n_2}} = \sqrt{\frac{p_1(1 - p_1)}{n_1} + \frac{p_2(1 - p_2)}{n_2}}$$

If n_1 and n_2 are large, the distribution of $\hat{p}_1$ and the distribution of $\hat{p}_2$ are approximately normal so the distribution of $(\hat{p}_1 - \hat{p}_2)$ can also be considered to be normal and we can construct confidence intervals in the same manner as previously.

To evaluate the standard error it is acceptable to use $\hat{p}_1$ and $\hat{p}_2$ in place of p_1 and p_2 provided n_1 and n_2 are large. If there is reason to suppose that p_1 and p_2 are equal, then the pooled estimate:

$$\hat{p}_0 = \frac{x_1 + x_2}{n_1 + n_2}$$

should be used in place of both p_1 and p_2.

Suppose that a sample survey in region A showed that 1100 out of 4000 randomly selected shoppers said that they regularly bought Wosh detergent powder, whereas in a sample of 2000 similar shoppers in region B only 200 stated that they bought this powder regularly. What are the 99% confidence limits for the difference in the proportion of all shoppers who buy Wosh regularly in the two regions?

We need to evaluate, for 99% confidence limits:

$$(\hat{p}_1 - \hat{p}_2) \pm 2.58 \text{ standard errors}$$

i.e.
$$(\hat{p}_1 - \hat{p}_2) \pm 2.58\sqrt{\frac{p_1(1-p_1)}{n_1} + \frac{p_2(1-p_2)}{n_2}}$$

From the samples we have:

$$\hat{p}_1 = \frac{1100}{4000} = 0.275 \qquad n_1 = 4000$$

$$\hat{p}_2 = \frac{200}{2000} = 0.1 \qquad n_2 = 2000$$

As n_1 and n_2 are large, we may use $\hat{p}_1$ in place of p_1 and $\hat{p}_2$ in place of p_2 when evaluating the standard error. Thus the 99% confidence limits are:

$$(0.275 - 0.1) \pm 2.58\sqrt{\frac{0.275 \times 0.725}{4000} + \frac{0.1 \times 0.9}{2000}}$$

$$= 0.175 \pm 2.58\sqrt{0.00004984 + 0.000045}$$

$$= 0.175 \pm 2.58\sqrt{0.00009484}$$

$$= 0.175 \pm 2.58 \times 0.009739$$

$$= 0.175 \pm 0.025 \quad \text{to 3 decimal places}$$

Thus, with 99% confidence, the true difference in the proportions buying Wosh regularly in regions A and B is 0.175 ± 0.025 or, in percentage terms, $17.5\% + 2.5\%$.

14.6 Exercises

1. If $\bar{x} = 40$ and $\sigma = 8$ in a sample of 36, find (*a*) the 95% confidence limits for μ and (*b*) the 99% confidence limits for μ.

2. A random sample of 100 sales invoices taken from a company's file of sales invoices has an average value of £42.80 with a standard deviation of £7.32. Find: (*a*) the 95% confidence limits for the true average value of invoices; (*b*) the 99% confidence limits for the true average value of invoices.

3. A company employs 56 sales representatives and wants to be sure that each one is allowed an adequate travel allowance each month based on a rate of 20p per kilometre. During the last month the average distance travelled by them was 484 kilometres with a standard deviation of 12 kilometres. What would be an adequate allowance? (An adequate allowance would cover, say, 95% of monthly travel of all sales representatives.)

4. A supplier of cables must quote the average breaking strain of each batch of cables produced. The value quoted must be within ± 10 kg of the true mean for 99% of the output. Records show that the breaking strains have a standard deviation of 80 kg. How big should each batch be for the supplier to estimate accurately?

5. (*a*) Candidates' marks in an examination in statistics were found to have a mean of 50% and a standard deviation of 10%. The marks were normally distributed. What proportion of candidates had marks (i) above 70%, (ii) below 40%?

   (*b*) It the marks gained referred to the results of 100 candidates, what is the reliability of the mean mark of 50% (use 95% confidence limits)?

   (The Institute of Chartered Secretaries and Administrators)

6. A transport company purchases tyres from a certain manufacturer. From each tyre a life of 24 000 kilometres is required. A new manufacturer tenders to supply tyres and a sample of 50 of its tyres has an average life of 25 000 kilometres with a standard deviation of 700 kilometres when tested.

   (*a*) Estimate the proportion of tyres supplied by this company that will have an average life of less than 24 000 kilometres even if the most optimistic estimate of the mean life at 95% confidence is assumed.

   (*b*) Explain how you reached this estimate.

7. (*a*) The central limit theorem states that, as n increases, the distribution of sample means is normal with mean μ and standard error $\sigma/\sqrt{n}$. Explain briefly the meaning of the terms: (i) distribution of sample means, (ii) standard error.

   (*b*) Items are manufactured to a mean weight of 3 kg and a standard deviation of 0.05 kg. Cartons each containing nine items are sold on the understanding that the mean weight of items in a carton is not less than 2.97 kg. Cartons not meeting this requirement are rejected.

(i) Calculate the proportion of cartons which are rejected. (ii) Suppose now that the proportion rejected can be reduced by adjusting the process so that the mean weight of all items is increased to 3.01 kg. This adjustment costs £0.36 per full carton. If the cost of a rejected carton is £10, calculate whether the adjustment is economically desirable.

8. A survey of 100 drivers shows that 35 drive cars less than one year old.

   (*a*) Estimate at the 95% confidence level the proportion of drivers in the population from which the survey was taken that drive cars less than one year old.

   (*b*) How big a sample would you need to take to be able to find the proportion to within ±1% at the same confidence level?

9. Explain what is meant by a confidence interval. A recent market survey for your company reveals that in the Greater London area 54% of people questioned had purchased your company's product within the last four weeks. Given that this percentage was based upon a sample of 800 respondents, calculate the 95% confidence interval for the proportion of people in Greater London who have bought your company's product.
(*Institute of Chartered Accountants: part question*)

10. The standard procedure used by a company for detecting defects in a product has been shown, from the experience of examining 5000 defects, to detect 90% of defects present. Calculate the standard error of the proportion of defects detected out of 100 examined and find the probability that, out of 100 defects, 15 or more are undetected by the procedure. An alternative procedure is suggested which during trials detects 368 out of 400 defects. What are the 95% confidence limits for the difference in proportions detected by these two methods?

(Answers at the end of the book.)

UNIT FIFTEEN

Significance testing

15.1 Introduction

In this Unit we shall be concerned with comparing observations with expectations and deciding whether the differences between them are what is termed *statistically significant*. In Unit 13 we looked at methods that can be used to find expected outcomes of a statistical experiment, and acknowledged that there will be some differences between observed and expected outcomes purely because of random factors. Now we consider how to determine whether the differences are so large that it is unlikely that purely random factors could have caused them. If this is so, then we must either find out why the differences have occurred or change the assumptions we have been making about the population under investigation.

When testing for significance, we always adopt a formal approach. We state the assumption that we are making about the population; this assumption is called the *null hypothesis* and referred to as H_0. Examples of null hypotheses are:

H_0: The mean lifetime, μ, of a population of light bulbs is 1 600 hr.
H_0: The proportion, p, favouring the Government in a parliamentary constituency is 0.6.
H_0: The observations come from a Poisson distribution.

The next step is to state an *alternative hypothesis*, H_1, and this is usually the opposite of the null hypothesis. Examples of alternative hypotheses are:

$H_1: \mu \neq 1\,600$ hr ($\neq$ means 'not equal to')
or $H_1: \mu < 1\,600$ hr
$H_1: p \neq 0.6$
H_1: The distribution is not Poisson.

Tests for significance can be classified as *one-tailed* or *two-tailed* depending on the form of the alternative hypothesis. For example, when testing:

$$H_0: \mu = 1\,600 \text{ hr}$$

against the alternative:

$$H_1: \mu \neq 1\,600 \text{ hr}$$

we are interested in whether the population mean is *different from* 1 600 hours either on the high side or on the low side and this would involve a two-tailed test. When testing:

$$H_0: \mu = 1\,600 \text{ hr}$$

against the alternative:

$$H_1: \mu < 1\,600 \text{ hr}$$

we are interested only in whether the mean is *less than* 1600 hours and this involves a one-tailed test. This could in fact be the case when we are dealing with the lifetimes of light bulbs as we would only be concerned if the mean lifetime was less than 1600 hours. If the mean was greater than 1600 hours, we would be getting a longer burning lifetime than we had anticipated and we would not wish to complain.

The test for:

$$H_0: \mu = 1\,600 \text{ hr}$$

against:

$$H_1: \mu > 1\,600 \text{ hr}$$

would also involve a one-tailed test. In an examination it is usually clear from the wording of a question whether a one-tailed or a two-tailed test is appropriate.

The null hypothesis could of course be completely verified if we examined the whole population. Using statistical methods, however, we wish to be able to test the hypothesis using the results of one sample taken from the population. We can never be sure that any conclusion we draw about the population from the sample will be correct because of the random variation between different samples from the same population. There are two ways in which we may be in error. We commit a *Type I error*, when we reject a null hypothesis that is in fact true. The maximum probability of committing a Type I error in a particular test is called the *significance level* of the test and we specify at the outset of a test what level we intend to use. Once the significance level has been specified, we can set up a decision criterion based on the assumption that the null hypothesis is correct. The 0.05 or 5% level is customary but the 0.01 (1%) and 0.001 (0.1%) levels are also commonly used. Using a 5% level of significance we are saying that only 5 times in 100 will we have rejected the null hypothesis when we should have accepted it. A *Type II error* is committed when we accept a null hypothesis when it is in fact false. The probability of committing this type of error can be investigated once the test procedure has been set up. If you progress to more advanced statistical work, you will investigate Type II errors in detail but we shall not consider them further here.

To set up the decision criterion for a test we need to use the sampling distributions discussed in Unit 14 and we shall give the specific procedures for various tests in later sections of this Unit. Having laid down the decision criterion, observations are made on a sample of the population. If the test statistic calculated from this sample falls in the rejection area as specified by the decision criterion, the null hypothesis is rejected and the alternative hypothesis is accepted.

Significance testing is one of the most important techniques in statistical method. It is the cornerstone of many statistical applications in business, particularly in such areas as market research and quality control.

15.2 Tests concerning means

(a) Two-tailed test for a single mean

Suppose, for example, that a machine fills packets with bird-seed, putting on average a weight of 1 kg in each packet. It has been found that the weight delivered varies from packet to packet, with a standard deviation of 0.04 kg. Unfortunately the machine does tend to drift away from the average on occasions and it then needs to be reset. In order to test for this drift, a sample of 50 packets is taken from the production line and the arithmetic mean weight of the sample is found to be 1.01 kg. Is this evidence that the machine needs resetting?

In this example, we set:

H_0: the mean weight, μ, of all packets is 1 kg: $\mu = 1$ kg
$H_1: \mu \neq 1$ kg

We use a two-sided alternative here as the manufacturer would need to reset the machine whether it drifted above or below the 1 kg setting. We shall use the 0.05 level of significance.

The sample size, n, is large (50) so we may assume that the means of samples drawn from the population of weights of all the packets follow a normal distribution with mean, μ, and standard error $\sigma/\sqrt{n}$ where σ is the standard deviation of the population of packet weights. A sample mean, $\bar{x}$, can take any value and assuming H_0 is true, has the distribution shown in Fig. 15.1. As we are performing a two-tailed test at the 0.05 level of significance we intend not to reject H_0 if $\bar{x}$ lies beneath the unshaded region in Fig. 15.1 but we shall reject H_0 if $\bar{x}$ lies beneath either of the two shaded areas because there is only a chance of 0.05 that $\bar{x}$ lies here if the true population mean is 1 kg. We need to know the values of the sample mean, $\bar{x}_a$ and $\bar{x}_b$, at the boundaries of the shaded regions. From the normal distribution tables, Table D, and from Unit 14.2(*c*), we know that these values correspond to Z scores of ± 1.96 and it is the custom to set up the significance test in terms of Z values rather than in terms of values of $\bar{x}$. Our decision criterion, therefore, is:

$$\text{if } -1.96 < Z < 1.96, \text{ do not reject } H_0;$$

$$\text{if } Z > 1.96 \text{ or } Z < -1.96, \text{ reject } H_0.$$

-1.96 to $+1.96$ is termed the *acceptance region* for Z. Values of Z leading to the rejection of H_0, that is, $Z > 1.96$ or $Z < -1.96$, are said to fall in the *rejection* or *critical region*. The values $Z = \pm 1.96$ are called the *critical values* of Z.

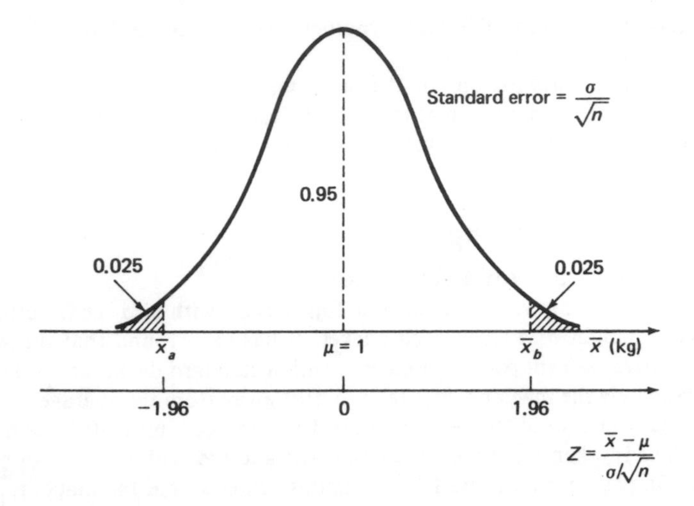

Fig. 15.1 Rejection regions for a two-tailed test of a mean

We now consider the information obtained by the manufacturer:

$$n = 50 \qquad \bar{x} = 1.01 \text{ kg} \qquad \sigma = 0.04 \text{ kg}$$

and we work out the *test statistic*, Z,

$$Z = \frac{\bar{x} - \mu}{\sigma/\sqrt{n}}$$

$$= \frac{1.01 - 1}{\frac{0.04}{\sqrt{50}}}, \text{ assuming } H_0 \text{ is true,}$$

$$= \frac{0.01}{0.04} \times \sqrt{50}$$

$$= 1.78$$

The value 1.78 lies in the region −1.96 to +1.96 which is the acceptance region. We conclude that there is not sufficient evidence to suggest that the machine needs resetting. The observed difference between the sample mean and 1 kg is likely to have occurred purely by chance.

Suppose that later in the day a second sample of size 50 was taken and found to have a mean weight, $\bar{x}$, of 1.02 kg. Would we alter our conclusions?

The null hypothesis, the alternative hypothesis, the significance level and hence the decision criterion remain unchanged. We need to work out the value of the test statistic, Z, on the basis of the second sample:

$$n = 50 \qquad \bar{x} = 1.02\ \text{kg} \qquad \sigma = 0.04\ \text{kg}$$

and we work out the test statistic, Z:

$$Z = \frac{\bar{x} - \mu}{\sigma/\sqrt{n}}$$

$$= \frac{1.02 - 1}{\frac{0.04}{\sqrt{50}}}, \text{ assuming } H_0 \text{ is true,}$$

$$= \frac{0.02}{0.04} \times \sqrt{50}$$

$$= 3.54$$

3.54 is greater than 1.96 so the value of the test statistic is now in the rejection region and we reject H_0. We conclude that the mean of the population appears to be significantly different from 1 kg and there is evidence that the machine needs resetting.

(b) One-tailed test for a single mean

Suppose that a firm manufactures light bulbs and the sales manager claims that tests have proved that the average life of these bulbs is 1000 hours. We wish to buy some of these light bulbs but before doing so we wish to check that the claim for their average life is correct. We therefore take a sample of 36 bulbs and test them; we find that the average life of the bulbs in the sample is 940 hours with a standard deviation of 126 hours. Is the manufacturer's claim justified?

In this example we shall perform a one-tailed test as we are only concerned as to whether the sample provides evidence that the true mean lifetime is *below* 1000 hours. We thus have:

H_0: the mean life, μ, of all the bulbs manufactured by the firm is 1000 hr: $\mu = 1\,000$ hr
$H_1: \mu < 1000$ hr

and we shall use the 0.05 level of significance.

Once again we have a large sample ($n = 36$) so the distribution of sample means can be considered to be normal and has mean, μ, and standard error $\sigma/\sqrt{n}$. Looking at Fig. 15.2 we will reject H_0 only if $\bar{x}$ lies under the shaded area of the normal curve because we are now performing a *one*-tailed test at the 0.05 level of significance. Using Table D, we find the Z score corresponding to an area of 0.05 beyond Z is 1.645 so, as we are concerned with values of $\bar{x}$ below the mean, the critical value of Z is -1.645. The decision criterion is:

$$\text{if } Z > -1.645, \text{ do not reject } H_0;$$

$$\text{if } Z < -1.645, \text{ reject } H_0.$$

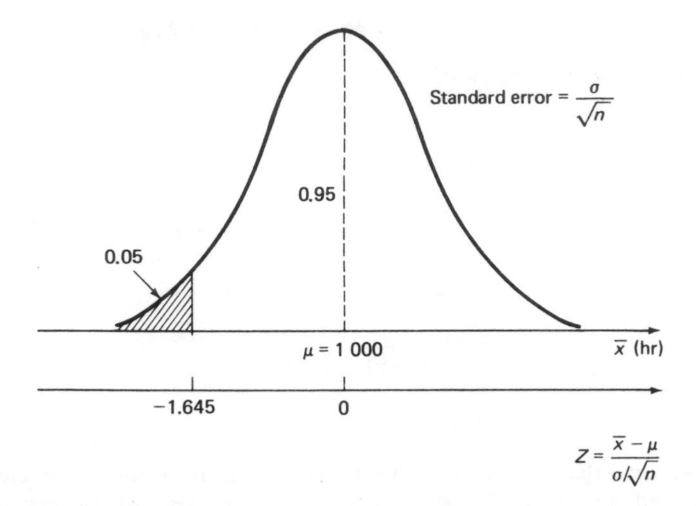

Fig. 15.2 Rejection region for a one-tailed test of a mean

From the sample we have:

$$n = 36 \qquad \bar{x} = 940 \text{ hr} \qquad s = 126 \text{ hr}$$

As the sample size is large, we may use s as an estimate of σ. The test statistic, Z, is given by:

$$Z = \frac{\bar{x} - \mu}{\sigma/\sqrt{n}}$$

$$= \frac{940 - 1000}{126/\sqrt{36}}, \text{ assuming } H_0 \text{ is true}$$

$$= \frac{-60}{126} \times 6$$

$$= -2.86$$

The value −2.86 is less than −1.645 so we reject H_0 and we conclude that the sales manager's claim that the mean lifetime of the firm's bulbs is 1000 hours is not justified.

(c) Testing for a difference between two means

From Unit 14.4 we know that, if we take a sample of size n_1, with mean $\bar{x}_1$ and standard deviation s_1, from a population with mean μ_1 and standard deviation σ_1, and a second, independent, sample size n_2, with mean $\bar{x}_2$ and standard deviation s_2, from a population with mean μ_2 and standard deviation σ_2, the distribution of the difference between the sample means $(\bar{x}_1 - \bar{x}_2)$

has mean $(\mu_1 - \mu_2)$ and standard deviation $\sqrt{\frac{\sigma_1^2}{n_1} + \frac{\sigma_2^2}{n_2}}$ and that this distribution will be normal provided the parent populations are normal or the sample sizes are large. We can use this result to set up a significance test for the difference between the means of two populations.

Suppose, for example, that there are two machines producing articles by different methods. The works manager wishes to compare the output of the two machines to determine whether they produce the same volume of articles on average. The output from the two machines is observed and the following results are recorded:

	Machine 1	Machine 2
Number of hours observed	35	38
Mean volume of output per hour	108 items	106 items
Standard deviation of output	5 items	6 items

Do these results show that there is a difference between the mean outputs of the two machines?

We have:

H_0: there is no overall difference between the mean outputs of the two machines: $\mu_1 = \mu_2$ or $\mu_1 - \mu_2 = 0$

H_1: there is an overall difference between the mean outputs of the two machines: $\mu_1 \neq \mu_2$ or $\mu_1 - \mu_2 \neq 0$

We are performing a two-tailed test because we are concerned whether there is any difference in mean output of the two machines, not simply whether one produces more than the other. We shall use a significance level of 0.05.

Our significance test will be in terms of the test statistic:

$$Z = \frac{(\bar{x}_1 - \bar{x}_2) - (\mu_1 - \mu_2)}{\sqrt{\frac{\sigma_1^2}{n_1} + \frac{\sigma_2^2}{n_2}}}$$

Looking at Fig. 15.3, we set that the critical values of Z are ± 1.96.

The decision criterion is:

$$\text{if } -1.96 < Z < 1.96, \text{ do not reject } H_0;$$

$$\text{if } Z > 1.96 \text{ or } Z < -1.96, \text{ reject } H_0.$$

Using the sample evidence we find:

$$Z = \frac{(108 - 106) - 0}{\sqrt{\frac{\sigma_1^2}{35} + \frac{\sigma_2^2}{38}}}, \text{ since } \mu_1 = \mu_2 \text{ assuming } H_0 \text{ is true}$$

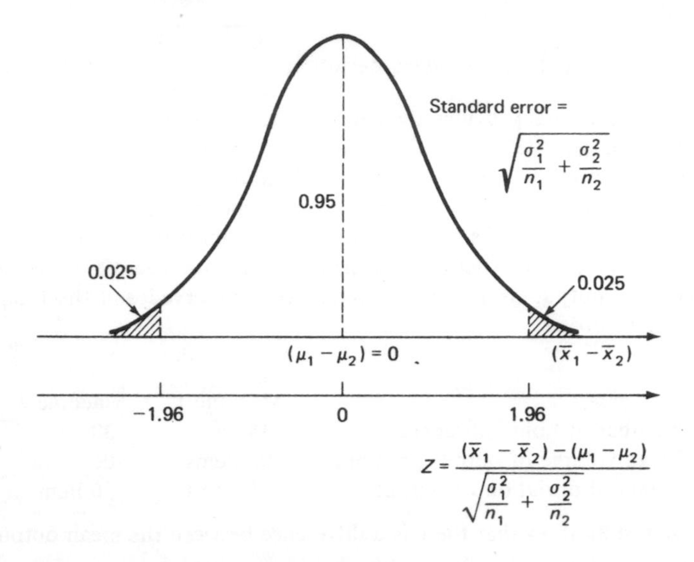

Fig. 15.3 Two-tailed test for a difference between means

$$= \frac{2}{\sqrt{\frac{5^2}{35} + \frac{6^2}{38}}}$$, estimating $\sigma_1{}^2$ by $s_1{}^2$ and $\sigma_2{}^2$ since n_1 and n_2 are large

$$= \frac{2}{\sqrt{1.662}}$$

$$= \frac{2}{1.29}$$

$$= 1.55$$

1.55 lies between −1.96 and 1.96 so Z lies in the acceptance region. We conclude that these samples do not provide evidence of a significant difference between the mean outputs of the two machines.

(d) Significance levels

In our tests we have used a level of significance of 0.05 (5%) and in any examination question which does not specify the significance level to be used, you should assume that the 5% level is appropriate. A sample result which leads to the rejection of the null hypothesis at the 5% level is termed a *significant* result. Using the 0.01 (1%) level of significance you should check using Table D that the critical values of Z for a two-tailed test are ±2.58 and for a one-tailed test −2.33 or 2.33. Sample results leading to rejection of the null hypothesis at the 1% level are referred to as *highly significant*. At the 0.001

(0.1%) level, the critical values of Z are ± 3.29 for a two-tailed test and 3.09 or -3.09 for a one-tailed test. Sample results leading to rejection of the null hypothesis at the 0.1% level are said to be *very highly significant*. In the presentation and interpretation of statistical results you should remember that the word 'significant' is used in a technical sense and has a precise meaning.

15.3 Tests concerning proportions

In the same way that we could use the distribution of the sample means to test hypotheses concerning means of populations, we may use the results of Unit 14.5 to perform tests of population proportions.

(a) Tests of a single proportion

Suppose a company manufacturing a washing powder claims that 40% of housewives use its brand. A random sample of 200 housewives is interviewed and they are asked if they use this powder; 68 reveal that they do. Does this sample support the manufacturer's claim?

We need to use a one-tailed test here since we are concerned only whether the market share is *less* great than claimed. We have:

$H_0: p = 0.4$ where p is the true proportion of all housewives using the powder

$H_1: p < 0.4$

We use the 5% significance level.

From Unit 14.5 we know that the test statistic:

$$Z = \frac{\text{Sample proportion} - \text{Population proportion}}{\text{Standard error of proportions}}$$

can be considered to have a standard normal distribution provided the sample size is large (Fig. 15.4).

The critical value of Z for a one-tailed test at the 5% level is -1.645. The decision criterion is:

$$\text{if } Z > -1.645, \text{ do not reject } H_0;$$

$$\text{if } Z < 1.645, \text{ reject } H_0.$$

The sample proportion, $\hat{p}$, equals 68/200 or 0.34. Assuming H_0 is true, the standard error of proportions (see Unit 14.5) is:

$$\sqrt{\frac{pq}{n}} = \sqrt{\frac{0.4 \times 0.6}{200}} = 0.03464$$

Thus the test statistic is:

$$Z = \frac{0.34 - 0.4}{0.03464}$$

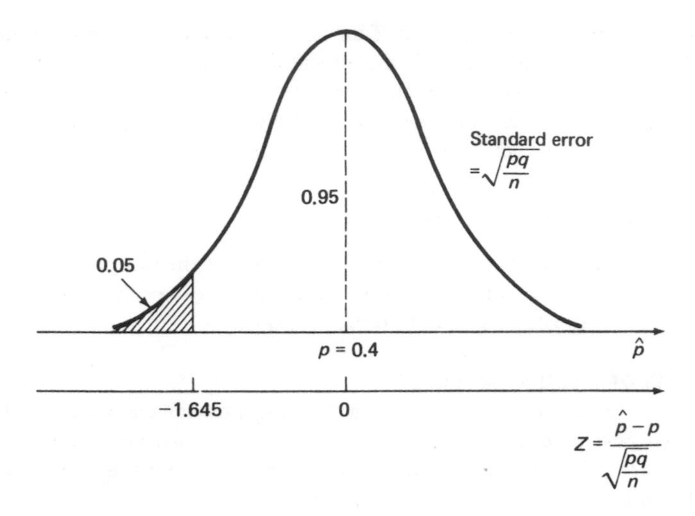

Fig. 15.4 One-tailed test of a proportion

$$= \frac{-0.06}{0.03464}$$

$$= -1.732$$

The value -1.732 is less than -1.645 so we reject H_0. We conclude that significantly fewer than 40% of all housewives use the brand.

(b) Tests for differences between proportions

We may wish to know whether differences between population proportions are significant. Suppose that a random sample of 100 people from one constituency shows 60% in favour of the Government, while an independent random sample of 144 from another constituency shows 55% in favour. Does the 5% difference indicate a *real* difference of opinion between the constituencies?

We have a two-tailed test here as we are dealing with a *difference* of opinion. Thus:

H_0: the proportions p_1 and p_2 of Government supporters in the two constituencies are the same: $p_1 = p_2$ or $p_1 - p_2 = 0$

H_1: the proportions p_1 and p_2 are different: $p_1 \neq p_2$ or $p_1 - p_2 \neq 0$

We use the 5% significance level.

From Unit 14.5 we know that the difference in the sample proportions $(\hat{p}_1 - \hat{p}_2)$ has approximately a normal distribution with mean $(p_1 - p_2)$ and standard error $\sqrt{\frac{p_1 q_1}{n_1} + \frac{p_1 q_2}{n_2}}$, provided n_1 and n_2 are large.

The test statistic is therefore:

$$Z = \frac{(\hat{p}_1 - \hat{p}_2) - (p_1 - p_2)}{\sqrt{\dfrac{p_1 q_2}{n_1} + \dfrac{p_1 q_2}{n_2}}}$$

The critical values for Z in a two-tailed test are ± 1.96 and the decision criterion is:

$$\text{if } -1.96 < Z < 1.96, \text{ do not reject } H_0;$$

$$\text{if } Z > 1.96 \text{ or } Z < -1.96, \text{ reject } H_0.$$

If H_0 is true, $(p_1 - p_2) = 0$ and our best estimate of both p_1 and p_2 is given by the pooled estimate, $\hat{p}_0$ (see Unit 14.5) where:

$$\hat{p}_0 = \frac{\text{Total number of Government supporters in both samples}}{\text{Total number of people in both samples}}$$

$$= \frac{n_1 \hat{p}_1 + n_2 \hat{p}_2}{n_1 + n_2}$$

$$= \frac{60 + 79.2}{244}$$

$$= 0.570 \text{ to 3 decimal places}$$

The standard error is:

$$\sqrt{\frac{0.570 \times 0.430}{100} + \frac{0.570 \times 0.430}{144}}$$

$$= \sqrt{0.00415}$$

$$= 0.0644 \text{ to 3 decimal places}$$

Assuming H_0 is true, the test statistic is given by:

$$Z = \frac{(0.60 - 0.55) - 0}{0.0644}$$

$$= \frac{0.05}{0.0644}$$

$$= 0.776 \text{ to 3 decimal places}$$

0.776 lies between -1.96 and $+1.96$ so we cannot reject H_0. We conclude that the difference between the two proportions is not statistically significant. The difference between the samples could easily have occurred by chance, even if the two constituencies actually have the same proportion of Government supporters.

15.4 The *t* distribution

(a) Introduction

When we used the sampling distribution of the mean to test hypotheses concerning means in Unit 15.2, if the population standard deviation, σ, was unknown, we used the sample standard deviation, s, as an estimate of σ. This approximation is only valid for large samples and cannot be used for small samples. The problem was studied by W. S. Gosset who wrote under the pen-name 'Student'.

Instead of working with the standard normal variate:

$$Z = \frac{\bar{x} - \mu}{\sigma/\sqrt{n}}$$

Gosset worked with the variate t where:

$$t = \frac{\bar{x} - \mu}{s/\sqrt{n}}$$

Here $\bar{x}$ is the mean and s is the standard deviation of a sample size n taken from a *normal* population with mean μ. (The t distribution should strictly be applied only to samples taken from a normal population although many examination questions leave this unstated and you have to assume that a given population is normal.) The sampling distribution of t, called *Student's t distribution*, is similar in shape to a standard normal distribution, ($\mu = 0$, $\sigma = 1$), being symmetrical about $t = 0$, but it becomes increasingly more spread out as the sample size n decreases. This is as we would expect since less information is contained in small samples and a greater variation in results is possible. The shape of the t distribution depends on n but when you consult tables of the t distribution (see Table F), you will find the values of t tabulated according to the *number of degrees of freedom* (*df*) rather than the sample size. The symbol ν (the Greek letter nu), is used to denote the number of degrees of freedom. The symbol ν is related to n but the relationship between the two depends on the particular problem under consideration and we shall indicate its value in each type of problem we deal with.

The t distribution table (Table F) gives us less information than the normal distribution tables. Table F gives the values of t corresponding to only a few areas under the t distribution curve. To demonstrate the information tabulated let us look at Table F when the number of degrees of freedom, ν, is 10. We find:

$$t_{0.100} = 1.372 \qquad t_{0.050} = 1.812 \qquad t_{0.025} = 2.228$$
$$t_{0.010} = 2.764 \qquad t_{0.005} = 3.169$$

These values tell us that:

an area of 0.100 lies beyond $t = 1.372$;
an area of 0.050 lies beyond $t = 1.812$;
an area of 0.025 lies beyond $t = 2.228$;

an area of 0.010 lies beyond $t = 2.764$;
an area of 0.005 lies beyond $t = 3.619$.

By symmetry we can also deduce for example that an area of 0.100 lies beyond $t = -1.372$. The t distribution for $\nu = 10$ is shown in Fig. 15.5(*a*). How do the above values of t compare with the corresponding Z scores? We said that if n is large, then t approaches the normal variable Z and if you look at the foot of Table F you can read off the values corresponding to $\nu = \infty$. You should recognize some of the values here such as $t_{0.05} = 1.645$ and $t_{0.025} = 1.960$ from our tests using the normal distribution. (See Fig. 15.5(*b*).)

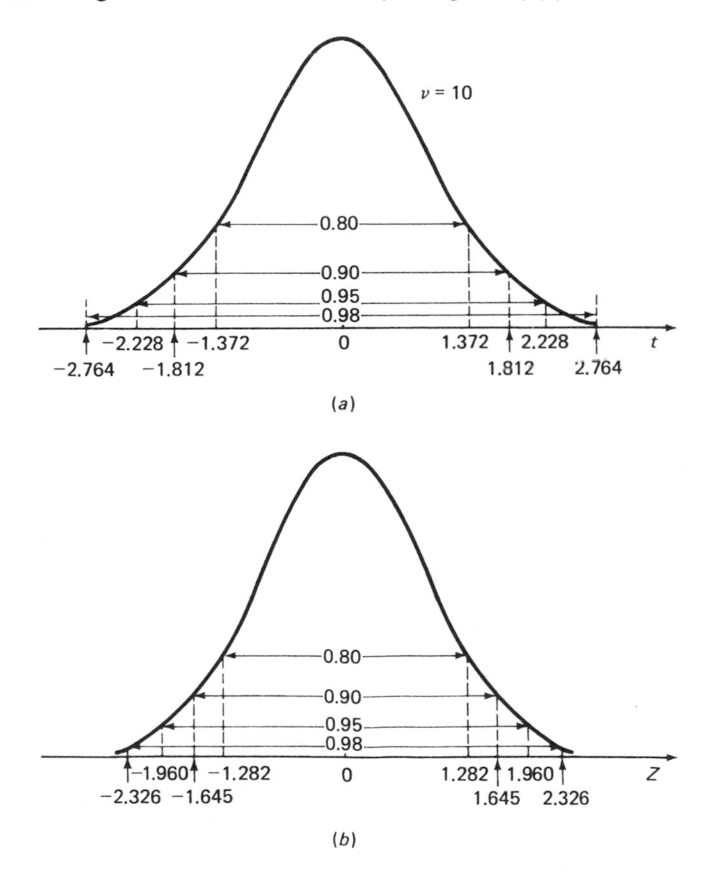

Fig. 15.5 Comparison of Student's t distribution ($\nu = 10$) and the standard normal distribution

(b) Tests concerning a single mean

Suppose, for example, that a company wishes to buy tyres for the vehicles in its transport fleet and requires tyres that last on average 20000 kilometres. A tyre company claims that it can supply tyres that meet this requirement. The buying company decides to test that claim and buys 16 tyres, tries them and finds that they last an average of 19 500 kilometres with a standard deviation of 1 200 kilometres. Can the supplier's claim be accepted?

We have a one-tailed test here because the buyer would be worried only if the tyres performed less well than claimed. Thus:

H_0: the mean life (μ) of all tyres supplied by the company is 20 000 kilometres: $\mu = 20000$ km
$H_1: \mu < 20000$ km

We use the 0.05 significance level. Thus to set up the decision criterion we need to find the critical value of t such that an area of 0.05 lies beyond t in the lower tail. Hence we require the value of $-t_{0.05}$ at the appropriate number of degrees of freedom. In tests concerning a single mean, the number of degrees of freedom, ν, is always one less than the sample size:

$$\nu = n - 1 = 16 - 1 = 15$$

From Table F:

$$t_{0.05} = 1.753$$

The decision criterion is:

$$\text{if } t > -1.753, \text{ do not reject } H_0;$$
$$\text{if } t < -1.753, \text{ reject } H_0.$$

The test statistic, t, is given by:

$$= \frac{\bar{x} - \mu}{s/\sqrt{n}}$$

$$= \frac{19500 - 20000}{1200/\sqrt{16}}, \text{ assuming } H_0 \text{ is true}$$

$$= \frac{-500}{1200/4} = \frac{-500}{300} = -1.667$$

The value -1.667 is greater than -1.753 so we cannot reject H_0. There is not sufficient evidence to suggest that the mean life of all tyres is significantly less than 20 000 km. Note that we are not concluding that the mean lifetime *is* 20 000 km – it might be worth while to carry out more trials before making any financial commitments.

(c) One-tailed and two-tailed tests using the *t* distribution

If we need to perform a two-tailed test concerning a mean at the 5% level, how do we find the critical values of t? Looking at Fig. 15.6, you can see that 0.025 of the area under the curve lies beyond $t_{0.025}$, so an area of 0.95 lies between

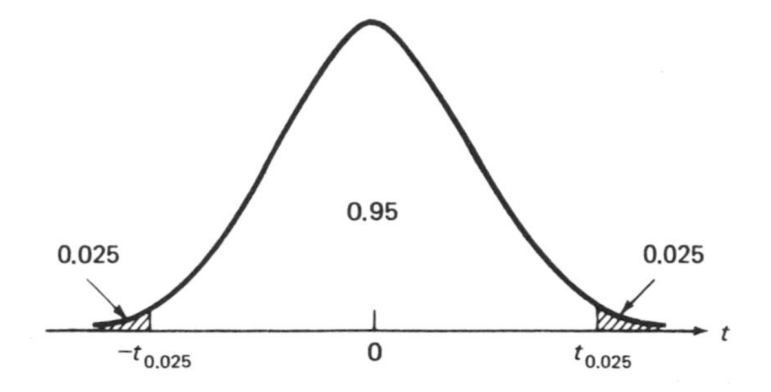

Fig. 15.6 Rejection regions for a two-tailed t test

$-t_{0.025}$ and $+t_{0.025}$. Thus $\pm t_{0.025}$ are the appropriate critical values for a two-tailed test at the 5% level. For $\nu = 15$, for example, we can see from Table F that these critical values are ± 2.13.

Not all tables of the t distribution are laid out in the same way. Table F is appropriate for one tailed tests because the entry under $t_{0.05}$ gives us the value of t with an area of 0.05 beyond it. Some tables in both books and examinations give values of t appropriate to two-tailed tests and the entry under $t_{0.05}$ would give the value of t with an area of 0.025 beyond it. It is advisable to learn a few values of t so that under examination conditions you know whether you have been presented with a one-tailed or a two-tailed table. The easiest values to learn are the ones corresponding to an infinite number of degrees of freedom, because these are the ones you are familiar with from the normal distribution.

In a one-tailed table you will find:

$$t_{0.05} = 1.645 \quad \text{or} \quad t_{5\%} = 1.645$$
$$t_{0.025} = 1.96 \quad \text{or} \quad t_{2.5\%} = 1.96$$

Note that the areas are sometimes quoted as percentages rather than proportions.

In a two-tailed table you will find:

$$t_{0.100} = 1.645 \quad \text{or} \quad t_{10\%} = 1.645$$
$$t_{0.050} = 1.96 \quad \text{or} \quad t_{5\%} = 1.96$$

(d) Testing a correlation coefficient for significance

The t distribution may also be used to test whether a correlation coefficient is significantly different from zero.

If you look back to Unit 9.5, you will find that in our example relating sales to advertising expenditure, we worked out that Pearson's product moment correlation coefficient had the value, $r = 0.83$. This coefficient was based on a sample of ten observations, $n = 10$. What we want to know is whether on the basis of this value for r we are justified in asserting that there is always a linear relationship between advertising expenditure (x) and sales (y). To carry out this test we always put the null hypothesis in the form that there is *no* correlation between the two variables x and y so that the population correlation coefficient, ρ, is zero.

$$H_0: \rho = 0$$
$$H_1: \rho \neq 0 \text{ (a two-tailed test)}$$

We use a significance level of 0.05.

Provided we assume that x and y follow a bivariate normal distribution (see Unit 9.6), then it can be proved that the statistic:

$$t = \frac{r\sqrt{n-2}}{\sqrt{1-r^2}}$$

follows a t distribution with $(n - 2)$ degrees of freedom.

In our example, $n = 10$, so we have 8 degrees of freedom. From Table F, the critical values of t for a two-tailed test at the 5% level are ± 2.306.

The decision criterion is:

if $-2.306 < t < 2.306$, do not reject H_0;

if $t > 2.306$ or $t < -2.306$, reject H_0.

Using our sample value of r, we have:

$$t = \frac{r\sqrt{n-2}}{\sqrt{1-r^2}}$$

$$= \frac{0.83 \times \sqrt{8}}{\sqrt{1-0.83^2}}$$

$$= 4.21$$

The value 4.21 is greater than 2.306 so we reject H_0. We conclude that ρ is significantly different from zero and that we would be justified in assuming a linear relationship between advertising expenditure and sales.

Some tables supplied for your use in examinations may give the critical values of r directly for different sample sizes. Fig. 15.7 shows the type of information that is given in such tables and it enables one to decide whether a value of r is significant at the 5% level without needing to set up a significance test. All values of r under the curve are not significant for the stated number of observations, while values of r above the curve are significant. In our example, $n = 10$ and $r = 0.83$. This point lies above the curve in Fig. 15.7 and indicates, as before, that there is significant correlation between advertising expenditure and sales. You can also see, however, that a value of $r = 0.8$ is *not* significant for only five observations of the variables. With so few observations a value of r as high as 0.8 is likely to occur by chance in a sample of observations even when there is no true linear relationship between the two variables.

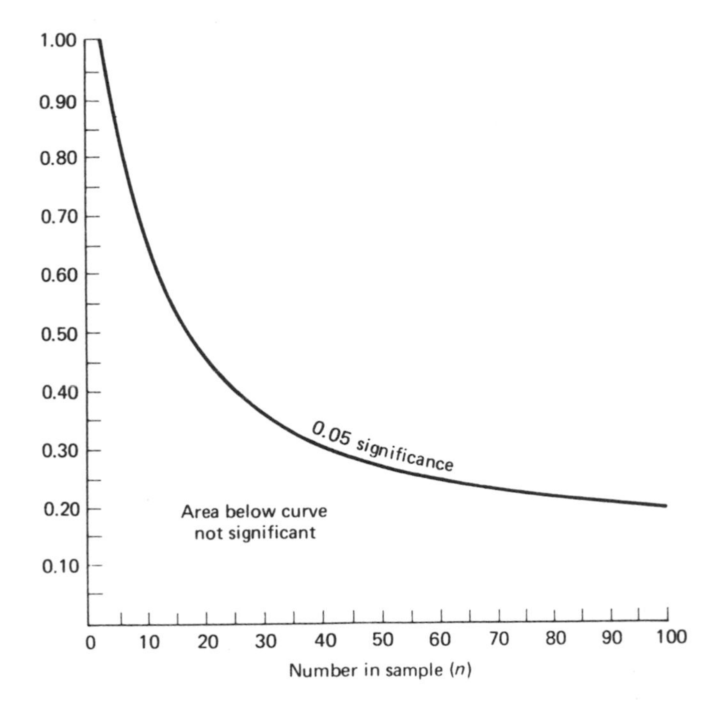

Fig. 15.7 Significance of Pearson's correlation coefficient r

(e) Other uses of the t distribution

We have by no means demonstrated all the uses of the t distribution and if you proceed further with your statistical studies you will certainly learn much more about it. Confidence intervals for a population mean may be found using a small sample when the population standard deviation is unknown. The procedure for finding a 95% confidence interval is broadly the same as in Unit 14.2(*c*) but, whereas the Z score was always 1.96 using the normal distribution, for the t distribution the value of $t_{0.025}$ depends on the sample size. $n - 1$ degrees of freedom are appropriate.

It is also possible to consider the differences between two population means using a t test. There are two basic types of test – one when we use two independent samples from the populations and the other when the samples are 'paired' in some way. For example, when comparing the mean heartbeat rate in men before and after heavy exercise, we could take a random sample of men and measure their heartbeats when they are resting. We could also take an independent random sample of men who have just done some heavy exercise and measure their heartbeats. Alternatively, we could take a random sample of men and measure their heartbeats while they are resting and then measure the heartbeats of these *same men* after they have exercised. This latter pairing procedure is of great value as it reduces the variation present in the investigation because we know that the only difference between the two samples is that they have exercised, whereas using independent samples there will be other varying characteristics between the two groups of men, for example, weight and smoking habits. When we wish to compare means of more than two populations, Fisher's F distribution which is related to the t distribution, may be used to set up an *analysis of variance*. This is a widely used technique. References to books containing further information on these topics are to be found at the end of this volume.

The t distribution may also be used in regression analysis to test whether the regression coefficient, b (see Unit 9.3(*b*)), is significantly different from zero. Confidence intervals for the parameters of the regression line and for the estimates made using the regression equation may also be constructed.

15.5 Chi-squared (χ^2) tests

(b) Introduction

The symbol χ (pronounced *kigh*) is the Greek lower-case letter chi and gives its name to some important significance tests. Chi-squared tests are used to compare an observed frequency distribution or table with the distribution or table of frequencies expected, assuming some null hypothesis is true.

The basic approach to a chi-squared test is to find the differences between the corresponding expected and observed frequencies. These differences are then squared. Each difference is divided by its expected frequency and these quotients are added together to give a single statistic, X^2, that is tested for significance. The calculation is expressed in the formula:

$$X^2 = \sum \frac{(F_O - F_E)^2}{F_E}$$

where F_O is the *observed* frequency, F_E the *expected* frequency and the sum is taken over all groups in the frequency distribution or table. It can be proved mathematically that the statistic, X^2, approximates to a continuous theoretical distribution known as the *chi-squared distribution*. Chi-squared cannot take negative values and its shape depends (as did the t distribution) on the number of degrees of freedom, ν. Its shape for $\nu > 2$ is shown in Fig. 15.8—note that the curve is not symmetrical and that its peak is at $\chi^2 = \nu - 2$. Table G at the end of the book gives values of χ^2 corresponding to various degrees of freedom. For example, corresponding to $\nu = 10$ we find:

$$\chi^2_{0.05} = 18.307 \quad \chi^2_{0.025} = 20.483 \quad \chi^2_{0.01} = 23.209 \quad \chi^2_{0.005} = 25.188$$

These values tell us that:

an area of 0.05 lies under the χ^2 curve to the right of $\chi^2 = 18.307$
an area of 0.025 lies under the χ^2 curve to the right of $\chi^2 = 20.483$
an area of 0.01 lies under the χ^2 curve to the right of $\chi^2 = 23.209$
an area of 0.005 lies under the χ^2 curve to the right of $\chi^2 = 25.188$

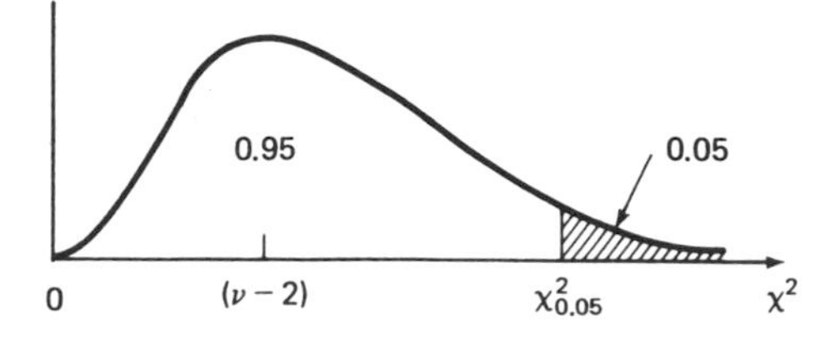

Fig. 15.8 The χ^2 distribution ($\nu > 2$)

A chi-squared test is always a one-tailed test. If the observed and expected frequencies exactly agreed, the test statistic, X^2, would be zero so we are always testing to see whether the differences between the frequencies are so large as to be unlikely to have occurred by chance, in other words, we are testing whether X^2 is significantly *greater than* zero.

(b) Goodness of fit test

To carry out a test for 'goodness of fit' using the χ^2 distribution, we proceed in the following way. We are given an observed frequency distribution. We construct a theoretical expected frequency distribution for comparison, assuming some null hypothesis is true. For example, we might postulate that the data follow a binomial distribution with parameter $p = 0.3$, or that the data follow a Poisson distribution, or that the data follow a normal distribution with a given mean and standard deviation. We next calculate the value of the test statistic, X^2, and if this value is large enough to lie in the critical region of our test we *reject the null hypothesis* and conclude that the postulated theoretical distribution is *not a good fit* to the observed distribution.

For example, let us use our calculations of Unit 13.5 where we found the probabilities of a range of chest sizes in an adult male population assuming a normal distribution with mean chest size 98 cm and standard deviation 5 cm. Suppose the manufacturer wants to check that this normal distribution is in fact a good description of actual chest sizes found. It commissions a market research firm to take a sample of 10000 men to find out their chest sizes. The results are shown in Table 15.1. The manufacturer now wants to compare these observed frequencies with the frequencies it would have expected assuming a normal distribution with mean 98 cm, standard deviation 5 cm. We have:

H_0: A normal distribution with mean 98 cm, standard deviation 5 cm, fits the data;

H_1: A normal distribution with mean 98 cm, standard deviation 5 cm, does not fit the data.

We use a significance level of 0.05.

The normal probabilities were calculated in Unit 13.5 and are given again in Table 15.2. Two extra groups—'less than 80 cm' and 'more than 115 cm'—have been added to those of Unit 13.5 because the normal distribution allows

Table 15.1 Chest sizes found in a survey of 10000 male adults

Size range (cm)	Number in size range
80–85	41
85–90	511
90–95	2173
95–100	3831
100–105	2643
105–110	740
110–115	61
TOTAL	10000

Table 15.2 Comparison of expected and observed frequencies of chest sizes

Size range (cm)	Theoretical probability	Expected frequency (F_E)		Observed frequency (F_O)		Difference $F_O - F_E$
Less than 80	0.0002	2	47	0	41	−6
80–85	0.0045	45		41		
85–90	0.0501	501		511		10
90–95	0.2195	2195		2173		−22
95–100	0.3811	3811		3831		20
100–105	0.2638	2638		2643		5
105–110	0.0726	726		740		14
110–115	0.0079	79	82	61	61	−21
115 and over	0.0003	3		0		
	1.0000	10000		10000		

for chest sizes in these regions even though the manufacturer is not prepared to make garments outside the range 80–115 cm. From these probabilities, we can work out the *number of men* we expect to be in each of the size ranges assuming H_0 is true. Our observations covered 10000 men so we must multiply each theoretical probability by 10000 to obtain the expected frequency. When working out an expected frequency, you will find that it does not always exactly equal a whole number. This is of no consequence but it is usually a waste of effort to work to more than one decimal place in the expected frequency.

It can be shown that the approximation of the distribution of χ^2 to a chi-squared distribution is not good when any of the expected frequencies is small. There is no precise rule as to what is 'small' but expected frequencies should not be less than 5 and preferably not less than 10. When such small frequencies do occur, these classes should be combined with adjacent classes to obtain a sufficiently large expected frequency before carrying out the χ^2 test. In our example we have one such small class at each end of the distribution so we combine them with adjacent classes to give 47 men in the range 'less than 85 cm' and 82 men in the range '110 cm and over'.

We can now work out the test statistic:

$$X^2 = \sum \frac{(F_O - F_E)^2}{F_E}$$

$$= \frac{6^2}{47} + \frac{10^2}{501} + \frac{22^2}{2195} + \frac{20^2}{3831} + \frac{5^2}{2638} + \frac{14^2}{726} + \frac{21^2}{84}$$

$$= 0.766 + 0.200 + 0.221 + 0.104 + 0.009 + 0.270 + 5.25$$

$$= 6.819$$

The number of degrees of freedom in a goodness of fit test is one less than the number of cells or categories of classification. Also, one degree of freedom must be subtracted for each parameter such as a mean or a standard deviation that has to be estimated, using the observed data, in order to work out the expected frequencies. In our example, there are 7 classes (we amalgamated the two small classes at each end). A mean of 98 cm and a standard deviation of 5 cm were specified and we did not work these out from the observed data. We thus have $(7 - 1) = 6$ degrees of freedom. From Table G, we find that with 6 degrees of freedom, $\chi^2_{0.05} = 12.592$. Our decision criterion is:

if $X^2 < 12.592$, do not reject H_0;
if $X^2 > 12.592$, reject H_0.

We found that X^2 equalled 6.819, which is less than 12.592. We conclude that there is no evidence to suggest that the observed data are not fitted by a normal distribution with mean 98 cm and standard deviation 5 cm. The manufacturer can now happily go ahead and make the numbers of sweaters as predicted in the different size ranges.

(e) Contingency tables

Another important application of the χ^2 test is to frequencies which are classified in two ways rather than one. A *contingency table* shows the frequencies in each classificatory cell. Suppose, for instance, a medical researcher is investigating a possible link between smoking and bronchitis. A survey is done to find out how many smokers and non-smokers suffer from bronchitis and the results of the survey are set out in a contingency table (Table 15.3). Each individual is classified in two ways, firstly, as a smoker or non-smoker and,

Table 15.3 Contingency table of observed frequencies of smokers and non-smokers who suffer from bronchitis

	Smokers	Non-smokers	Totals
Bronchitis sufferers	200	200	400
Non-sufferers	150	450	600
TOTALS	350	650	1000

secondly, as one who does or does not suffer from bronchitis. We next construct, for comparison, the contingency table we would expect assuming some null hypothesis is true. When analysing data in this way, we are usually trying to see whether there is some link or association between the factors of classification. Here we are interested in whether smokers are more likely to suffer from bronchitis than non-smokers. The null hypothesis, *when testing for association*, is always that there is *no* association between the two variables. If the test statistic, X^2, is large enough to be in the critical region, the null hypothesis is rejected and we conclude that there *is* an association between the variables.

We have:

H_0: There is no association between smoking and bronchitis;
H_1: There is an association between smoking and bronchitis.

We use the 0.05 significance level. The number of degrees of freedom, ν, for a chi-squared test of this type equals:

$$(n_r - 1)(n_c - 1) \cdot$$

where n_r is the number of rows in the contingency table and n_c is the number of columns (excluding the totals). For our example, we have two rows and two columns so:

$$\nu = (2 - 1)(2 - 1) = 1$$

Looking in the chi-squared table, Table G, for one degree of freedom we find:

$$\chi^2_{0.05} = 3.841$$

The decision criterion is:

if $X^2 < 3.841$, do not reject H_0;
if $X^2 > 3.841$, reject H_0.

We must now work out the value of X^2. To do this we must evaluate the frequency we would expect in each category of classification assuming the null hypothesis is true. If there is no association between bronchitis and smoking, we would expect bronchitis to be distributed among both smokers and non-smokers in the same proportion as it is distributed in the entire sample. Out of the total of 1000 people, 400 suffered from bronchitis so assuming H_0 is true, 400/1000 of the 350 smokers, 140, should suffer from bronchitis and 400/1000 of the 650 non-smokers, 260, should also suffer from bronchitis. Similarly 600/1000 of the 350 smokers, 210, and 600/1000 of the 650 non-smokers, 390, should not suffer from bronchitis. The expected frequencies are shown in Table 15.4.

Table 15.4 Expected frequencies of smokers and non-smokers who suffer from bronchitis, assuming no association

	Smokers	Non-smokers	Totals
Bronchitis sufferers	140	260	400
Non-sufferers	210	390	600
TOTALS	350	650	1000

The expected frequency in each cell can always be calculated using the expression:

$$\text{Expected frequency} = \frac{\text{Column total} \times \text{Row total}}{\text{Grand total}}$$

Thus, for column 1, row 1, this gives:

$$\frac{350 \times 400}{1000} = 140$$

and similarly for the other entries.

The test statistic, X^2, is:

$$X^2 = \sum \frac{(F_O - F_E)^2}{F_E}$$

where the sum is taken over all the cells. Thus:

$$X^2 = \frac{(200 - 400)^2}{140} + \frac{(200 - 260)^2}{260} + \frac{(150 - 210)^2}{210} + \frac{(450 - 390)^2}{390}$$

$$= 25.71 + 13.85 + 17.14 + 9.23$$

$$= 65.9 \text{ to 1 decimal place}$$

At this point we must take note of a correction that should be made when we are analysing a contingency table with only one degree of freedom, as here. We stated that the chi-squared distribution is a continuous distribution but the test statistic, X^2, is a discrete variable because the observed frequencies can only be whole numbers. We should strictly apply a continuity correction every time we use the chi-squared test but in practice we need use it only in situations when we have just one degree of freedom. This continuity correction is called *Yates' correction* after the statistician who first showed its importance. All that we need to do is subtract $\frac{1}{2}$ from the *absolute* differences between the observed and expected frequencies before those differences are squared. The test statistic is then:

$$\sum \frac{(|F_O - F_E| - \frac{1}{2})^2}{F_E}$$

and in our example has the value:

$$\frac{(|200 - 140| - \frac{1}{2})^2}{140} + \frac{(|200 - 260| - \frac{1}{2})^2}{260} + \frac{(|150 - 210| - \frac{1}{2})^2}{210}$$

$$+ \frac{(|450 - 390| - \frac{1}{2})^2}{390}$$

$$= 25.29 + 13.62 + 16.86 + 9.08$$

$$= 64.8 \text{ to 1 decimal place}$$

64.8 is greater than the critical value of 3.841. We thus reject H_0 and conclude that there *is* an association between bronchitis and smoking.

15.6 Statistical quality control

(a) Introduction

Statistical quality control is a system in which sampling theory is applied to the practical problems of routinely controlling the quality of manufactured or packaged goods. The term 'quality' denotes something about the product which can be measured such as its length, weight or impurity content. Alternatively, it may only be possible to classify a product as 'good' or 'bad'. In the first case we are dealing with *variables*, while in the second case we are dealing with *attributes*.

The basic principles underlying control by variables are:

(i) Even when a manufacturing or packaging process is working in a satisfactory and stable manner, there is certain to be random variation in the quality of the product and this can usually be approximated by a normal distribution.

(ii) Under these conditions, the means of small samples, of the same size, taken from the output, will approximate to a normal distribution.

(iii) Any sample mean which is significantly different from the overall average is an indication that the process is no longer working in a satisfactory manner.

These principles also apply to quality control by attributes except that the proportion of defective items in a sample is measured rather than the sample mean.

(b) Control charts

Suppose, for example, we have a machine making bolts of nominal length 2 cm. Even when the machine is working satisfactorily, not every bolt is exactly of length 2 cm but by observation it has been found that the length of bolt has approximately a normal distribution with a mean, μ, of 2 cm and a standard deviation, σ, of 0.010 cm.

In order to keep a check on the quality of the product, we decide to take a sample of four bolts every quarter of an hour. We measure the lengths of each of these four bolts and find the mean length, $\bar{x}$, of the sample. The means of all these samples should follow a normal distribution with mean μ and standard error $0.010/\sqrt{4}$, i.e. 0.005 cm using the theoretical results of Unit 14.2(*b*) and assuming that the machine is still in stable operation.

Using the normal distribution table, Table D, we know that 95% of the area under a normal curve lies within 1.96 standard deviations of the mean and also that 99.8% of the area under a normal curve lies within 3.09 standard deviations of the mean. So 95% of the sample means should lie between $(2 - 1.96 \times 0.005)$ cm and $(2 + 1.96 \times 0.005)$ cm, that is between 1.9902 and 2.0098 cm. Similarly 99.8% of the sample means should lie between $(2 - 3.09 \times 0.005)$ cm and $(2 + 3.09 \times 0.005)$ cm, that is between 1.98455 and 2.01545 cm. We set up a chart, called a *control chart*, with these limits drawn in and we can then plot the observed sample means on the chart in time order (Fig. 15.9). These limits are referred to as *control limits* or *control lines*. The limits at 1.96 standard errors are called the *inner* limits or the *warning*

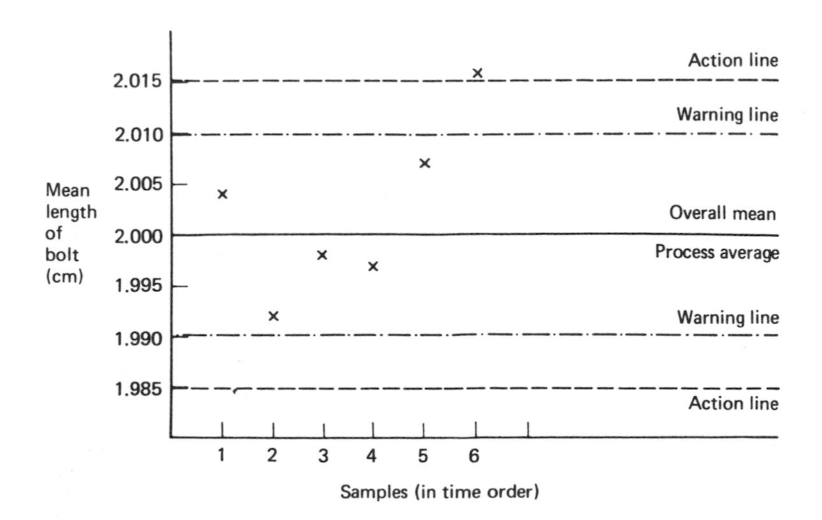

Fig. 15.9 A control chart for sample means (sample size 4)

limits. If the machine is running satisfactorily, 'in control', $2\frac{1}{2}$% or 1 in 40 observations of the sample mean can be expected to be outside one of these warning lines. The limits at 3.09 standard errors are called the *outer* control limits or the *action* lines. If the machine is in control, we expect only 0.1% or 1 in 1000 readings to be outside either of these lines. Action lines are usually drawn in red on a control chart.

Table 15.5 displays some results for samples taken from the process and the means of these samples are plotted on the control chart (Fig. 15.9). Samples 1 to 5 have means within the inner control lines and are taken as evidence that the machine is behaving satisfactorily. The mean of the sixth sample, however, is outside the outer control limit. The mean for this sample, therefore, is *highly significantly different* from the overall mean of 2 cm since if the machine is behaving as expected, only 0.1% or 1 in 1000 samples should have a mean as high as this. Consequently this sample is taken as evidence that the machine has gone 'out of control' and is tending to make longer bolts.

Table 15.5 Quality control data; bolt lengths in cm

Sample	1	2	3	4	5	6
Bolt lengths	2.002	1.988	2.006	2.003	2.016	2.023
(cm)	2.011	2.001	2.002	1.987	2.000	2.021
	2.011	1.996	1.991	1.996	1.999	2.006
	1.992	1.983	1.993	2.002	2.013	2.014
TOTALS	8.016	7.968	7.992	7.988	8.028	8.064
Mean	2.004	1.992	1.998	1.997	2.007	2.016

The term 'out of control' does not imply that the machine is behaving dangerously or that it is necessarily producing unacceptable bolts. It simply implies that some factor other than the inherent variability of the machine has started to operate—maybe a component part has worked loose, leading to a change in the setting of the machine or perhaps there has been some fluctuation in the supply of electricity to the machine. The quality control chart indicates that the process should be stopped whilst an investigation into the possible cause of the increase in mean is made.

The statistical control limits have nothing at all to do with the engineering tolerance limits which may have been specified. Such tolerance limits can also be put on the chart and before any quality control scheme is set up, a manufacturer should check that the tolerance limits are both sufficiently far apart and suitably positioned relative to the control limits so that it is possible for the process both to be in statistical control and to produce individual items not outside the tolerance limits.

(c) Use of the sample range in quality control

As well as monitoring the sample means, it is advisable, at least in the early stages of a quality control scheme, to measure the variability of the readings in a sample. This is necessary because the mean averages out the fluctuations between readings and can mask the fact that there are unacceptable variations within the sample. It is possible, particularly with newer electronic calculators, to calculate routinely the standard deviation of each sample as a measure of the variation. It is likely that the use of the standard deviation in quality control work will increase in the future. It is more common at present, however, to use the sample *range* as the measure of variation. This is much easier to calculate and is more readily understood by those, not usually statistically qualified, who have to carry out the routine quality control observations.

The range of each sample can be worked out and plotted on a *range* chart. The inner and outer control lines for these charts are not easily calculated and in an examination you will be provided with the appropriate constants to enable you to construct the chart. Note that, as a sample range cannot be negative, a range chart always starts at zero. It is not usual to put lower control lines on a range chart as we are concerned primarily with detecting large variations within a sample. It can be useful to put these lines in as a check on the quality control staff who may be tempted either to 'invent' data or not to record what they consider to be 'rogue' readings.

(d) Setting up a quality control scheme

Before we could draw in our control lines in Fig. 15.9 we needed to have estimates of μ, the overall mean length of bolts and σ, the standard deviation of the bolt lengths. How do we obtain these estimates? First we must have the process working in good order from a technical point of view with any teething troubles overcome. Next a large number of bolts should be taken off the production line as a series of small samples and the lengths of the bolts in these samples measured. A sample size of between 3 and 12 is appropriate and about 400 lengths should be measured in all. Using these results we then carry out the following steps:

(i) Calculate the mean, $\bar{x}$, of each sample.

(ii) Calculate the mean of the sample means, $\bar{\bar{x}}$. This is called the *grand mean* and is used as the estimate of μ.

(iii) Calculate the range, w, of each sample.

(iv) Calculate the mean of the sample ranges, $\bar{w}$.

(v) Estimate the standard deviation of the length of bolts, σ, by multiplying $\bar{w}$ by the constant a_n, where a_n depends on the size of sample. For $n = 4$, for example, $a_n = 0.4857$. Other values of a_n are given in Table J. This relationship between the range and the standard deviation assumes that the population of bolt lengths is normal.

(vi) Decide on the size of sample, n_1, to be used in the quality control scheme. This does not have to be the same sample size as used in (i)–(v) above. Set up a control chart for the sample mean with limits at:

$$\bar{\bar{x}} \pm 1.96 \frac{a_n \bar{w}}{\sqrt{n_1}}$$

and:

$$\bar{\bar{x}} \pm 3.09 \frac{a_n \bar{w}}{\sqrt{n_1}}$$

(vii) Set up a range chart, if desired.

(viii) Take samples of size n_1 regularly from the output and enter the sample means (and ranges) on the chart(s).

(ix) Interpret the control charts as follows.

A point outside an outer control limit is taken as evidence that something in the process has changed and requires investigation. Such a point will occur because of the inherent random variation in the process only once in a thousand samples.

Sometimes the trend of points will indicate a drift of the process even before any points fall beyond the action line. It is usual to look upon the following occurrences as signals that some interfering factor is influencing the process and requires investigation:

one point outside the outer control line,
two successive points between the inner and outer control lines,
five successive points in control but on the same side of the mean line.

(e) Attribute sampling

In some manufacturing and packaging processes it is more convenient to classify items as 'good' or 'bad' and to measure the number or the proportion defective in a sample, rather than making any measurements. Using the normal approximation to the binomial distribution (see Unit 13.6(*a*)) we can set up control limits either for the *number of defectives* in a sample of size n at:

$$np \pm 1.96\sqrt{npq}$$

and

$$np \pm 3.09\sqrt{npq}$$

or for the proportion of defectives in a sample of size n at:

$$p \pm 1.96\sqrt{\frac{pq}{n}}$$

and

$$p \pm 3.09\sqrt{\frac{pq}{n}}$$

and a control chart can be constructed as before. The value of p, the true proportion of defectives produced by the process, is estimated from the results of the first few samples. The lower limits are not always drawn in on the chart since a low number or proportion of defectives implies good quality production. Samples are taken regularly from the production process and the results entered on the chart.

Control charts are not the only method of assessing the quality of production. Sometimes it is necessary or more convenient to deal with a *batch* of completed goods, as, for instance, when a delivery of components coming into a factory is to be checked to see if it is of acceptable quality. Sampling methods are still employed as it would be too costly or impracticable to test every item. One sample is taken from the batch and the batch is accepted if the number of defectives in the sample is less than or equal to some prescribed number, c, called the acceptance number. The batch is rejected if the number of defectives in the sample is greater than c. This type of control is known as *acceptance sampling* and tables (for example, *BS 6001*, see Further Reading) detailing the characteristics of such sampling schemes are available. Obviously, the consumer desires there to be only a small chance of accepting a batch when the proportion defective is large, but the producer wants there to be a large chance of acceptance when the proportion defective is small. From a study of *BS 6001* a suitable scheme balancing the consumer's and producer's risks can be chosen.

References to specialist texts on statistical quality control will be found at the end of this book.

15.7 Exercises

1. A manufacturer produces cables with a mean breaking strain of 1 000 kg and a standard deviation of 50 kg. A sample of 50 cables has a mean breaking strain of 984 kg. Does this indicate that the mean breaking strain of cables produced is less than 1 000 kg? Test at the 0.05 level of significance.

2. The expected life of light-bulbs produced in a factory is 1 500 hours. A sample of 36 light-bulbs showed a mean life of 1 430 hours with a standard deviation of 100 hours. Test at the 0.01 significance level to see if the mean life is less than 1 500 hours.

3. There are two methods of packing a product. The weights packed by each method were recorded over a period of time as follows:

	Method A	Method B
Mean weight (g) packed	756	735
Standard deviation (g)	12	14
Number of packets observed	100	125

   Test to see if the difference between the mean weights packed is significant at the 0.05 level.

4. An opinion poll reveals that of 100 voters sampled, 56 say that they will vote for a particular party at the next election. Is the party assured of getting significantly more than 50% of the vote at the election? Test at the 5% level.

5. (*a*) A printing firm which manufactures coloured labels for the packing trades has found over past years that the fraction defective on its main production line has been 3%. A random sample of coloured labels containing 500 items has just been checked and is found to have 25 defective labels. Is this evidence of a significant increase in defectiveness?

(*b*) A random sample of 1 000 labels from a different firm has been found to contain 60 defective items – is this sample significantly different in its level of defectiveness from the sample of 500 items with 25 defectives?

(*Institute of Chartered Secretaries and Administrators*)

6. Repeat Exercise 2 for a sample of 10 bulbs. Why is your approach different for this sample size?

7. A machine is used to fill bags of sugar so that the net weight of sugar in each bag is 1 kg. The machine can be adjusted and a randomly chosen sample of nine bags has a mean weight of 1.06 kg with a standard deviation of 0.08 kg. Test the hypothesis that the machine needs to be adjusted.

8. A machine producing delicate components for the electronics industry has been suffering a rejection rate of 65%. The machine is adjusted and nine test runs after adjustment give the following rejection rates:

61 63 70 58 62 62 65 57 61

Test at the 0.05 significance level to see if the rate has been reduced.

9. (*a*) State the two principal types of use of the chi-squared test. Describe clearly in each application what conclusions are to be drawn if the calculated value of chi-squared exceeds the tabular value.

(*b*) A brand of matches is claimed on the box to have average contents of 40 matches. A single box bought by a friend of yours is found to contain 36 matches and he is considering sending a complaint. However, someone has advised him to examine a few more boxes and have a *t* test carried out if he is still not satisfied.

Explain to your friend, who has little statistical knowledge, why this is good advice. Also explain to him clearly what conclusions the *t* test allows to be drawn in this situation. (You are not intended to carry out a *t* test or describe the calculations involved.)

(*Association of Certified Accountants*)

10. An analysis was carried out of the incidence of traffic accidents involving drivers working 'round-the-clock' for an organization in terms of time of day and the results were as shown below:

Time of day (24-hr clock)	00.01 -04.00	04.01 -0800	08.01 -12.00	12.01 -16.00	16.01 -20.00	20.01 -24.00	TOTAL
Number of accidents	14	16	24	22	24	20	120

Are these results consistent with the assumption that the rate at which accidents occur is uniform throughout the 24 hours?

(*Institute of Personnel Management*)

11. Using the results of Table 13.4 (see Unit 13.2), test whether the student's pair of dice was unbiased. Use the 0.05 level of significance.

12. The following data show the ownership of video cassette recorders (VCRs) by social class in a random sample of 150 households in a town. Test to see if there is an association between ownership of VCRs and social class in that town.

Class of head of household	VCR owned	VCR not owned
Executive	15	10
Managerial	23	8
Working	54	40

13. In an office 200 workers were asked if they were satisfied with their jobs. The results of the survey were:

	Yes	No
Executive officers	60	40
Clerical officers	50	50

Is there a significant relationship between grade of job and job satisfaction? Test at the 0.05 level of significance and use Yates' correction.

14. (*a*) Discuss the advantages of using sampling procedures to control the quality of manufactured items.

(*b*) Samples of 4 items are taken periodically from a production line to provide a check on quality.

The component dimension is set at a mean value of 4.0 cm and the standard deviation is known to be 0.08 cm. Sketch a control chart for sample means.

Eight samples taken sequentially had the following mean values:

3.97 4.02 3.94 4.01 4.09 4.08 4.15 4.14

Plot these means on your chart and comment on what your chart shows.
(*Institution of Industrial Managers, Certificate of Industrial Management*)

15. Explain briefly how control charts are used to control the 'quality' of the product in industrial processes.

1000 ohm resistors are being mass produced. Random samples, each consisting of five resistors, are taken at intervals at the beginning of the production process and their values in excess of 990 ohms are tabulated below:

(1)	(2)	(3)	(4)	(5)	(6)	(7)	(8)	(9)	(10)
10	11	16	5	10	19	19	15	9	17
15	18	10	8	15	16	10	8	16	9
9	12	14	10	8	10	6	2	10	4
13	11	7	6	13	9	13	10	9	12
17	3	4	12	16	14	12	6	2	13

Design a control chart based on the above data using 1 in 40 and I in 1000 control limits. Consider the control of the process by showing on the charts the control point corresponding to each sample. Comment on the control of the process.
(*Institution of Industrial Managers, Certificate of Industrial Management*)

(Answers at the end of the book.)

UNIT SIXTEEN

Interest and investment appraisal

16.1 Introduction

In this unit we shall be looking at an area where the statistician is able to help the accountant and economist. We shall be examining techniques that are not concerned with making estimates or decisions on the basis of samples or past events but are concerned with making decisions about taking actions based on current conditions and the expectations of future events. This area is known as *investment appraisal* and involves simple calculations that are useful when deciding if an investment project should be undertaken, or, if there is a choice of investment projects, which, if any, of them should be undertaken. We must also look at simple and compound interest calculations.

The subjects of *interest* and *investment appraisal* go together because interest is important to investment projects, as many are financed by borrowed funds on which interest must be paid. Alternatively, a firm may finance an investment from reserves of cash, in which case the interest that could be earned by lending those reserves outside the firm is lost. The discerning management will take these factors into consideration when appraising the viability of an investment project.

16.2 Simple interest

Simple interest means that the interest in each time period is calculated as a percentage of the *original* capital sum. The capital sum grows by the same amount in each period. Its growth follows an *arithmetic progression*, which means that it grows by the same amount in each time period. Suppose £100 was invested at a simple interest rate of 5% per annum for 5 years the total amount that the sum would grow to is

$$£100 + £5 + £5 + £5 + £5 + £5 = £125$$

so the arithmetic progression is, in £:

$$100, 105, 110, 115, 120, 125$$

Simple interest, I, can be calculated by applying the formula:

$$I = \frac{PRT}{100}$$

where P is the principal amount invested, R the percentage rate of interest per annum and T the number of years for which the investment is undertaken. For the example above the calculation is:

$$I = £\frac{100 \times 5 \times 5}{100} = £25$$

The formula can be rearranged to enable us to calculate what rate of interest is required to make the capital sum grow to a certain amount over a period of time. Say, for example, we wanted to invest £400 so that the sum grows to £520 over 6 years. What rate of simple interest would be needed? The formula is:

$$R = \frac{100I}{PT}$$

and the value of I we require is £120, P is £400 and T is 6 years, so

$$R = \frac{100 \times 120}{400 \times 6} = 5\% \text{ per annum}$$

We can confirm this by simple addition:

$$£400 + £20 + £20 + £20 + £20 + £20 + £20 = £520$$

The arithmetic progression is, in £:

400, 420, 440, 460, 480, 500, 520

We can also calculate the amount of principal to be invested at 5% per annum for the sum to grow to £520 earning simple interest for 6 years. The formula is rearranged:

$$P = \frac{100I}{RT}$$

In the example:

$$P = \frac{100 \times 120}{5 \times 6} = £400$$

The principal of £400 is known as the *present value* of £520 at a simple interest rate of 5% per annum in 6 years from now.

16.3 Compound interest

Compound interest means that the interest is added to the capital sum so that the next amount of interest is calculated on the original capital sum *plus* the first amount of interest. Suppose, for example, £100 was invested at a compound interest rate of 5% per annum, at the end of the first year the interest earned would be £5, 5% of £100. At the end of the second year the interest earned would be 5% of £105, that is £5.25. At the end of the third year the interest earned would be 5% of £110.25, that is £5.5125, and so on.

The growth of a capital sum invested at a rate of compound interest follows what is known as a *geometric progression*. Unlike an arithmetic progression, each successive term in a geometric progression does not change by a constant amount: in this instance each term increases by *more* than a constant amount. Each successive term is found by *multiplying* the preceding term by a fixed number. In the example of £100 invested at a compound interest rate of 5% per annum, each successive amount in the series is found by multiplying the previous amount by 1.05. The geometric progression is, in £ correct to the nearest penny:

$$100,\ 105,\ 110.25,\ 115.76,\ 121.55,\ \ldots$$

You can see that a capital sum grows more quickly with compound interest than with simple interest. At the end of four years the capital sum has earned £1.55 more than it would have earned at a rate of simple interest.

The formula for calculating the amount, A, that a sum will grow to at a rate of compound interest is:

$$A = P \times \left(1 + \frac{R}{100}\right)^{T}$$

where P is the principal amount invested, R the percentage rate of interest per annum, and T the number of years for which the investment is made.

So if we invested £100 at a rate of 5% per annum compound interest for 5 years, the amount obtained would be, correct to the nearest penny:

$$A = £100 \times \left(1 + \frac{5}{100}\right)^{5} = £127.63$$

The best way of tackling the calculation is with a calculator although it is possible to use logarithm tables, as follows:

$$\begin{aligned}\log A &= \log 100 + (\log 1.05 \times 5)\\ &= 2.0 + (0.0212 \times 5)\\ &= 2.106\end{aligned}$$

The antilogarithm of 2.106 is 127.6, and A is therefore £127.6.

The problem with using logarithms is that if you have only four-figure tables available the results are somewhat approximate. Using four-figure logarithm tables for the above calculation gave a result (£127.6) slightly different from that obtained using a calculator (£127.63). For calculations of compound interest four-figure logarithms are not to be recommended. If it is essential that a precise result be obtained, then it is best to use seven-figure logarithm tables if a calculator is not available.

To calculate the amount of principal to be invested at a given rate of interest for the sum to grow to a particular sum over a period of time earning compound interest the formula is:

$$P = \frac{A}{\left(1 + \frac{R}{100}\right)^T}$$

Suppose we wanted a sum of £600 after 4 years at a compound interest rate of 6% per annum, how much would we need to invest? The calculation is:

$$P = \frac{600}{\left(1 + \frac{6}{100}\right)^4}$$

$$= £475 \text{ to the nearest } £$$

As with simple interest this amount is the *present value* of £600 at a compound interest rate of 6% per annum in 4 years' time. This method of calculating P is known as *discounting* the future amount.

Another useful technique which follows from these compound interest calculations is a method of calculating a *constant repayment*, the amount that must be paid at regular intervals to clear a loan or mortgage. This is of great importance to finance houses, banks and building societies. For example, suppose you take out a mortgage of £10000 repayable over 25 years at an interest rate of 12% per annum. What annual repayments are required to clear this loan, assuming the interest rate does not change?

It can be proved that the amount, Y, of the repayment is given by:

$$Y = PK^T\left(\frac{K-1}{K^T-1}\right)$$

where P is the amount borrowed, K is $\left(1 + \frac{R}{100}\right)$, R is the percentage rate of interest per annum and T is the term, in years, of the loan.

In the example of the £10000 mortgage:

$$K = 1 + \frac{12}{100} = 1.12$$

$$Y = £10000 \times 1.12^{25}\left(\frac{1.12 - 1}{1.12^{25} - 1}\right)$$

$$= £1275.00 \text{ per annum}$$

The above arithmetic is tedious and is best performed with a calculator or seven-figure logarithm tables. The use of four-figure logarithm tables may give results that are somewhat inaccurate.

16.4 Investment appraisal

In this section we will look at methods that can be used to decide whether an investment should be undertaken, or which out of a choice of possible investments should be undertaken. When appraising investment projects, the principal factors to be taken into consideration are:

(*a*) the initial cost of the investment;
(*b*) the amount of income resulting from the investment, or the amount of cost the investment will save;
(*c*) the estimated life of the investment;
(*d*) the timing of the income or cost savings;
(*e*) the interest to be paid on money borrowed to finance the investment, or, if the investment is to be funded from reserves of cash, the interest that could be earned on that money if it were lent elsewhere;
(*f*) the amount of capital available—it might be easier to finance two £10000 projects in consecutive years rather than one £20000 project over two years.

The techniques of appraising investments can be separated into two main divisions:

(i) Conventional methods These can be further subdivided into:

(1) the pay-back method;
(2) the average rate of return method.

These methods are fairly unscientific, but nevertheless very important because they are the ones most commonly used. A survey in the journal *Economics* (Autumn 1978), 'The Investment Decision: Theory and Practice', showed that of the 42 firms that took part in the survey, 22 used the pay-back method of investment appraisal and 6 used the average rate of return method.

(ii) Discounting methods These can be subdivided into:

(1) the net present value method;
(2) the internal rate of return method.

These methods provide a more scientific approach to investment appraisal. They are of particular interest to the economist because they take into account the *opportunity cost* of the use of capital funds. Money used to finance a project itself costs money. If it has to be borrowed, the cost is the interest that has to be paid. If the project is financed from the cash reserves of the firm then the cost is the interest that could have been earned on those funds if they had been lent out. Both methods borrow from the *present value* concept of compound interest using the principle that £1 at the present time is better than £1 some time in the future.

Table 16.1 gives details of five prospective projects that a company is considering.

Table 16.1 A company's prospective projects

Project	A	B	C	D	E
Capital cost (£)	10000	15000	25000	40000	50000
Net proceeds (£)					
Year 1	2000	5000	25000	–	15000
Year 2	3000	10000	–	–	15000
Year 3	5000	5000	–	20000	15000
Year 4	5000	–	–	30000	15000
Year 5	5000	–	–	10000	15000
Year 6	–	–	–	–	–
Total income (£)	20000	20000	25000	60000	75000

In real life most investment projects would be more costly than these and would stretch over longer periods of time, but for demonstrating the methods used these examples will serve the purpose. We shall compare these projects using the different methods of appraisal.

16.5 The pay-back method

This method ranks investments in order of preference by considering the time taken to pay back the initial cost of the project. For projects that earn the *same amount* of net proceeds each year, the *pay-back period* can be determined by dividing the initial cost by the net proceeds in each year. Unfortunately, not all our projects earn the same amount of proceeds each year, and this is true of many investment projects. The pay-back periods are easily determined, however, by adding up the proceeds expected in successive years, until the total equals the cash outlay:

Project	Pay-back period (years)
A	3
B	2
C	1
D	$3\frac{2}{3}$
E	$3\frac{1}{3}$

The drawbacks of this method can clearly be seen. The most favourable project is C which just covers the original cost at the end of the first year of the life of the project. This method takes no account of the earnings after the pay-back period for the full life of the project. The timing of receipts is ignored.

Money expected in the future should not be taken at its face value.

The method is suitable for investment projects where:

(*a*) the cost is low and the project is completed in a short time;
(*b*) the investment is productive as soon as the initial cost is incurred;
(*c*) the net proceeds are easily determined.

16.6 The average rate of return method

Using this method the investment which shows the highest average rate of return is considered to be the most profitable. Unfortunately, there is more than one way in which the *average rate of return* can be determined. The simplest method is to divide the gross (or total) proceeds from the project by the life of the investment and to express this as a percentage of the initial outlay. Using this method for project A the average rate of return would be:

$$\frac{20000 \div 5}{10000} \times 100 = 40\%$$

Capital equipment, however, becomes worn out with use and its value is *depreciated* year by year to allow for this. Depreciation is allowed for in the calculation by using the formula:

$$\text{Average rate of return} = \frac{\text{Average annual proceeds} - \text{Average annual depreciation}}{\text{Initial cash investment}}$$

Again using project A as an example, assuming that the capital equipment depreciates over five years so that the average annual depreciation is the initial cash investment divided by five, the average rate of return is:

$$\frac{(20000 \div 5) - (10000 \div 5)}{10000} \times 100 = 20\%$$

Evaluating each project in the same way over its lifetime gives:

Project	Average rate of return (%)
A	20 over 5 years
B	11.1 over 3 years
C	0 over 1 year
D	10 over 5 years
E	10 over 5 years

Using this method project A gives the greatest average rate of return and would be preferred on those grounds. A weakness of this method is that comparable results can only be obtained when the lifespans of the projects are the same.

16.7 The net present value method

This method *discounts* the expected future proceeds from a project to the *present value* of those incomes; the discounted present value is then subtracted from the initial cost of the project to see if the net present value of the proceeds exceeds the initial cost. If it does, the investment will be profitable.

In using this method the *discounting rate* to be applied must first be determined. The discounting rate that is often applied when making net present value calculations is the current rate of interest that must be paid on borrowed funds. This, of course, will also be the rate of interest that could be earned elsewhere if internal funds are used to finance the project. This may not be the correct discounting rate, however. Many would argue that a company should aim to earn at least its current rate of return on capital when undertaking new investment. On the other hand, others might argue that firms should always be seeking to employ new capital even if it reduces the average rate of return on capital employed.

To evaluate the projects we will use a discounting rate of 15% per annum. Table I at the back of the book gives *discounting factors* for £1 at various rates of interest over periods of time. The column headed '15%' shows us the present value of £1 for any time period up to 18 years in the future if the rate of interest that could be earned on that £1 lent elsewhere were 15% per annum. It is by these *present value factors* that we discount our future investment proceeds. From Table I we can see that the present value of £1 at 15% per annum over 5 years is:

	£
in 1 year	0.8696
in 2 years	0.7561
in 3 years	0.6575
in 4 years	0.5718
in 5 years	0.4972

Present value tables are not always supplied in an examination but it is easy enough to work out the true present value of a future sum of money provided you have a calculator or seven-figure logarithm tables. From Unit 16.3 we know that the present value (P) of an amount of money (A) at a percentage annual rate of compound interest (R) payable in T years' time is:

$$P = \frac{A}{\left(1 + \frac{R}{100}\right)^T}$$

When A = £1 and R = 15%:

$$P = £\frac{1}{\left(1 + \frac{15}{100}\right)^T}$$

Taking values of T in succession as 1, 2, 3, . . ., you will find that the resulting values of P agree with those extracted from Table I. If you need to calculate half-yearly present values, remember that $1.15^{\frac{1}{2}}$ is $\sqrt{1.15}$ so, for example:

$$\pounds\frac{1}{1.15^{\frac{3}{2}}} = \pounds\frac{1}{1.15\sqrt{1.15}} = \pounds 0.8109 \text{ to 4 significant figures}$$

To evaluate a project we work out its *discounted present value* by totalling the successive terms of the present value formula for each year of the project:

$$P = \frac{A_1}{\left(1 + \frac{R}{100}\right)^1} + \frac{A_2}{\left(1 + \frac{R}{100}\right)^2} + \frac{A_3}{\left(1 + \frac{R}{100}\right)^3} + \ldots + \frac{A_n}{\left(1 + \frac{R}{100}\right)^n}$$

For project A the discounted present value is:

$$\frac{\pounds 2000}{(1 + 0.15)^1} + \frac{\pounds 3000}{(1 + 0.15)^2} + \frac{\pounds 5000}{(1 + 0.15)^3} + \frac{\pounds 5000}{(1 + 0.15)^4} + \frac{\pounds 5000}{(1 + 0.15)^5}$$

$$= \pounds 1739.13 + \pounds 2268.43 + \pounds 3287.58 + \pounds 2858.77 + \pounds 2485.88$$

$$= \pounds 12640 \text{ to the nearest pound}$$

Alternatively we can use the table of discounting factors, Table I, to reach the same result and this calculation is laid out in Table 16.2.

Table 16.2 Calculation of the discounted present value of project A

Year	Proceeds (£)	Discounting factor at 15% per annum	Discounted proceeds (£)
1	2000	0.8969	1739.20
2	3000	0.7561	2268.30
3	5000	0.6575	3287.50
4	5000	0.5718	2859.00
5	5000	0.4972	2486.00
Discounted present value of project A			12640.00

The net present value of project A is found by subtracting the initial cost:

$$\text{Net present value} = \pounds 12640 - \pounds 10000$$

$$= \pounds 2640$$

The result is positive, indicating a profitable investment. Project A is compared with the other projects in Table 16.3.

Table 16.3 Comparison of the profitability of projects A–E

Year	Discounted proceeds, to the nearest £, assuming a 15% discounting rate				
	A	B	C	D	E
1	1739	4348	21739	–	13043
2	2268	7561	–	–	11342
3	3288	3288	–	13150	9863
4	2859	–	–	17153	8576
5	2486	–	–	4972	7458
TOTALS	12640	15197	21739	35275	50282
Less initial cost	10000	15000	25000	40000	50000
Net present value of future profits	2640	197	−3261	−4725	282

Project A shows the highest net present value so it is the project to be preferred. Some of the results obtained by this method may seem surprising. For example, returns of £75000 on an initial investment of £50000 may seem attractive at first glance, but discounting greatly reduces the value of the proceeds.

16.8 The internal rate of return method

This method is similar to the net present value method but instead of reducing proceeds to a value to compare with the original cost of the project at a particular rate of interest, the *internal rate of return* method, also known as the *yield* method, seeks to find the rate of interest that will reduce the proceeds from the project to the original cost. If this rate of interest is acceptable—that is, if it equates with a reasonable return on capital invested or at least is more than the cost of borrowing the funds to finance the investment—then the project will be profitable. Where the proceeds from the investment are not regular, as with projects A to E, then the only way to find the internal rate of return is by trial until the correct rate is found, using a calculator or present value tables.

We shall use project A as an example and try a rate of 20% per annum first (see Table 16.4).

The initial cost was £10000 so a discounting rate of 20% is too low. The higher the rate of interest the lower the net present value of the proceeds. Next we try a rate of 25% (see Table 16.5).

The rate is too high this time, so we will try 24% (see Table 16.6).

Table 16.4 Project A: internal rate of return of 20% per annum

Year	Proceeds (£)	Discounting factor at 20% per annum	Discounted proceeds (£)
1	2000	0.8333	1667
2	3000	0.6944	2083
3	5000	0.5787	2894
4	5000	0.4823	2412
5	5000	0.4019	2010
TOTAL			11066

Table 16.5 Project A: internal rate of return of 25% per annum

Year	Proceeds (£)	Discounting factor at 25% per annum	Discounted proceeds (£)
1	2000	0.8000	1600
2	3000	0.6400	1920
3	5000	0.5120	2560
4	5000	0.4096	2048
5	5000	0.3277	1639
TOTAL			9767

Table 16.6 Project A: internal rate of return of 24% per annum

Year	Proceeds (£)	Discounting factor at 24% per annum	Discounted proceeds (£)
1	2000	0.8065	1613
2	3000	0.6504	1951
3	5000	0.5245	2623
4	5000	0.4230	2115
5	5000	0.3411	1706
TOTAL			10008

The total discounted proceeds in this instance are £10008 which is very close to the initial cost. The internal rate of return for project A is therefore 24% per annum.

If we repeat this process for the other projects we can compare their internal rates of return (see Table 16.7). You should check these results for yourself.

Table 16.7 Projects A–E: internal rates of return compared

Project	Internal rate of return (to nearest 1%) per annum
A	24
B	16
C	–
D	11
E	15

Yet again project A is the preferred project, since it gives the highest rate of return. There is no internal rate of return for project C because there is no interest rate low enough to reduce the proceeds to a present value that equates with the initial outlay. This project would certainly not be undertaken if profit was the motive for the investment!

16.9 Other considerations

It should be remembered that it may not be only the most profitable investment that is undertaken. Firms will probably study many possible investment projects at any time and decide to go ahead with several; they may even choose to undertake projects that do not appear to be very profitable, perhaps under pressure from the Government, environmentalists or trade unions. A project may be undertaken purely on an entrepreneur's intuition.

There are additional factors that may prevent accurate investment appraisal.

(i) It may sometimes be difficult to estimate the future income from an investment project. A firm, for example, may introduce a computer which improves through-put of data, increases the amount and scope of information available to the management, and greatly reduces the time-span before information is available. This will improve management control which will undoubtedly raise the revenue of the firm, but quantifying the amount by which revenue will be raised may be extremely difficult.

(ii) It may be difficult to quantify the risk element in an investment project.

(iii) Unforeseen circumstances may invalidate the appraisal: we have all seen the effects of a rise in the rate of inflation on investment projects.

(iv) The life of the project may have been wrongly estimated. Returns may start coming in much later than expected, perhaps because of labour problems during the development stage of the project.
(v) Government intervention may affect the project: new legislation may cause plans to be changed; increased or new taxes may increase costs.

Nevertheless investment projects should be carefully appraised before being undertaken, and all information available at the time taken into consideration. The discounting methods of appraisal are preferable because they tend to take into account more of the facts than the other methods. For an important project a useful compromise is to present to management three estimates of net present value:

(*a*) the most likely estimate of the outcome based on facts available at the time;
(*b*) an optimistic estimate of the likely outcome;
(*c*) a pessimistic estimate of the likely outcome if plans go badly astray.

While this approach does not eliminate uncertainty it does attempt to take it into account.

16.10 Exercises

1. Find by how much the amount of compound interest exceeds the amount of simple interest on a sum of £1 000 invested at $12\frac{1}{2}$% per annum for 4 years.

2. Find the amount at compound interest on a sum of £2 500 invested for 2 years at 10% per annum when:
   (*a*) interest is paid annually;
   (*b*) interest is paid half yearly;
   (*c*) interest is paid quarterly.

3. Draw a graph showing over a period of 15 years the growth in the value (i.e. principal plus interest) of £100 if invested at:
   (*a*) simple interest of 10% per annum;
   (*b*) compound interest of 7% per annum.
   From the graph determine the year in which the total value will be equal under both (*a*) and (*b*) above.

4. Calculate the present value of the following cash flows at the given rate of interest:
   (*a*) £500 one year away when R = 10% per annum;
   (*b*) £1 750 three years away when R = 25% per annum;
   (*c*) £265 two years away when R = 11% per annum;
   (*d*) £2000 nine years away when R = 8% per annum.

5. If you borrow £1200 repayable over five years at an interest rate of 15% per annum, what is the annual repayment?

6. What is the pay-back method of investment appraisal and what are its advantages and disadvantages?

7. Explain what is meant by discounting methods of investment appraisal. How are discounting factors calculated?

8. Compare the net present values of the following two investment projects if the current rate of interest is 14% per annum.

	A	B
Initial cost	£1 000	£1 000
Yield		
Year 1	£400	£400
Year 2	£500	£400
Year 3	£350	£400
Year 4	£300	£400

9. How should a firm using discounting methods of investment appraisal select the discounting rate to be used? What other aspects should be taken into consideration?

10. If a firm will not consider any investment project unless it earns a return of at least 12% per annum will it undertake a project that costs £400 initially and yields one receipt of £1 000 at the end of the tenth year?

11. What is the rate of return on a project that yields £1 200 000 at the end of the tenth year and costs £450 000 initially?

12. A firm's yield on capital employed is 15% per annum. An investment project requires an initial outlay of £6 500 and is expected to yield receipts of £1 200 for 10 years. Should the project be undertaken?

13. Contrast the methods of investment appraisal that can be used by companies and say which method you would recommend and why.

(Answers at the end of the book.)

UNIT SEVENTEEN

Population statistics

17.1 Introduction

The statistical study of human populations is called *demography* and specialists in this field are termed *demographers*. *Vital statistics*, in this context, are the 'statistics of life' such as the numbers of births, marriages and deaths. The primary results of demographic studies are statistical tables classifying a community by such variables as age, sex, marital condition and geographical distribution. Secondary statistics such as birth rates, fertility rates and death rates are also calculated.

Demographic studies are of great importance to Government planners at both national and local level and are of commercial value to marketing organizations.

The United Nations makes population studies on an international basis and publishes a *Demographic Year Book*. The UN Population Commission was formed after the Second World World War and is particularly concerned with improving the collection of population statistics in developing countries. The World Health Organization publishes periodic reports concerning birth and death rates and the distribution of sickness.

In the United Kingdom, population statistics are gathered for the most part in two ways: by *vital registration* and by the *decennial census*.

17.2 Vital registration

There is a Registrar General for England and Wales and another for Scotland. They are ultimately responsible for ensuring that all births, marriages and deaths are registered. In England and Wales, the Church kept lists of births, marriages and deaths from early times but prior to the setting up of the General Register Office by an Act of Parliament of 1836, registration was not centralized. Since then various Acts relating to vital registration have been passed. In 1874 the requirement for a death certificate signed by a doctor and giving the cause of death was instituted. This enables the causes of death to be analysed. It was not until 1926 that still births had to be registered.

In 1936 an Act was passed requiring that, whenever a birth is registered, the mother's age and date of marriage and the number of children born to her, both living and dead, in this and in any previous marriages must be disclosed. In addition, when a death is registered, the marital condition of the deceased and the age of the surviving spouse, if any, must be given. When a married woman dies, the year and duration of marriage and the number of children born in that marriage are required.

In Scotland it was not until 1854 that the first Act concerning registration was passed but the present-day requirements concerning the registration of births, marriages and deaths are almost identical to those for England and Wales.

The Registrars General are also concerned with statistics of morbidity (sickness), abortions and divorces. Some infectious diseases are notifiable but it is obviously impossible to ensure that every illness is recorded and much work on the distribution of sickness is done by sampling from the sickness benefit claim forms submitted to the Department of Social Security. Further studies can be made by analysing hospital records and the records of general practitioners. The Registrars General publish their findings on the size and age distribution of the population and their analyses of births, marriages and deaths in their Annual Reports. Since 1974 these have taken. the form of a series of small reference volumes, each dealing with one topic or a number of closely related topics. Preliminary tables make some salient information available earlier than the reference series. Office of Population Censuses and Surveys (OPCS) monitors give quick release of selected information as it becomes available; for instance, the Registrars General weekly and quarterly returns of births and deaths are published in this form. You will find copies of these publications in any good reference library.

17.3 The decennial population census

(a) History of the census

The Office of Population Censuses and Surveys is responsible for carrying out the Census of the UK Population which is taken every ten years. The first complete census of Great Britain as a whole was taken in 1801 and censuses have been taken at ten yearly intervals ever since, with the exception of 1941 when no census was taken because of the Second World War. In 1939, however, a modified census (the National Registration) was taken.

You should study some of the recent census reports, relating to the 1981 and 1991 censuses, in your nearest library.

The latest census was taken on 21 April 1991. Complete publication of the results takes several years. There are several constraints when choosing a suitable census date. It is a requirement of an Act of Parliament that a census be taken every ten years. In addition the United Nations requested that countries should take a census at the beginning of the decade. The member states of the EC also decided to synchronize their censuses. It is inadvisable to choose a time of year or day of the week when people are likely to be away from home, so the summer and Bank Holiday periods are ruled out. In winter, bad weather may make it difficult for the enumerators to deliver and collect the census forms and it is preferable for the date to be in British Summer Time to give more evening daylight for the collection of forms. For all these reasons, a Sunday in late March or early April is usually chosen but it is not the same Sunday in every census, because of the variability in the date of Easter.

There is provision in the Act for a census to be taken after an interval of five years and in 1966 a 'sample' census was taken. Ten per cent of the population was asked to fill in a census-type form. It was intended to take another such census in 1976 but this was cancelled as a result of cuts in public spending.

(b) The 1991 census forms

There were separate 12-page forms for England, Wales and Scotland, with slight variations in the questions between the forms. The Welsh form, for example, asked questions about knowledge of the Welsh language. The Scottish form asked similar questions about Scottish Gaelic. There were also separate forms for persons who were spending census night in communal establishments, for instance at sea. For England, the census form consisted of questions on the following topics:

General population information

1. Full name
2. Sex
3. Date of birth
4. Marital status
5. Relationship in household
6. Whereabouts on census night
7. Usual address
8. Term-time address of students and schoolchildren

Migration

9. Address one year ago
10. Country of birth
11. Ethnic group
12. Long-term illnesses (including problems due to old age)

Employment

13. Activity last week
14. Hours worked per week
15. Occupation
16. Name and business of employer

Workplace and transport to work

17. Address of place of work
18. Main means of transport used on journey to work

Higher qualifications

19. Higher educational, professional and vocational qualifications attained after age 18

In addition questions were asked about the housing and amenities of each household:

H1. Number of rooms
H2. Type of accommodation

H3. Tenure
H4. Availability of amenities: bath or shower, inside or outside WC, central heating
H5. Number of cars or vans available for use by members of the household

Questions 1, 2, 3, 6, 7 are basic to the population count. Question 1 enables the persons on the form to be distinguished and this makes it easier for the householder to complete the form and for census staff to apply to the right person when information is missing. Names and addresses are not entered in the computer records to ensure confidentiality. The actual census returns are kept under strict security for 100 years after which time they will be made available for general inspection by the public.

Question 2 enables statistics to be quoted separately for men and women and provides basic knowledge about the structure of households and families.

Question 3 asks for date of birth rather than age. In the 1971 census, age was requested but it was felt that this question was not always answered accurately. This question provides what the OPCS calls a 'benchmark' for central and local administration. The allocation of very large sums of money such as the Rate Support Grant and the corresponding scheme for the health services is affected by the age distribution of the population. Many plans for social services and for estimating future income and expenditure of the social security system use these figures. Insurance companies also use them for the construction of National Life Tables.

Question 4 on marital status provides basic knowledge about the structure of households and families, nationally and locally, including the various categories of one-parent families. It also yields basic statistics on rates of marriage, divorce, re-marriage and fertility; and economic activity rates for single and married women. The statistics can also help in assessment of housing demand and enable the effects of alternative tax and social security systems to be evaluated.

Question 5 enables the number of families, rather than households, to be counted. The different types of families can be identified and this is of particular use to local social services.

Questions 6, 7 and 8 are needed to count those people who are absent from home on census night. The form asked for a return of people who are usually resident at the address and are present there on census night, of people usually resident but absent on census night and also of visitors. For some purposes, for example resource allocation, it is the numbers usually resident in an area that are needed whether or not they actually happen to be there on census night. The 1981 Census did not ask about the term-time address of students and schoolchildren.

Question 9 provides information on population movement within the country and is vital to the future planning of local authorities with regard to housing, employment, the elderly and the overseas-born.

Questions 10 and 11 allow areas with concentrations of immigrants to be identified. It enables the social and economic conditions of immigrants to be compared with those of the native population. A test census was carried out in Haringey in 1979 and because of the hostile reaction and poor response to these questions, the Government decided to omit them from the 1981 census form as it was felt that the census as a whole might be jeopardized and that the answers given were unreliable. The ethnic group question was new to the 1991 Census.

Question 13 finds out how many people of each age are in the working population, how many are potentially available for work and how many are not available because they are disabled or retired. From these statistics the proportion working or seeking work in each age group in each area can be calculated. This is the only comprehensive source of statistics about the working population. The regular statistics collected by the Department of Employment do not cover the self-employed. The regular statistics do not include those who are not registered as unemployed. Household surveys exclude people who are resident in institutions.

Question 15 is the basis for the estimates of social classes which are used extensively, for example, in market research. In conjunction with Question 16 it shows the deployment of skilled workers and enables future labour demand to be estimated. Medical research on occupational mortality and morbidity as well as industrial mortality is based on these questions. The statistics are also used by insurance companies and others concerned with occupational pension schemes.

Question 16 determines the industry in which a person is employed as distinct from his or her occupation. Regular employment statistics are gathered on an industry basis without reference to occupation.

Question 17 enables the Department of Employment to determine 'travel-to-work' areas. These are the smallest areas for which unemployment rates are calculated. These travel-to-work areas are used by the Department of Industry for designating assisted areas. The then Manpower Services Commission used these areas in organizing its services and giving advice on labour availability to prospective employers. Travel-to-work statistics in conjunction with Question 18 are used by local authorities and local transport authorities. The Department of Transport uses census statistics when planning trunk road schemes. The Department of the Environment uses information on journeys-to-work when planning the provision of new jobs and homes and the related transport services.

Question 19 asks for detailed information of any higher educational, professional and vocational qualifications obtained after the age of 18 so that studies of the future demand and supply of highly qualified manpower can be made. These form the basis of future decisions on the scale and balance of higher education.

Question H1 in conjunction with Question H2 allows overcrowding and under-occupancy to be measured. Areas which characteristically contain overcrowded accommodation can be identified for possible special action.

Question H2 allows an estimate of the number of dwellings to be made, dependent on information about shared accommodation. Potential housing demand can be estimated by the amount of inadequate housing and overcrowding when two or more households share accommodation.

Question H3 allows trends in owner-occupation to be assessed. The amount of furnished and unfurnished accommodation rented from private landlords can be measured and the effects of changes in the law assessed. Information on the numbers occupying local authority housing, their family structure and socio-economic characteristics help in developing housing policy at local level.

Question H4 provides a measure of inadequate housing. The answers help to identify areas for improvement under Housing Action Area and General Improvement Area programmes.

Question H5 provides information which is of use to public transport authorities when decisions on reductions of services are made. Before the 1981 Census, the Government had decided that this question was not sufficiently important to be included on the census form but Members of Parliament decided that it should be re-instated. Answers to questions on the Welsh and Scottish forms about the Welsh language and Scottish Gaelic provide a guide to policies on the development of the languages, of value particularly to broadcasting authorities.

(c) Census organization

The country was divided into around 130,000 *enumeration districts* each of which was the responsibility of one enumerator who delivered and collected a census form at each household and communal establishment within its clearly defined area. The enumeration districts covered the entire area of the country without omission or overlap and represented workloads that could be performed in the time available. The boundaries of enumeration districts were chosen to coincide with national, county, local government ward and civil parish boundaries as they existed on Census day.

A four-tier field force was established to deliver and collect the census forms. This was based on the method used successfully in 1981. 135 Census Area Managers (CAM) were each responsible for an area containing up to half a million people. Each CAM recruited up to 25 Census Officers (CO) who controlled a local area of about 25 000 people. Those areas were known as Census Districts.

Each CO was responsible for the recruitment and training of some 7 800 assistants and 117 500 enumerators.

Care was taken to avoid, if at all possible, placing an enumerator in an area where he had personal or business contacts. If an enumerator was known to the householder, the census form could be returned in a sealed envelope to be opened by the Census Officer for the area. All forms were collected by the enumerators as the cost of organizing facilities for return by post was too great.

The enumerator had to check that the form was completely filled in and to give any help required, including calling in interpreters for those with a limited knowledge of English.

(d) Processing and publication of the results

Census forms from a little under 22 million households were processed. The first results were published in July 1991 in preliminary reports for England and Wales and for Scotland. These reports were derived from summary records made by field staff. They did not, therefore, tally precisely with subsequent reports.

The main processing of the Census began in June 1991, and the publication of reports was scheduled for completion in 1994. The statistics to be made available are more extensive than those produced from any previous population census of Great Britain; with modern computer technology this is only to be expected.

As in previous censuses, returns from the 1991 census were presented on a county to county basis. It was planned from the outset that the 1991 local reports should provide results from all the topics covered by the census. Initial proposals for local statistics were based on those for 1981. Comments were then invited from advisory groups representing government departments and local and health authorities. Meetings were held in various parts of the country with interested parties to discuss the proposals. The form of local statistics was developed and refined through these and further rounds of consultation.

Some topics—those where answers to the questions were more difficult and costly to process—were only processed for a 10% sample. These were questions that often required a written response rather than a simple tick or number in a box, such as occupation, name and business of employer and higher qualifications.

It was decided that an abbreviated set of statistics from the Census results for small areas throughout Great Britain—Small Area Statistics (SAS) and Local Base Statistics (LBS)—would be produced. The topics and cross-classifications for inclusion were decided prior to the Census in consultation with interested parties. The SAS and LBS tables are produced for each enumeration district and these are aggregated to local government wards and districts. When dealing with tables for one enumeration district only, it is possible that individuals possessing particular characteristics could be identified. SAS will not, therefore, be made available for any areas with fewer than 50 persons and 16 households, but an unmodified count of males, females and households is available for all enumeration districts.

Further information on all aspects of the Census can be found in *OPCS Census Monitors*, newsletters and reports.

17.4 Demographic rates

From the information provided by the population census and vital registration, many statistics can be calculated.

(a) Birth rates

(i) The crude birth rate is the simplest measure of the birth rate and is the number of live births in one year per 1000 of the population:

$$\text{Crude birth rate} = \frac{\text{Total number of live births in one year}}{\text{Total population at the middle of the year}} \times 1000$$

The crude birth rate depends on the age and sex distribution within the population and is consequently not of great use.

(ii) The legitimate and illegitimate fertility rates take into account the fact that almost all births are to women between the ages of 15 and 44 inclusive. Thus:

$$\text{Legitimate fertility rate} = \frac{\text{Number of live births to married women in one year}}{\text{Number of married women aged 15–44 at the middle of the year}} \times 1000$$

and:

$$\text{Illegitimate fertility rate} = \frac{\text{Number of illegitimate live births in one year}}{\text{Number of single women and widows aged 15–44 at the middle of the year}} \times 1000$$

(iii) The age-specific fertility rate takes account of the variation in fertility over the 15–44 age range. This age range is divided into smaller, five-year, groups and a fertility rate is calculated separately for each age group:

$$\text{Age-specific fertility rate} = \frac{\text{Number of live births in an age group in one year}}{\text{Number of women in that age group at the middle of the year}} \times 1000$$

Table 17.1 shows an example of the calculation of age-specific fertility rates.

Table 17.1 Calculation of age-specific fertility rates

Age group	Number of women at mid-year	Total number of live births in one year	Age-specific fertility rate per 1000 women
15–19	10500	200	19.0
20–24	10500	1800	171.4
25–29	8500	1100	129.4
30–34	8000	800	100.0
35–39	7500	350	46.7
40–44	7500	100	13.3
TOTALS	52500	4350	

(iv) The gross reproduction rate (GRR) tackles directly the problem of estimating how quickly a population is reproducing itself. The GRR is the ratio of the hypothetical total of *female* babies, who would be born to a group of women who start their child-bearing period together and who neither die nor migrate until they have reached the end of that period, to the total number of women considered.

We can work out the GRR from the information given in Table 17.1. The age-specific fertility rates give the numbers of children born each year to 1000 women in the different age groups. If 1000 women started their child-bearing periods together at age 15, between them they would produce 19.0 babies every year for the five years until they were 20, then 171.4 babies every year for the five years until they reached 25. In total they would produce:

$$\begin{aligned} & 5 \times \text{sum of the age-specific fertility rates in Table 17.1} \\ &= 5 \times 479.8 \\ &= 2399 \text{ babies} \end{aligned}$$

Not all these babies are female. From population studies the ratio of girls to boys is found to be approximately 0.485:0.515. Thus 0.485 × 2399, that is, 1163.515, of the babies are girls. Therefore, we have:

$$\text{GRR} = \frac{1163.515}{1000} = 1.16 \text{ to 2 decimal places}$$

A GRR of 1.00 indicates that births are just sufficient to maintain the present population. If the GRR is greater than 1.00 the births are more than enough to maintain the present population.

(v) The net reproduction rate (NRR) The GRR can be criticized because it takes no account of the deaths of females at all ages from 0–45. The Life Tables published by the Registrar General give the proportion of women who survive to each age. To adjust the GRR for mortality, we take the product of the sur-

Table 17.2 Adjusted age-specific fertility rates

Age group	Age-specific fertility rate	Survival factor	Adjusted fertility rate
15–19	19.0	0.94	17.9
20–24	171.4	0.93	159.4
25–29	129.4	0.91	117.8
30–34	100.0	0.90	90.0
35–39	46.7	0.88	41.1
40–44	13.3	0.86	11.4
TOTAL			437.6

vival factor for the midpoint of each age group and the age-specific fertility rate before calculating the reproduction rate. The calculation, using fictitious survival factors, is given in Table 17.2.

$$\text{NRR} = \frac{5 \times 437.6 \times 0.485}{1000}$$
$$= 1.06 \text{ to 2 decimal places}$$

The NRR was thought to provide a basis for reliable population projections, but it has some defects since it is dependent on the fertility and mortality rates assumed in its calculation. Fertility rates can change quickly as, for example, after the Second World War.

(b) Death rates

(i) The crude death rate is the simplest measure of mortality and is the number of deaths per 1000 of the population:

$$\text{Crude death rate} = \frac{\text{Number of deaths in one year}}{\text{Total number in the population at the middle of the year}} \times 1000$$

(ii) The standardized death rate overcomes the effects of differing age distributions on the value of the crude death rate. For instance, a seaside resort to which elderly people retire has a higher crude death rate than an industrial area, simply because there is a greater proportion of old people in the population. To compare two such areas more usefully, we calculate for each the standardized death rate. To do this we consider a standard population and calculate what the death rates in the areas would have been if each had had this standard population. The calculation of a standardized death rate for a certain region is shown in Table 17.3. The crude death rate is:

$$\frac{13\,270}{414\,000} \times 1000 = 32.05 \text{ per } 1000$$

and the standardized death rate for the region is:

$$\frac{7724}{360\,000} \times 1000 = 21.46 \text{ per } 1000$$

The standardized death rate is much lower in this example than the crude death rate because of the relatively large numbers of elderly people in the region.

The problem of defining a standard population is similar to that of fixing a suitable base year for an index number. Theoretically, any population would do, as all comparisons are made relative to the standard. There are advantages, however, in choosing one which is representative of the general situation as it actually exists. In UK official statistics, one standard used is the population of England and Wales in 1981, and the other is a synthetic 'world' standard population.

Table 17.3 Calculation of a standardized death rate

Age group	Actual population (thousands)	Actual number of deaths	Age-specific death rate per 1000	Standard population (thousands)	Expected number of deaths in the standard population
0–4	30	480	16	39	624
5–9	27	54	2	37	74
10–14	30	30	1	35	35
15–24	65	325	5	65	325
25–44	132	792	6	107	642
45–64	89	3471	39	58	2262
65–84	41	8118	198	19	3762
TOTALS	414	13270		360	7724

(iii) Other death rates can be calculated for sub-sections of the population of interest. Those of particular interest are the infantile mortality rate, which gives the death rate per 1000 live births, of children under one year old; separate rates for males and females and for particular age groups; occupational mortality rates; separate rates for different social classes and for different causes of death.

17.5 Exercises

1. Describe the population and vital statistics available in Britain and explain why the collection of such statistics is given high priority in most countries.

2. Periodic population censuses have been carried out in the UK for many years. Describe briefly the way in which the censuses are conducted, the information collected during the most recent census and the form in which the results are summarized.

   (*Institute of Personnel Management*)

3. An insurance company is interested in the structure and dispersion of the UK population. What official sources of statistics can be used to find such information? Why might the company be interested in this information?

4. (*a*) What basic data are essential for the production of accurate demographic statistics?

   (*b*) Define the following terms relating to demographic studies, and state how the rates are calculated:

   (i) birth rate;
   (ii) fertility rate;
   (iii) crude death rate;
   (iv) standardized death rate.

(*c*) 'The birth rate is not a good measure of the way the population is replacing itself!'
Comment on this statement.
(*Institute of Cost and Management Accountants*)

5. The table below shows the number of deaths and the population, according to age groups, in a certain town in a given year; the last column shows the corresponding percentages of the population of the country. Complete the table and hence calculate the standardized death rate per 1000.

Age group	Number of deaths	Population of town	Death rate per thousand	% population of country
0 and under 25	80	4000		25
25 and under 45	40	4000		45
Over 45	290	4000		30

6.

Age group	Town A Population	Town A Number of deaths	Town B Population	Town B Number of deaths	Standard population
0–10	1200	16	1100	17	1200
10–44	5800	17	4800	25	6000
45+	3000	36	4100	44	2800

Using the data given above, calculate

(i) the crude death rate for town A;
(ii) the standardized death rate for town A;
(iii) the crude death rate for town B;
(iv) the standardized death rate for town B.

Hence state, with reasons, which town offers the healthier environment.

7. The number of days' work lost due to sickness in a company's two factories were as given in the table:

Age	Oxford factory Employees	Oxford factory Days lost	Birmingham factory Employees	Birmingham factory Days lost
20–	113	904	84	704
30–	167	1168	126	792
40–	89	401	229	1082
50–	51	294	94	520

Compare the numbers of days lost, by age groups; and, taking the total work force as standard, compute the standardized rate of days lost for each factory. Comment on the similarities and differences between the days lost for the two factories.

(*Institute of Statisticians*)

8. In calculating the premiums payable for life assurance policies, insurance companies use information derived from mortality tables. Extracts from a mortality table are given below:

Males per 1 000 births

Age in years	Number living
0	1 000
20	958
40	905
60	680
80	413

(*a*) Find the probability that a male aged 20 will:
(i) attain the age of 40 years
(ii) attain the age of 40 years but not 80;
(iii) not attain the age of 60 years.

(*b*) On his 25th birthday a young man decides to invest £1 000 in a single payment policy which matures only if and when he attains the age of 60 years. The insurance company can invest at a rate which allows it to apply 5% compound interest to the £1 000 investment. Considering the possibility that the man may die before he reaches the age of 60 years and therefore will not collect anything, state the maximum amount the insurance company can promise to pay at the age of 60 years.

9. What is the distinction between gross reproduction rate and net reproduction rate? Calculate these rates for the following data:

Age group	Total number of women (thousands)	Total number of *female* births	Survival factor
15–19	2400	28 000	0.96
20–24	3000	227 000	0.96
25–29	2700	231 000	0.95
30–34	3400	196 000	0.94
35–39	2700	114 000	0.94
40–44	3000	33 000	0.93

(Answers at the end of the book.)

UNIT EIGHTEEN

Computers and statistics

18.1 Introduction

No modern textbook on statistics would be complete without a look at the computer as a tool for the collection and analysis of statistical data and the production of statistical information. In this unit we will see why the computer is a powerful and useful tool for the statistician. We will also look at some developments in computing that have been made particularly for the statistician.

18.2 The computer

The computer is an electronic tool, and the word 'tool' is important. Many people regard the computer with awe, as though it works by magic. In fact the computer is like any other tool: it does exactly as it is instructed to by its user.

There is a factor, however, that makes the computer different from other tools. With other tools the user operates to a given set of instructions every time he or she wants to do a job. With a computer the set of instructions to do a job is written down once, and is then stored within the machine. This set of instructions is then followed every time the same job needs to be done. The set of instructions is called a *program*.

This feature of computers is ideal for the statistician. Let us consider the correlation coefficient formula in Unit 9.5. You may have already experienced the problems of calculating r using an ordinary calculator. Even if you are proficient with the method, you may still make simple mistakes along the way. I have often found that my incorrect results are due to minor errors in copying figures or reading displays during a calculation. However, once the formula for calculating r has been embodied in a computer program, and that program has been tested to prove that it works, we can be certain that it will always produce the correct result. (This assumes, of course, that the data are put into the computer correctly, but more about that later.)

You can see through the Units of this book that there are awkward formulae to deal with when using many statistical techniques. These are easily written into computer programs. We will also see in Unit 18.7 that the latest spreadsheet packages contain the programs to do these calculations.

18.3 Computer hardware

The machinery of the computer is called *hardware*. It consists of several components.

(a) The processor

This is the brain of the machine. It carries out the instructions in a program, does the calculations that are needed and enables decisions to be made about data and the results of calculations. In large computers the processor is called the central processing unit (CPU).

(b) Memory

This is where the computer's programs are stored while they are running. Data that are needed by any program are also stored in the memory while they are being used for processing.

(c) Input

The data that are used by the computer programs have to be fed into the machine through an input device of some sort. Usually the data have be entered into the machine by means of a keyboard that looks like a typewriter. The data can be entered directly into the computer, or entered on to magnetic tape or disk to be read into the computer later. There are, however, different types of input devices, some of which are of particular interest to the statistician. We will look at those devices in more detail in Unit 18.5.

(d) Storage

The memory of the computer is limited. The amount of data that a computer user needs to store would not fit into memory. As you know, some statistical applications, such as sales forecasting, need large amounts of historical data. The computer writes data it needs to store from memory on to storage devices such as magnetic tapes or magnetic disks. These data can then be quickly read back into the computer when it is required in the future. The data keyed into the computer, as described in section (*c*) above, are often written on to a storage device until they are processed.

(e) Output

Obviously we need facilities that enable us to see the statistical information the computer produces. Modern computers offer a range of output devices. Information can be printed, drawn or displayed on screens. The statistician can have information presented in tables, or in the form of histograms, charts, graphs or pie charts. Output can be displayed in most of the methods of presentation shown in Unit 5.

18.4 Computer software

The programs that run on the computer are known collectively as *software*. The computer cannot operate without software, and there are different types of software.

(a) Systems software

The first thing the computer does when it is switched on is to load into its memory a special type of software. This is *systems* software, and is also known as an operating system. This software controls the operation of the computer. It looks after the users of the computer, making sure that they all are allocated a fair share of the computer's resources. It also looks after users' data and makes sure that the data are protected so that unauthorized people do not have access to them. Systems software also provides computer programs to do standard jobs, such as sorting and copying files of data.

(b) Applications software

Applications software can best be described as the user's software. It consists of computer programs that have been designed and written to solve particular user problems. In a company, for example, there will be applications software to do jobs such as producing the weekly payroll, keeping records of stocks of products and accounting for sales of goods to customers. Any programs written to produce statistics will be part of the applications software. For example, regression analysis (Unit 9.3) is an application of computing for which a computer program might be written to do the complex calculations needed. Regression analysis may be part of sales-forecasting applications software.

In Unit 18.6 we will look at some applications software that has been developed purely for the use of the statistician.

18.5 The computer and the statistician

We can now look at all the aspects of the computer that make it an ideal tool for the statistician. In some sections we will be repeating and expanding upon some items that we have already discussed.

(a) Data collection

In earlier Units we said that large samples are necessary in order to get meaningful statistics. The larger the sample, the more effort is needed to read forms, categorize data and count responses before the analysis of the data can even begin. Data can easily be entered into the computer. Forms can be displayed on the computer's input screen, and the computer can be programmed to skip through the data items so that an operator can quickly key-in responses.

There is also available special input hardware that the statistician can make use of. *Optical character recognition* (OCR) enables forms to be read directly by the computer so that data can be entered into the computer without keying. The use of OCR calls for careful form design. The data to be read by the input device have to be in particular positions on the form. The data that are entered on to the form must be typed or printed very carefully for the device to read them properly.

Optical mark recognition (OMR) is probably even more useful for the statistician. The input devices for this technique read marks such as bold lines, ticks or crosses made on forms and interpret them as data to be fed into the computer. Forms used in statistical surveys often require the person entering data on the form to respond to a question by just making a mark to indicate a yes or no answer. Sometimes a question requires the respondent to make a selection by putting a mark such as a thick stroke in a box. (See Unit 2.7.) An example of a form that is designed to be read by an optical mark reader is shown in Fig. 18.1.

The use of OCR and OMR saves time and effort in reading, categorizing and counting data. OMR makes form completion quicker and easier when 'multiple-choice' type questions (see Unit 2.7) can be used.

(b) Speed of analysis

The computer works at electronic speeds. The speeds at which it does calculations and processes data are measured in millionths of a second. It can provide almost instant analysis of masses of data. It can complete a large number of very complex calculations in fractions of seconds once the data have been input. It trivializes the calculations involved in statistical analysis.

(c) Presentation of statistical information

Computer output devices can present statistics in the form of tables, charts, graphs or most of the formats shown in Unit 5. The output can be printed on a printer, drawn on a graph plotter or printer, or displayed on a VDU screen.

(d) Storage

A vast amount of data can be stored in a very small space. It can be quickly retrieved and analysed over and over again. Data can be stored over a long period of time and be constantly added to. This makes for very large samples of data over time. The storage of large amounts of historical data is essential for statistical applications such as sales-forecasting using regression analysis (see Unit 9.3).

(e) Accuracy

As has been said before, once a computer program has been written to perform a statistical calculation, and that program has been thoroughly tested to make sure that it works correctly, we can be certain that it will always give us an accurate result for the data it is working on. That last phrase is important, because obviously we will not get the correct result if we work on incorrect data. We know that some statistical calculations are complex, even forbidding. Having computer programs to deal with that complexity is a boon for the statistician.

BLOOMSBURY HEALTH AUTHORITY

Bloomsbury Health Authority want to know how satisfied patients are with their care in hospital. Could you please help us by completing this questionnaire? Look for the box under the face which best expresses your views and fill in the white space as shown in the example:

EXAMPLE:

Your answers are not seen by staff caring for you; they are read directly by a machine so it is best to use either a pencil or a blue pen. We would welcome any additional comments, but could you please write them on a separate piece of paper.

WHAT DO YOU THINK ABOUT THE FOLLOWING:

	very satisfied 4	satisfied 3	somewhat dissatisfied 2	very dissatisfied 1
doctors				
nurses				
the clinical treatment you receive				
the information that is given to you about your treatment				
the control of any pain				
atmosphere on the ward				
noise on the ward				
your surroundings (layout, furnishings, decor, etc.)				
ward cleanliness				
the way your day is organised on the ward				
bathroom and toilets				
food				
radio, TV, dayroom				
telephones				
overall, how satisfied are you with your hospital stay				

BHA3U

Thank You for Your Help

Fig. 18.1 A form designed to be read by OMR

(f) Saving of calculations

It follows from what was said in the above section that once we have programs written for all statistical calculations there is no need for the statistician to worry about them. It is only necessary to enter the data and run the program. It can be argued that the teaching of statistics should now be changed—that there is no need to learn any formulae. All that needs to be learned is the statistical technique that is needed to solve a particular problem, and how to run the program for that technique.

(g) Standard software

We will see in the next section that most of the statistical programs that are needed have already been written and are available for the statistician in the form of *software packages*. The statistician only needs to obtain a package and learn how to use it.

18.6 Statistical packages

We have seen that the computer takes the hard work out of statistical calculations, and that once a program has been written to do such calculations it can be used over and over again, always giving correct results. Obviously, as for other computer applications, it was not long before an enterprising person realized that a system of programs to perform statistical calculations would be very useful—that people might even be prepared to pay for such a system. We refer to programming systems that have been written for particular applications as *packages*. There are now many statistical packages available—packages such as SPSS, BMD and SAS.

Packages save people the hard work of writing their own programs. It means that they do not have to learn a programming language, or pay expensive programmers to write them for them. It means that they do not have to test their own programs to make them work. Instead they can obtain a set of programs that have already been tried and tested by other users. It means that they do not have to wait for programs to be written before they can start to use them for their statistics. They can be operational very quickly.

The use of packages can change the way in which we approach statistics. Some people give up on the subject very quickly because they are put off by having to master the mathematics and formulae. By using a package, the user of statistics does not need to learn the formulae. All that is required is to find out how to input data into the package and select the commands that produce the required results. It also means that the teacher of statistics can concentrate on teaching the techniques of statistics and how to use the results of calculations rather than how to perform the calculations.

We will now take a closer look at one particular package called *Minitab*. It is an easy-to-use general-purpose statistical package designed for those with no previous experience with computers. The use of the package follows the same pattern as doing statistical calculations by hand. The computer is instructed to read in data and put it into columns. Once the data are stored in columns, powerful commands are available to work on the data. For example, the command AVERAGE, followed by a column number, will result in the computer finding the average of the data in that column. The command STANDARD DEVIATION, followed by a column number, will result in the computer finding the standard deviation of the data in the column. There are commands that enable the user to print data as histograms or graphs as well as commands to perform most statistical calculations.

Fig. 18.2 shows an example of the use of Minitab for a regression calculation using the data in Table 9.3 in Unit 9.3. The print-out shows first an instruction

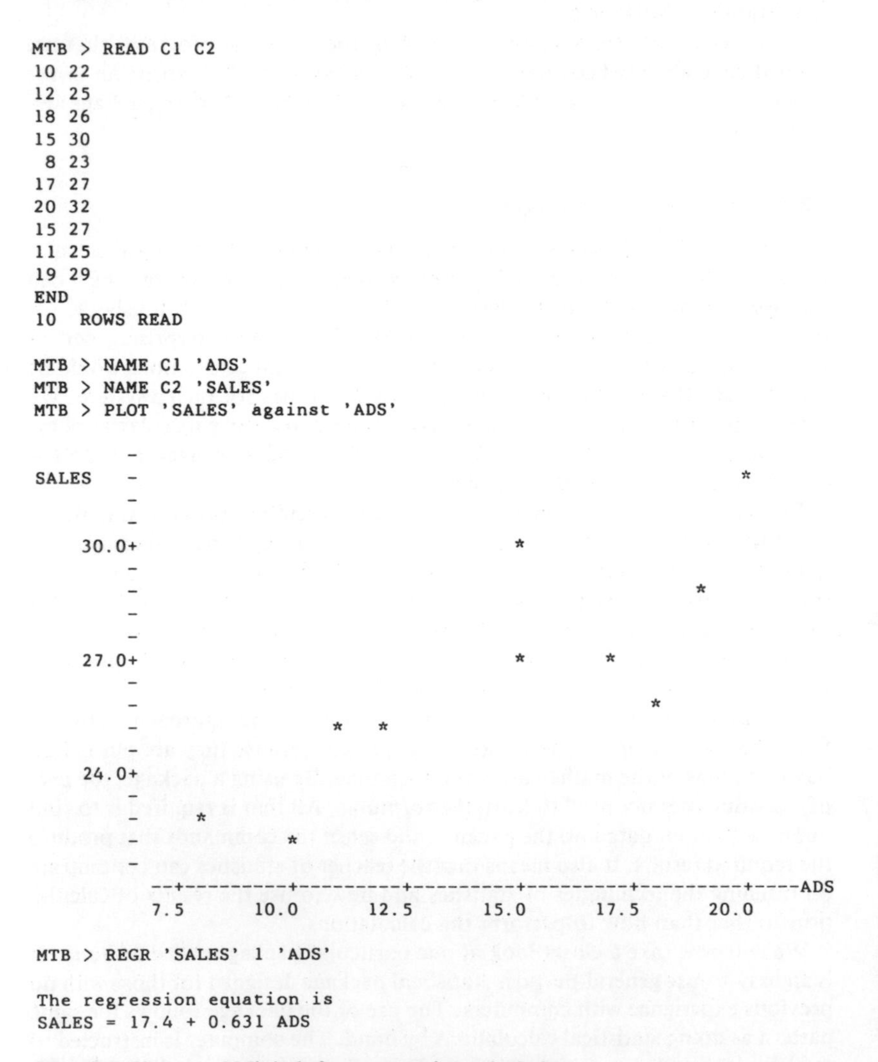

```
MTB > READ C1 C2
10 22
12 25
18 26
15 30
 8 23
17 27
20 32
15 27
11 25
19 29
END
10  ROWS READ

MTB > NAME C1 'ADS'
MTB > NAME C2 'SALES'
MTB > PLOT 'SALES' against 'ADS'

           -
SALES      -                                                    *
           -
           -
      30.0+                                   *
           -
           -                                                *
           -
           -
      27.0+                                   *       *
           -
           -                                              *
           -                      *   *
           -
      24.0+
           -
           -       *
           -               *
           -
         --+---------+---------+---------+---------+---------+----ADS
          7.5      10.0      12.5      15.0      17.5      20.0

MTB > REGR 'SALES' 1 'ADS'

The regression equation is
SALES = 17.4 + 0.631 ADS
```

Fig. 18.2 The use of Minitab for a regression calculation

to read data into two columns, C1 and C2. The two columns are then named, 'ADS' and 'SALES'. The next instruction—PLOT 'SALES' against 'ADS'—causes a scatter graph of the data to be printed, as can be seen. Minitab is then told to produce the regression analysis of column 2 on column 1 by this instruction: REGR 'SALES' 1 'ADS'. You can see the resulting equation.

Another example of the use of Minitab is shown in Fig. 18.3. It is the standard deviation example from Unit 8.7, the data for which is shown in Table 8.6. You can see how Minitab makes light of the problem. The first command tells Minitab that a set of data is to be entered. The frequency distribution is then entered, as you can see, in the form of each frequency followed by the value of x that has that frequency. Once the set of data, c1, has been entered, Minitab can then be instructed to produce statistics from the data set.

The command 'aver c1' causes the mean to be printed. The command 'stan c1' causes the standard deviation to be printed, while the command 'desc c1' causes a host of statistics to be printed, including a trimmed mean that is calculated by leaving out the extremes of the distribution. The command 'hist c1' causes a histogram to be printed, but as you will see Minitab prints histograms with the bars horizontal.

As well as statistical packages there are also computer software packages available for applications, such as simulation or operational research, where statistical techniques are needed as part of the application.

```
MTB > set c1
8(4) 12(5) 15(6) 25(7) 17(8) 13(9) 10(10)
MTB > end
MTB > aver c1
   MEAN    =      7.1000
MTB > stan c1
   ST.DEV. =      1.7203
MTB > desc c1

                   N      MEAN    MEDIAN    TRMEAN     STDEV    SEMEAN
C1               100     7.100     7.000     7.111     1.720     0.172

                 MIN       MAX        Q1        Q3
C1             4.000    10.000     6.000     8.000

MTB > hist c1

Histogram of C1   N = 100

Midpoint   Count
       4       8  ********
       5      12  ************
       6      15  ***************
       7      25  *************************
       8      17  *****************
       9      13  *************
      10      10  **********

MTB > stop
```

Fig. 18.3 The use of Minitab to process the data shown in Table 8.6

18.7 Spreadsheets

Spreadsheet packages are widely used software systems. They are often associated with financial analysis, but in fact can be applied to many other problems. For example, most colleges these days will use a spreadsheet to produce lists of students' examination results, showing as well as individual marks such things as average marks for subjects and students and even perhaps standard deviations of marks for each subject. Spreadsheets are particularly useful for doing 'what if?' exercises. For example, what if we changed the pass mark for the examination. How would it affect the number of passes or failures?

There are many spreadsheet programs available, such as Lotus 123 and Microsoft ® Excel, but all spreadsheets do basically the same thing: they provide a large worksheet or table with a great number of rows and columns into which data can be entered and stored. The rows within the worksheet are identified by numbers, and the columns are identified by letters (or sometimes numbers). Fig. 18.4 shows an example of a section of a Microsoft ® Excel worksheet with the data of methods of travel to work, as shown in Fig. 18.5, entered into it.

You can see how each row/column can be identified by its number/letter, so each entry in the table has an individual row/column address. Each address in the worksheet is known as a cell. Any type of data can be entered into each cell, such as alphabetic characters for labels and headings for when a table is printed out and numeric data for items that need a numeric value, such as a

Microsoft Excel

File Edit Formula Format Data Options Macro Window Help

Normal

E12 =SUM(E6:E11)

Sheet1

	A	B	C	D	E	F	G	H	I
1									
2									
3	Age group		Own	Public	Totals				
4			Means	transport					
5									
6	Under 18		12	6	18				
7	18-25		15	10	25				
8	25-40		25	18	43				
9	40-55		13	32	45				
10	Over 55		6	12	18				
11									
12	Totals		71	78	149				
13									
14									
15									
16									

Ready NUM

Fig. 18.4 A section of a spreadsheet

student's mark for a particular examination. Formulae can also be entered into a cell. In the cell at the foot of a column in the table, for example, the formula for finding the average of all the values in that column could be entered. When values are then entered into the column, the average would automatically be calculated from the formula and stored in that cell.

The overall size of the worksheet provided in a spreadsheet is very large, typically 63 columns and 254 rows. Obviously this is more than can be seen on a VDU screen, so only part of the worksheet is shown on the screen at any time. The user can use arrow keys on the keyboard to move through the worksheet and display the parts that are wanted on the screen.

We already know that statisticians make use of tables, so a spreadsheet that enables us to store data in this way is useful. The ability to include formulae in the table and to have results of the formulae calculations automatically generated and stored greatly increases the spreadsheet's value.

With the latest spreadsheet technology it is not necessary to enter many of the formulae. Statistical tools are included as part of the spreadsheet package. Once the data have been entered into a worksheet in the form of a table the appropriate statistical calculation can be performed on the data by selecting the necessary tool from the spreadsheet menus. The variety of statistical tools available in a modern spreadsheet package is comprehensive. Microsoft Excel, for example, provides some 80 statistical routines. Fig. 18.5 shows a Microsoft Excel screen with part of the list of statistical routines available.

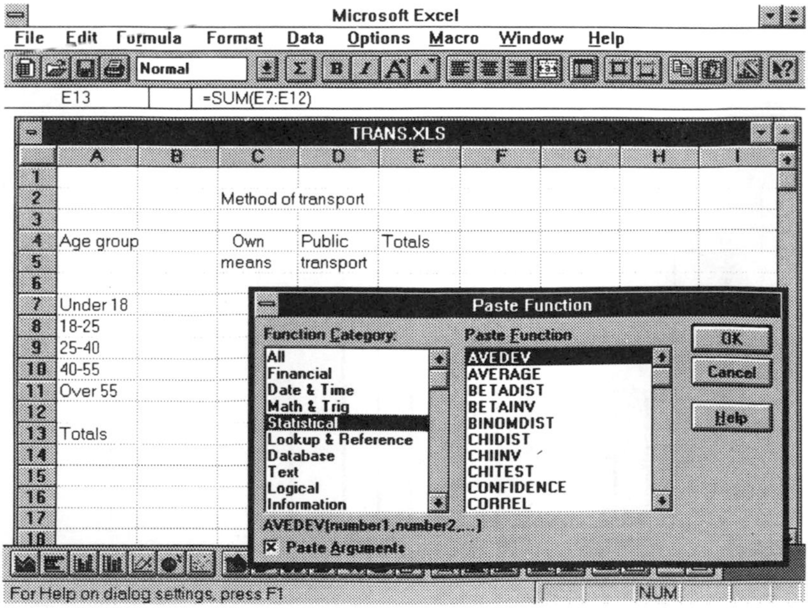

Fig. 18.5 Microsoft statistical routines

It is only necessary to enter the data into the cells of the spreadsheet, highlight the cells containing the data from which the statistic is to be calculated and select the appropriate routine from the list. Microsoft Excel has been used to recalculate most of the exercises where data have been changed for the third edition of this book. As can be imagined, this saved a great deal of time, and there was certainty that the answers would be correct.

Spread-sheet packages also provide the facility for the data in them to be displayed or printed in the form of charts, such as bar charts or pie charts, and graphs. Fig. 18.6 shows the contents of a small spreadsheet printed in the form of a bar chart. It shows the data from Table 5.1 in Unit 5.3. as entered in Microsoft Excel, and that table printed as a bar chart by Microsoft Excel.

This, however, is a very simple example. The facilities for producing graphical output from statistics in the latest spreadsheet technology, coupled with modern laser and colour printers, are again comprehensive. The ability to print sophisticated and professional charts and diagrams for inclusion in

	Method of transport		
Age group	Own means	Public transport	Totals
Under 18	12	6	18
18-25	15	10	25
25-40	25	18	43
40-55	13	32	45
Over 55	6	12	18
Totals	71	78	149

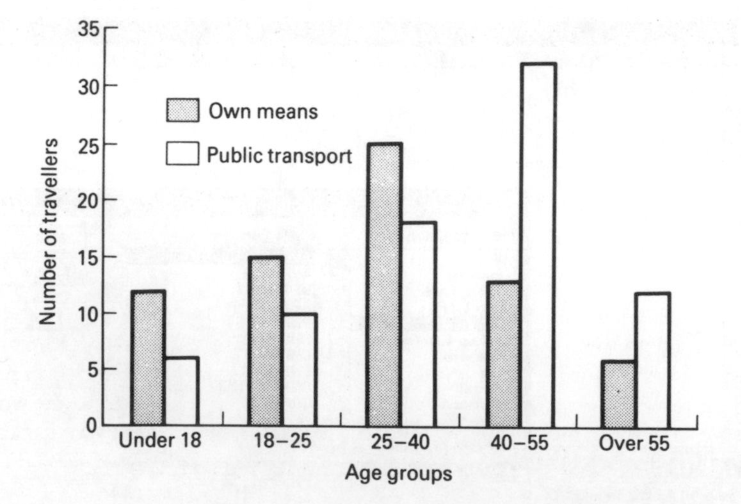

Fig. 18.6 The data from Table 5.1 entered in a spreadsheet (top) and printed as a bar chart (bottom)

reports, accounts and other documents is brought within the scope of the non-specialist.

Using Microsoft Excel, Fig. 18.7(a)–(f) shows the same data displayed in different formats. Each chart was obtained by a simple click of a mouse button. A user can click through the formats to find the best display. Fig. 18.8 shows the range of diagrams available in Microsoft Excel.

Spreadsheets are useful aids in decision-making where statistics are concerned. We mentioned 'what if?' exercises earlier in this section. Here is another example of their use. The spreadsheet layout in Fig. 18.9 shows the analysis of the capital investment project from Table 16.2 in Unit 16.7 at an interest rate of 15%. The discounted proceeds column has been entered into the spreadsheet using the net present value formula from page 284. The user can easily ask the question 'what if the interest rate was 12%?' by simply changing the interest rate in the formula to the new rate. The results of the change are then immediately displayed, as shown in Fig. 18.10. All values in the table have been immediately recalculated. The usefulness of the spreadsheet as an aid to decision-making in this way is often neglected.

There is no doubt that the spreadsheet is now an indispensable tool for the statistician, or for the non-statistician who needs to include statistical information in reports and other documents, or as visual aids for talks, lectures or presentations. Not only does the spreadsheet software perform the complex calculations, and provide excellent presentation tools, it also recalculates automatically when the data in the underlying worksheet tables are changed. Any such changes are also, of course, reflected automatically in any diagrams that are drawn from the data.

Data can also be linked to data in other spreadsheet tables, and merged into a new worksheet. Cross-tabulation with data held in databases (see 18.8) is also possible. The power of the latest spreadsheet technology is invaluable to those who work with statistics in any capacity.

18.8 Database packages

Database packages have been developed to enable non-computer specialists to write computer systems. They are sometimes referred to as *applications generators*, which implies that they are concerned with *what* the user wants the computer to do rather than *how* it is done. The aim of a database package is to allow a user of a computer who is not a computer expert to develop systems by telling the package what is required in terms of creating and keeping records, inputting data and producing outputs of various kinds. Users should be able to do this without needing to know how to write computer programs. In practice, however, it is questionable whether the complete computer novice would be able to develop more than the most simple of systems using such a package.

Examples of database packages are DBASE III+ and DBASEIV, FOXBASE, Paradox and Microsoft Access.

For the statistician who wants to keep fairly basic data records, and to produce summaries and statistical reports from those records, a database package

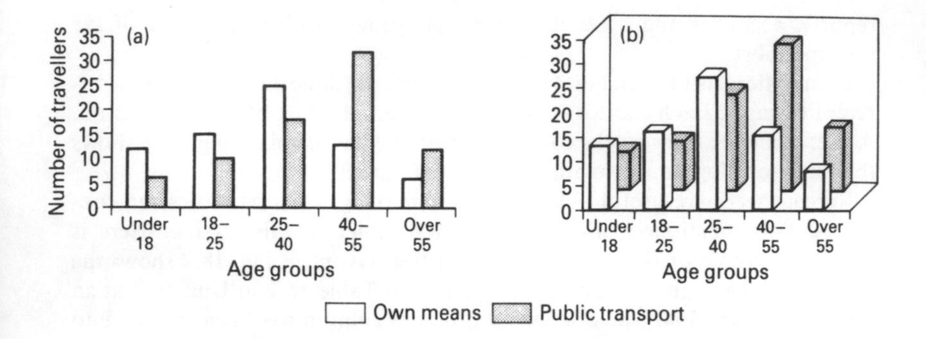

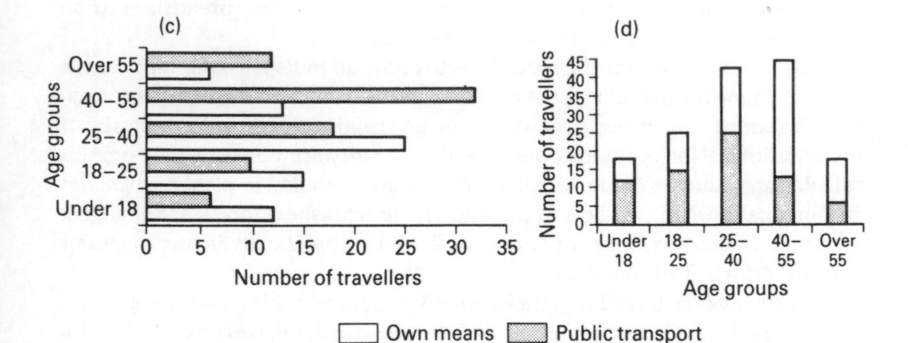

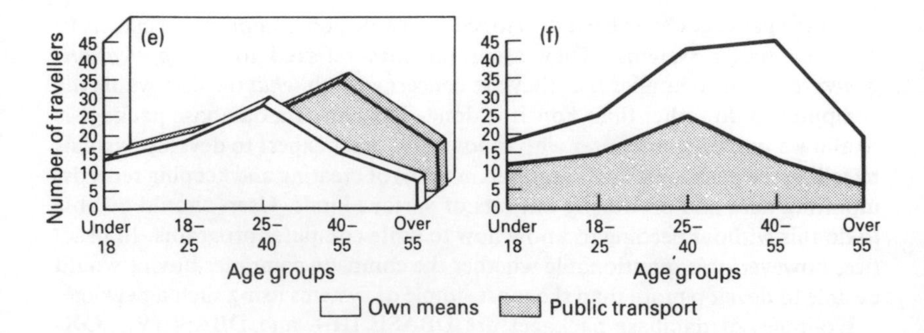

Fig. 18.7 Alternative display formats

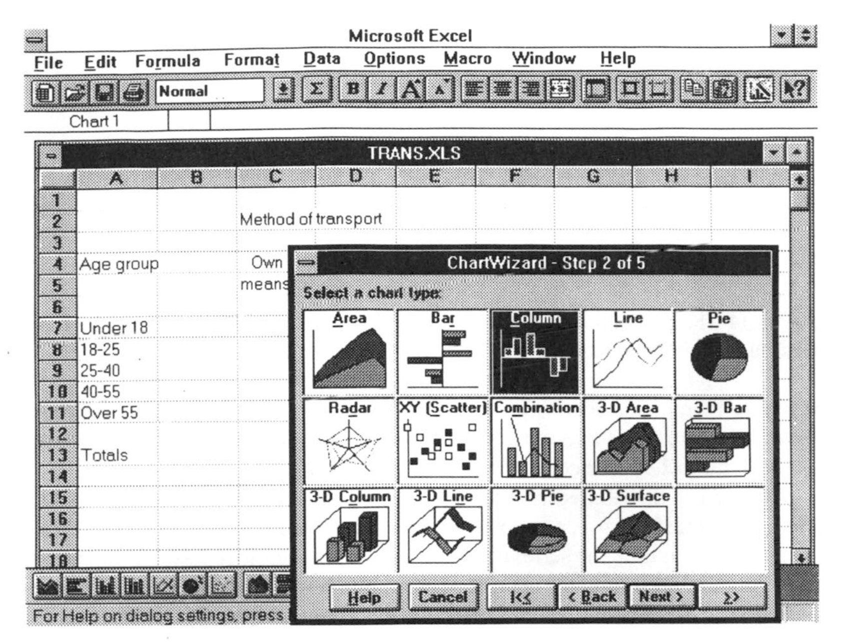

Fig. 18.8 Microsoft Excel diagram options

Project A

Year	Proceeds	Discount factor % 15	Discounted proceeds
1	2000	0.8696	1739.13
2	3000	0.7561	2268.43
3	5000	0.6575	3287.58
4	5000	0.5718	2858.77
5	5000	0.4972	2485.88
			12639.79

Fig. 18.9 The capital investment project of Table 16.2 entered in a spreadsheet with an interest rate of 15%

could be very useful. Using such software packages, databases and data-entry screens to go with them are very easy to create. Most database packages have facilities for collating data from databases, summarizing data and producing output in tabular and graphical format, as well as for importing and exporting data into or out of other databases and spreadsheets.

18.9 Computer access to official statistics

The Central Statistical Office (CSO) now provides a service to users of macroeconomic time series who wish to have the information in computer-

Project A

Year	Proceeds	Discount factor % 12	Discounted proceeds
1	2000	0.8929	1785.71
2	3000	0.7972	2391.58
3	5000	0.7118	3558.90
4	5000	0.6355	3177.59
5	5000	0.5674	2837.13
			13750.92

Fig. 18.10 The capital investment project of Table 16.2 entered in a spreadsheet and recalculated with an interest rate of 12%

readable form. This service is the *CSO Databank*. The data available cover the main economic indicators and most other major economic indicators, such as those published in *Economic Trends* (see Unit 3.3).

Databank is distributed on magnetic tape in a standard format that can be used, or adapted for use, on most medium to large computers. The CSO issues a new tape each time the series are updated. Users therefore have timely access to series of interest, and the service offers a much wider coverage than is possible with the paper publications. Information about Databank can be obtained from the CSO.

Eurostats also are available in various electronic forms, such as magnetic diskette, and can be accessed on-line. Information about the services available can be obtained from the Office for Official Publications of the European Community (see Unit 3.4).

Suggestions for projects

The following projects are particularly appropriate for use as a basis for assignments on BTEC courses with a statistical content, but whatever your reason for studying statistics, you will find it advantageous to attempt some of the projects. They are designed to help you acquire and demonstrate an understanding of the techniques of statistics and their application to real-life situations. You will collect data, design documents and write reports analysing and interpreting the data. Some of the projects require research into published statistics. Although you will need to be able to handle the necessary formulae and calculations, the emphasis is on the use of statistical techniques and the explanation of their use. Those units of particular relevance to each project are indicated.

Many of the projects are designed for individual work; others are group projects on which a number of students can work together. The projects are not necessarily self-contained and can be adapted to suit requirements and time available. You may well find that part of a project will be sufficient; some may be returned to and completed over a period of time.

Please remember that if you require the help of firms or individuals when undertaking a project it is essential that when asking for their help, you explain fully what you are doing and make sure that they know you are engaged on an educational exercise.

Individual projects

1. Obtain some company reports and accounts. (These are published annually by limited companies and can be obtained directly by writing to the companies' registered offices. You may also obtain some by visiting the office of a stockbroker, where you may be allowed to have some that are not required.)

   Look carefully at the reports and study the information they contain: they are useful sources of statistical data.

   (*a*) Use the accounts to draw dual and component bar charts to show the make-up of capital employed, that is share capital and loan capital, for different companies.

   (*b*) Calculate share capital and loan capital as percentages of capital employed and show those percentages on percentage component bar charts or pie charts. Which do you think display the facts best?

(*c*) Do you notice any significant differences in these percentages for different types of companies – that is, do manufacturing companies raise higher proportions of their capital through the issuing of shares than do commercial or property companies? If there are significant differences, can you explain them? (*Units 3, 5, 15*)

2. The Central Gas Company charged domestic consumers of gas as follows: each residence paid a standing charge per quarter of £8.00 and gas used was charged at a rate of 39p per cubic foot.

(*a*) Draw a graph that a clerk could have used to read off the total gas bill according to the gas that had been used.

(*b*) Write instructions for the clerk on how to use the graph.

(*c*) Explain in your own words what is meant by the statement 'the total bill is a function of the gas used'.

(*d*) If y is the total gas bill, x the cubic feet of gas used and $y = a + bx$ what is a and what is b?

(*e*) Find some further examples of similar charging systems, state their functions as above and draw their graphs. (*Unit 9*)

3. In Project 1 you were asked to obtain some company annual reports and accounts. Use them again in this project.

(*a*) From the profit and loss account of each company draw a scatter diagram with 'turnover' on the horizontal axis and 'profit before tax' on the vertical axis.

(*b*) Draw a similar scatter diagram with 'capital employed' from the balance sheet on the horizontal axis and 'profit before tax' on the vertical axis.

(*c*) From your diagrams, is there any evidence of a relationship between the two variables in either case? (In other words, is profit related to turnover or capital employed?)

(*d*) See if any reports show past records. Draw a scatter diagram showing the relationship between capital employed and profits before tax for a company for several years. Is there an evident relationship? If there is wide variation from year to year try to explain the variation on economic grounds.

(*e*) Measure the degree of relationship between the two variables. What difficulties did you encounter when making the calculation and how did you get round them? (*Unit 9*)

4. This project is concerned with the measurement of correlation using official statistics, namely the *Annual Abstract of Statistics* and *Regional Trends*.

See if you can measure the correlations between:

(*a*) the Index of Output of the Production Industries for industries and the number of stoppages in those industries;

(*b*) the Index of Output of the Production Industries and the number of unfilled vacancies in those industries;
(*c*) The percentage unemployment in each region and the amount spent on: (i) alcoholic drink; (ii) tobacco; (iii) services and miscellaneous; (iv) food.

Write up your results and say what the measurements you have made show. Have they any implications for industry and the economy? Do they say anything about the current state of our society? Are there any other correlations you could measure that may be revealing and interesting?

(*Units 3, 9, 10*)

5. The Sales Manager of your company has sent you the following details of sales of some different product lines for previous years.

	1990		1991		1992		1993	
Product	Tonnes sold	Price per tonne (£)	Tonnes sold	Price per tonne (£)	Tonnes sold	Price per tonne (£)	Tonnes sold	Price per tonne (£)
A	1000	50	1000	52	1000	56	1200	62
B	2000	16	2500	20	2500	25	2800	28
C	4500	28	4000	30	3600	33	3000	38
D	1500	15	1500	18	1500	22	1200	25
E	3000	10	3000	12	2800	15	2400	19
F	2200	12	2500	15	2700	20	2700	24
G	3000	22	4000	20	5000	18	5500	16
H	4000	8	4500	10	4500	13	4800	18

He is interested in using an index number based on 1990 to show how sales are performing compared with that year. While this is easily calculated using the total sales figure, he has a sneaking feeling that there should be a 'weighting' applied to reflect the changing pattern of sales. He asks you to confirm this and explain to him exactly what 'weighting' means.

(*a*) Send him a memorandum explaining to him what is meant by an aggregative index and why it is used.
(*b*) Let him know how a weighting should be applied to these figures and what type of indices can be used.
(*c*) Recommend what index you think should be used, giving reasons why.
(*d*) Calculate the index numbers for the years subsequent to 1986 using the method you are recommending and send them to the Sales Manager in your memo.

(*Unit 10*)

6. Following the work that you did for the Sales Manager on the price index in Project 5, he sends you a set of sales figures for 1988:

Product	Tonnes sold	Price per tonne (£)
A	1000	50
B	2100	15
C	4400	27
D	1500	15
E	2900	10
F	2200	11
G	3000	20
H	3800	8

He has looked at sales in the past and has reached the conclusion that they were fairly stable until 1988 and that prices rose more quickly after that year. He wants the price index recalculated, based on 1988.

(*a*) Change the base of the index to 1988 and write a memorandum to the Sales Manager to show what you have done and that complete recalculation is not necessary.

(*b*) In the same memorandum show the Sales Manager how a chain base can be used, and explain why this may sometimes be necessary.

(*Unit 10*)

7. The company for which you work is considering where to locate its new production unit. The directors have decided that it would be preferable to locate it where there is a considerable local market for high-class durables that will appeal to those enjoying better incomes. The directors reason that if much of the production can be sold locally then there will be a substantial reduction in transport costs, thus improving profits and enabling the products to be price competitive.

You, as the accepted expert in statistics in the company, are asked to investigate.

(*a*) From official statistics, find the weightings for the main groups in the Retail Prices Index (RPI) for last year.

(*b*) Compare these weightings with the *Regional Trends* published by the Central Statistical Office to see if the spending patterns in some regions differ. By looking at the price indices for the main groups of the RPI and taking into account the differences between regional and national weighting, can you determine if some regions may be better off than others?

(*c*) Compare your findings with personal incomes in the regions to see if that will give you any indications.

(*d*) Write a brief report on your findings, mentioning in it any other economic factors, such as regional unemployment, that should be taken into account. (*Units 3, 10*)

8. In the most recently available copy of the *Annual Abstract of Statistics* or *Economic Trends*, find the table giving the Index of Output of the Production Industries for the past seven years.
   (*a*) Which industries would you advise your firm to diversify into and why?
   (*b*) Which would you not advise the firm to expand into and why?
   (*c*) Pick out the industrial heading where you think further analysis is most needed.
   (*d*) If possible, discuss the matter with other students. Discuss particularly the use of the present industrial classification. Is, for example, the structure of business in this country changing to the extent that more concentration should be made on a breakdown of the various service industries rather than the traditional industries? Is this important for future business and economic planning? (*Units 3, 10*)

   This project can also be done by a group of students.

9. The table on page 324 compares household incomes in 1987 and 1992.
   (*a*) From the figures calculate the arithmetic mean of the total weekly household income using: (i) the frequency distribution of weekly household income; (ii) the average total weekly household income for each class in 1987, and the gross normal weekly household income in 1992.
   (*b*) Compare your results. Do they tell you anything about the calculation using the frequency distribution? What assumption about the data is made when making the calculation from the frequency distribution?
   (*c*) Calculate the median of the incomes for each year and compare them with the arithmetic means. Does this tell you anything?
   (*d*) Draw Lorenz curves to show how evenly (or unevenly) income is distributed in this population. Interpret the diagram you have drawn and show how you derived the necessary values for it.
   (*e*) Using the most recent copy of the *Family Expenditure Survey*, calculate the current mean and median incomes and compare them with the values obtained in (*a*) and (*c*). (*Units 3, 5, 7*)

10. The tables on pages 326–7 show expenditure on commodities and services as a percentage of total household expenditure in 1987 and 1992.
    (*a*) From the figures calculate the mean weekly household expenditure for both years, and the median weekly household expenditure for both years.
    (*b*) Compare your averages with average incomes (calculated in Project 9). If you were a personnel officer for a company and wanted to assess if wages were keeping up with household expenditure, what conclusions could you reach from these figures?

Project 9 Household incomes in 1987 and 1992 compared

1987			1992		
Household group	Number of households	Gross normal weekly household income (£)	Household group	Number of households	Gross normal weekly household income (£)
Households with gross normal weekly income:			**Household with gross normal weekly income:**		
Under £45	416	38.85	Under £60	405	48.71
£45 and under £60	428	52.00	£60 and under £80	504	68.33
£60 and under £80	576	69.67	£80 and under £100	432	90.30
£80 and under £100	468	89.60	£100 and under £130	517	115.09
£100 and under £125	475	112.91	£130 and under £160	453	144.10
£125 and under £150	409	137.72	£160 and under £200	502	179.21
£150 and under £175	362	161.71	£200 and under £240	456	218.99
£175 and under £200	379	187.17	£240 and under £280	449	260.32
£200 and under £225	427	212.49	£280 and under £320	502	299.93
£225 and under £250	362	237.07	£320 and under £370	518	344.06
£250 and under £275	358	262.08	£370 and under £420	443	393.96
£275 and under £325	676	299.18	£420 and under £470	449	444.77
£325 and under £375	534	350.12	£470 and under £540	464	504.37
£375 and under £450	569	410.51	£540 and under £640	451	589.01
£450 and under £550	430	469.42	£640 and under £800	415	712.78
£550 or more	527	814.26	£800 or more	458	1136.57

Source: *Family Expenditure Survey*

(*c*) Calculate the standard deviation of weekly household expenditure for both years. What does your result tell you about the distribution of expenditure for the two years?

(*d*) Using the most recent copy of the *Family Expenditure Survey*, calculate the current mean, median and standard deviation of weekly household expenditure and compare these values with those for 1987 and 1992.

(*e*) Your company is considering diversifying into new markets. From the percentages of expenditure spent on different commodities or services could you advise on the direction that the diversification should take? Give values and reasons to support your arguments.

(*f*) Suppose the Chancellor of the Exchequer seeks to impose new indirect taxes on goods and services to raise revenue. At the same time, the Chancellor wants the burden of these taxes to fall on those who can afford to pay them. From the tables on pages 326 and 327 suggest areas and items which could now be taxed; give reasoned arguments for your suggestions. (*Units 7, 8, 10*)

11. Collect copies of your local newspaper and take out the classified advertisements for used cars.

(*a*) From the advertisements obtain prices of two popular cars that have similar new prices. Find the average price of each of those cars for models from one year to nine years old.

(*b*) Measure the relationship between age and average price for each of these cars. What do your results seem to indicate?

(*c*) For each of the cars calculate the regression line of y upon x. Draw scatter graphs for each of the cars showing age and average prices and draw the regression lines on each of the scatter diagrams.

(*d*) Can you forecast the price of each of the cars for models that are ten years old? Explain any unrealistic results you get.

(*e*) Write a report on this project, the work you did and the difficulties you ran into when collecting the data and performing the calculations. (*Units 7, 9, 11*)

12. The Sales Department sends you the following quarterly tonnage sales of a product:

	Quarter 1	Quarter 2	Quarter 3	Quarter 4
1987	263	285	290	244
1988	272	285	287	250
1989	258	273	287	241
1990	256	263	270	243
1991	265	273	270	256
1992	282	290	294	273
1993	294	302	303	275

Project 10 Expenditure on commodity or service as a percentage of total household expenditure

(*a*) 1987

Household group	No. of households	Average total weekly household expenditure (£)	Commodity or service (% of total expenditure) Housing	Fuel light and power	Food	Alcoholic drink	Tobacco	Clothing and footwear	Household goods	Household services	Personal goods and services	Motoring expenditure	Fares other travel	Leisure goods	Leisure services	Miscellaneous
Households with gross normal weekly income																
Under £45	416	47.06	10.92	14.83	28.94	4.16	4.76	6.01	6.23	5.25	4.16	3.48	2.19	4.48	4.44	0.11
£45 and under £60	428	59.17	16.60	13.13	26.23	2.81	4.95	6.30	7.01	5.29	3.48	2.33	2.37	4.28	5.00	0.22
£60 and under £80	576	78.61	18.24	10.95	25.42	3.50	4.29	5.38	7.51	4.71	3.60	6.13	2.30	3.27	4.40	0.29
£80 and under £100	468	95.38	20.20	10.03	24.24	3.52	3.87	5.01	5.84	4.77	3.76	7.52	2.10	3.52	5.40	0.22
£100 and under £125	475	116.37	19.30	8.09	23.12	4.21	4.06	6.51	6.71	4.40	3.73	8.02	2.03	4.35	5.15	0.32
£125 and under £150	409	154.41	17.91	6.26	18.69	3.91	3.21	5.69	6.88	3.43	2.85	8.41	1.72	3.35	0.00	0.00
£150 and under £175	362	148.92	18.12	6.78	21.44	4.38	3.78	6.49	6.40	4.20	3.76	0.11	2.67	5.08	5.84	0.36
£175 and under £200	379	163.78	19.02	6.59	20.16	4.53	3.02	5.97	7.35	3.71	3.25	13.59	2.64	4.69	5.13	0.33
£200 and under £225	427	172.70	17.48	6.10	20.11	4.82	2.95	6.47	7.94	4.07	3.60	13.57	1.96	5.31	5.19	0.43
£225 and under £250	362	190.82	16.20	5.62	19.07	5.25	2.97	6.36	7.41	3.99	3.93	14.12	1.88	5.05	7.69	0.38
£250 and under £275	358	205.52	16.40	5.16	19.05	4.35	2.54	7.00	8.00	4.04	4.32	13.79	2.15	4.52	8.24	0.45
£275 and under £325	676	220.21	15.82	5.29	19.35	4.70	2.67	7.92	7.64	4.30	3.57	13.27	2.65	5.64	6.71	0.47
£325 and under £375	534	253.66	15.41	4.57	18.50	4.72	2.27	7.69	7.16	3.94	3.39	14.14	2.50	5.65	9.45	0.61
£375 and under £450	569	277.77	14.89	4.19	18.22	4.90	1.96	7.69	6.20	4.60	4.05	16.39	2.28	4.96	9.14	0.54
£450 and under £550	430	311.34	15.70	4.16	17.17	5.21	1.40	7.67	8.06	4.48	3.84	14.91	2.64	5.00	9.35	0.42
£550 or more	527	442.44	14.03	3.32	14.49	4.81	1.02	7.62	6.93	4.88	3.87	12.87	2.99	4.42	18.23	0.52

(*b*) 1992

Under £60	405	80.18	14.53	10.70	23.76	3.84	3.64	5.51	8.04	7.17	3.16	7.62	2.37	4.07	5.31	0.31
£60 and under £80	504	86.02	15.97	11.57	23.94	3.09	3.55	5.49	10.15	6.52	4.00	3.22	2.16	4.56	5.57	0.22
£80 and under £100	432	109.57	16.17	10.70	24.43	2.86	3.92	5.79	7.78	5.58	3.70	6.07	1.82	4.38	6.14	0.33
£100 and under £130	517	135.39	16.35	8.56	24.28	3.56	4.11	4.75	7.73	5.36	3.73	9.32	2.24	3.66	5.80	0.55
£130 and under £160	453	160.23	16.83	7.41	22.59	3.75	3.63	5.77	7.95	4.69	4.24	8.04	2.76	5.35	6.09	0.47
£160 and under £200	502	186.26	16.61	6.49	20.74	3.88	3.03	5.58	7.55	4.57	3.32	12.85	2.34	4.46	8.02	0.57
£200 and under £240	456	220.68	16.49	5.74	19.64	3.94	2.68	6.49	9.19	4.82	4.58	10.80	1.88	4.03	9.05	0.66
£240 and under £280	449	245.11	18.09	5.09	18.17	4.20	2.50	5.59	8.85	4.49	4.32	11.26	2.99	5.08	8.97	0.40
£280 and under £320	502	251.40	17.83	5.15	19.05	4.83	2.60	5.97	7.55	4.43	3.48	14.13	1.91	4.99	7.61	0.46
£320 and under £370	518	280.13	18.74	4.57	18.33	4.35	2.27	6.33	7.96	4.40	3.77	13.77	1.70	4.89	8.37	0.54
£370 and under £420	443	315.69	17.27	4.44	17.67	4.13	2.22	6.30	7.87	4.48	3.77	15.89	1.76	4.88	8.45	0.85
£420 and under £470	449	348.84	17.52	4.08	17.39	4.25	1.71	6.70	7.55	5.09	3.82	15.03	1.93	5.19	8.70	1.04
£470 and under £540	464	368.00	18.07	3.79	16.94	4.13	1.49	6.45	8.13	4.89	4.07	14.30	2.51	5.55	9.02	0.68
£540 and under £640	451	406.50	17.99	3.65	16.52	4.47	1.39	6.49	7.23	5.23	3.71	14.80	2.93	4.81	10.03	0.76
£640 and under £800	415	487.33	18.56	3.32	14.97	4.37	1.02	6.18	8.28	4.90	3.76	15.97	2.20	5.11	10.73	0.63
£800 or more	458	703.91	16.64	2.68	12.07	3.65	0.63	5.52	8.28	5.03	3.21	13.24	4.79	5.03	18.52	0.69

Source: *Family Expenditure Survey*

There is considerable seasonal variation in the figures and the Sales Manager wants to know how sales of the product are really going.

(*a*) Remove the variation from the figures and show the true trend of sales of the product and the original data on a graph.

(*b*) Explain in a memorandum what types of variations can occur in time series.

(*c*) Explain other methods that can be used to find the trend with examples that will serve to make the methods clear. (*Unit 11*)

13. Collect copies of your local newspaper and take out the classified advertisements for used cars. (If you have done Project 11 the advertisements collected then can be used again.) Select data relating to two popular cars.

(*a*) Calculate the arithmetic mean price of each of these cars for models that are four years old.

(*b*) Calculate the standard deviation of prices for models of these cars that are four years old.

(*c*) Compare the results of your calculations. Do they tell you anything about the comparative merits of each of the cars?

(*d*) You are asked by the Transport Manager of your company to recommend one of the above cars to be purchased in considerable numbers for the company's sales force. The cars will be kept for four years and then sold. The Transport Manager wants to purchase cars that will maintain their value while in use with the company. Write a report to the Transport Manager using the results of your survey as evidence and recommend one of the cars. In your report say what other points should be considered in making the decision and discuss any factors that may have affected your statistical results. (*Units 7, 8, 15*)

14. Obtain copies of your local newspaper and take out the classified advertisements for used cars. (If you have done Projects 11 and 13 the advertisements collected then can be used again.) Select data on four-year-old cars.

(*a*) State the conclusion you can draw about the mean price of all cars that are four years old and of popular cars that are four years old (at the 95% confidence level).

(*b*) Obtain a further copy of the local newspaper for another date and calculate the mean price of cars that are four years old, for all cars and for popular cars. Calculate the standard deviation of prices for all cars and for popular cars.

(*c*) Using your results test the hypotheses that the mean prices of both types of cars have changed between the dates of the two newspapers (at 0.05 significance level). (*Units 7, 8, 14, 15*)

15. A production process manufactures steel rods with a target diameter of 63 mm. As part of the company's efforts to maintain efficient production of these rods it is decided that a sample of five rods will be taken from the production line every hour and the mean diameter of those five rods

monitored to see if performance standards are being met. An unskilled clerk with no knowledge of statistical techniques will be employed to do this sampling job.

(*a*) List the measurements you would need to take to draw up a control chart to be used for the recording of the samples. Show how you would calculate the warning and action limits for the process.

(*b*) Write instructions for the clerk who will be taking the samples. These instructions must specify how to select the sample, how to measure and calculate the mean diameter of the five rods in the sample and how to record the results on the control chart. There must also be an instruction on how to tell if there could be something wrong with the production process and what to do in this situation. (*Unit 15*)

16. The Managing Director of the company for which you work finds out that you are taking a course in statistics, so you are presented with the following problems on which the directors have to reach decisions and asked whether there are statistical techniques that will help them in making the decisions.

For each of the problems write a short description of a statistical technique that will help the directors to make their decisions, giving an example of the use of the technique with some actual values.

Problem 1

There are problems of breakdowns on the production lines within the factory. The firm employs a team of fitters to deal with breakdowns, but a breakdown usually occupies a fitter for a complete day. From past records the cost of having to wait until the next day before a breakdown is rectified can be calculated, and past records also show the average number of breakdowns that occur daily. The cost of employing an extra fitter is also known. How can the directors decide whether the cost of employing a further fitter is likely to be less than the cost incurred by a production line failing when all fitters are already engaged on repair work?

Problem 2

The company is experimenting with new packaging for its product. In test areas it is selling the product in a variety of packages and recording the response of sales to the packages. How can the directors decide if some packages are selling better than others and if the differences are significant?

Problem 3

A machine is used to fill the packages with the product. The machine can be set to put a particular weight into the packages, but it does not always put the set weight in. The variation round that set weight can be measured. The firm will be liable to prosecution if too many of the packages contain less than the stated weight. How can they decide to what weight the filling machine should be set so that no more than, say, 5% of the packages produced will contain less than the stated weight? (*Units 2, 13, 15*)

17. Your company is considering two investment projects, both of which will cost £60000 in the year of installation. The net cash proceeds to be expected from the projects are as follows:

End of year	Project A (£)	Project B (£)
1	10000	20000
2	20000	30000
3	30000	20000
4	20000	10000
5	20000	10000

You are asked to appraise these projects by the approved methods of investment appraisal.

(*a*) Prepare a report showing the investment appraisal methods that you have used and the results obtained. Use tables to show the comparisons of results.

(*b*) Recommend to the Board of Directors the investment project that should be undertaken, if only one can be afforded. State your reasons for this recommendation.

(*c*) Say whether both investments should be undertaken if funds are available and state your reasons why.

(*d*) Include in your reasoning any economic considerations that should be taken into account at the time when the investment decisions are being made. *(Unit 16)*

18. The Managing Director of your company, very impressed by the assistance that you have been able to provide with the application of statistical techniques to decision-making in the past, asks again for your help. The company wants to expand by building a new factory. There is no land available at the present production site, so the company is considering new localities for the new production unit.

Before committing the company to the vast expense of constructing this new unit the board wants to be fairly certain that the returns will justify the expense, particularly as the funds for this project must be borrowed. In the past investment decisions within the company have been made more by the use of intuition than by any other method. The board now thinks that a more scientific method of appraising the investment should be attempted.

(*a*) Write a report to the Managing Director describing methods of investment appraisal that could be used to help the board come to their decision. It is important to give examples of the use of these methods so that it is clear to board members how these methods work.

(*b*) Prepare an outline of other social and economic factors that should be taken into consideration when deciding on the location of a new factory or other business premises. (*To expand this project into a group project prepare this outline as a group and then present the outline to another group acting as a questioning board.*)

(*c*) Write a brief report on any official statistics that would help the board in deciding on a new location for the production unit.

(*Units 3, 16, 17*)

19. Mr Wally is interested in expanding and opening new discos. He only wants to open discos, however, in places where success is reasonably certain. He asks you to suggest likely areas for new discos.

Use official statistics to try and find such areas. You may be interested, for example, in areas where:

(i) the population has a high proportion of young people;
(ii) the population is growing;
(iii) people spend a good part of their income on entertainment;
(iv) unemployment among the young is low.

These are just suggestions: you may be able to think of many other things to look at.

Report your findings, quoting your sources and using suitable methods of presentation for your information, and recommend six towns or cities for the location of six new discos.

(*Units 3, 5, 17*)

20. The many occasions on which you have been able to help the Board of Directors by suggesting statistical techniques that can aid their decision-making has finally borne fruit in that they are considering setting up a small department to advise on the use of statistics throughout the company. If such a department is created you will be asked to become its head. Unfortunately some board members need to be convinced of the need for such a department, despite the fact that the company is rapidly expanding on both the manufacturing and commercial fronts.

You are therefore asked by the Board of Directors to prepare an argument in favour of this new department outlining the areas where the department could serve the company.

(*a*) Prepare your argument in the form of a report for the board members.

(*b*) Prepare an oral argument to present to the board. You will be required to appear before the board at the meeting when they discuss your report.

(*To expand this into a group project a group of students can act as board members and question you on the need for the new department.*)

(*All Units*)

21. The Midville City Council is concerned about traffic and environmental problems within the city. Traffic has grown to such an extent that the city centre particularly seems to be coming to a standstill. They realize that something must be done.

Many schemes have been mooted, such as pedestrian-only paved areas, traffic restriction on inner city roads, park and ride facilities, strict limitations on parking with prohibitive car parking fees, cycle paths and so on. As always they meet strong opposition to any schemes to limit traffic from the motoring lobby and from city centre shops and businesses.

You have now been asked to conduct a survey to see how the general public feels about the problem. The council feels that strong public opinion in favour of limiting traffic may help them get schemes approved.

As a first stage, they have asked you to produce a report saying how the survey should be conducted, knowing that the resources are not available to ask the entire population of Midville and process their replies. You are also asked to design a draft questionnaire to be included with the report. (*Unit 2*)

22. The following report appeared in the *Wilshire Gazette*:

> **Survey shows parents' concern**
>
> Mr Jim Morrison, MP for South Wilshire, has found that people in his constituency want to see a change back to traditional methods of education. He has conducted a survey of members of the public in the South Wilshire district asking for people's views on progressive education.
>
> 'The replies I have received', he tells us 'show that 76% of people want an end to progressive education and a return to traditional classroom teaching. This is conclusive proof that the great majority of the population of this country are fed up with those educationalists who have forced progressive methods upon us.'

Write a critique of this report. Find other newspaper articles that use statistics in this way and write similar critiques of those. (*Unit 2*)

Group projects

23. With stop-watches and clip-boards, make observations of traffic travelling into the main commercial or industrial area in your locality. Work in groups with each group covering a different route.
    (*a*) Record the time interval between cars going into the area past the point where you are observing.
    (*b*) Record the number of occupants in each car.
    (*c*) Draw up a frequency distribution for the time intervals between 200 cars observed at each point.
    (*d*) Draw up a frequency distribution for the number of occupants of the 200 cars observed at each point. (Note that the problems involved in drawing up the distributions for (*c*) and (*d*) are different because one variable is continuous and one discrete.)
    (*e*) From the frequency distributions calculate the mean number of occupants in each car and the mean times between cars. Calculate the median times between cars.
    (*f*) Design a questionnaire that could be used in a survey to determine why

people use a car for their journey, and whether they have alternatives. (Points you may be interested in are cost, convenience, lack of alternatives, possibilities of sharing.)

(*g*) If it is possible, conduct the survey of people's use of their cars, write a report on how you collected and analysed the data and give your conclusions. (*Units 2,4, 7*)

24. Obtain copies of your local newspaper and take out the classified advertisements for used cars.
   (*a*) Distribute the pages among the members of the group and let each group member list the models and prices of cars that are four years old.
   (*b*) Bring together all the prices to form a frequency distribution.
   (*c*) Let each member of the group have a copy of the frequency distribution so that they can each calculate the arithmetic mean of the car prices. Find each other's errors until a result is agreed.
   (*d*) Discuss the results among the group and decide if the mean price is a meaningful average. If it is not, why?
   (*e*) Is there an average that is more typical of cars that are four years old? Calculate the mode and the median and discuss the results.
   (*f*) From the original data extract the prices of the popular cars that are four years old. Calculate the arithmetic mean price of those cars and compare it with the mean price of all cars that are four years old.
   (*g*) Compare the range, quartile deviation and standard deviation of the frequency distribution for all car prices and for the popular car prices. What do your results tell you about these frequency distributions? (*Units 4, 7, 8*)

25. The work force in your company wants the management to recognize their trade union. The management is reluctant to agree to this, in the fear that it will mean endless industrial disruption and excessive wage settlements that will force up labour costs within the company.

   You are asked to investigate and report whether available statistics support this view.
   (*a*) Using official statistics from the *Annual Abstract of Statistics* and any other appropriate sources, such as the *Employment Gazette*, see if you can measure any correlation between the number of disputes in industries and the average rates of pay in those industries.
   (*b*) Discuss the methods and statistics you have used with other members of the group doing the same exercise and, with the aid of the suggestions brought out during the discussion, see if you can improve upon your data.
   (*c*) Write a report to management telling them of your findings, showing your correlation coefficients, and stating your conclusions. Include in the report any benefits or drawbacks you can see in the recognition of a trade union.
   (*d*) Bearing in mind the company's plan to expand by building a new pro

duction unit, extend the assignment to include any correlation between stoppages and pay in the regions using *Regional Trends* produced by the Central Statistical Office.

(*e*) In a conclusion to your report say if you have any reservations about your statistical calculations, or rather the data you have used for them, and the difficulty you may have had in gathering the information. (*Units 3, 9*)

26. Use your local newspaper as a source of data on house prices by referring to the classified advertisements. Divide the locality into areas using a street map and assign an area to each member of the group.

(*a*) Draw up a frequency distribution of house prices in each area.

(*b*) Calculate the mean price of houses in each area.

(*c*) Discuss the variations in house prices between the areas. Can they be explained on social and economic grounds?

(*d*) Write a report on your findings using appropriate methods of presentation.

(*e*) Compare your findings with figures for previous years. (You may need to visit your local newspaper office to obtain these figures.) Using these figures as a base calculate a simple price index for present house prices in your locality. (*Units 4, 7, 10*)

27. This is a long-term project that can spread over a number of years. The aim is for you to prepare your own weighted price index to compare with the official Index of Retail Prices and to discuss at regular intervals the methods adopted and the results found. You should also compare your indices with the official index to discover and discuss the shortcomings and difficulties in preparing and using such an index.

(*a*) Over a period of time, record expenditure, to be used as weightings, under various headings. You may well use the groupings of the official Index of Retail Prices as a guide to your headings. Write a brief report outlining the groupings used and why, paying particular attention to any deficiencies you find with the official groupings.

(*b*) At regular intervals record the current prices of the items in your index groupings and calculate your individual index. Meet with the rest of the group and discuss your indices to see if you can explain any noticeable differences between them.

(*c*) At the end of the exercise write an essay describing the project, your findings both individually and as a group, and say what difficulties you encountered. Conclude by saying how useful you think these indices are as a guide to government, trade unions, firms and individuals in the economy.

As further reading for this project you are recommended to obtain a copy of *A Short Guide to the Retail Prices Index* (published by HMSO for the Department of Employment). (*Units 2, 10*)

28. Divide the group into two sections. One section will be a management team

from a company, one a trade union team involved in a pay negotiation. Each group must use official statistics, particularly the appropriate indices of pay and prices, to argue its case.

The management team will use its figures to argue that pay has kept ahead of the cost of living. The union team will argue that it has not and that the Retail Prices Index and other indices are inappropriate measures in the situation anyway with, of course, reasons why this is so.

After each section has prepared its data and the arguments supporting its case, the two sections will meet and negotiate. The negotiations should be watched by another group who will judge which section has best prepared and argued its case. *(Units 3, 10)*

29. The objective of this project is to demonstrate the concept of a sampling distribution of the means and to show that this is a frequency distribution which is symmetrical about the true population mean. It is necessary for your group to collect a considerable number of samples. The samples must therefore be as simple as possible and it is suggested that you collect data on only one variable. The samples must all come from the same population. One example of a data source for this project is the age of students in a college. Another is the length of partly used pencils.

    Each group member will collect random samples of the same size, say 30 to 40 items. About 100 samples are needed, which means that each member of your group will have to collect several samples.

    When the samples have been collected each group member must then calculate the arithmetic mean of his samples. Use the results as follows:

    (*a*) Form a frequency distribution of the means calculated from each sample.

    (*b*) Give a copy of this frequency distribution to each member of your group. Use the data to draw a histogram of the frequency distribution and then an estimated frequency curve on that histogram. Compare these diagrams and see what the shape of the distribution is.

    (*c*) Use one sample to calculate a standard deviation for the population. From that sample state the estimated true population mean within 95% confidence limits.

    (*d*) Show on the frequency curve drawn in (*b*) the limits within which 95% and 99.9% of the means of all samples of this size taken from the population will fall.

    (*e*) Each group member should compare the means of their samples against the true population mean as estimated from the frequency curve drawn in (*b*) and test to see if any that appear to be very different are significantly different at the 0.05 significance level.

    (Units 4, 7, 8, 14)

30. The Personnel Officer of the company for which you work has heard of your expertise with statistics and seeks your help. He wants to assess the apprentice scheme. When apprentices are recruited they undergo an aptitude test and if they pass that test they are then given an interview.

Critics of the selection procedure say that the aptitude test is serving no purpose and that apprentices should be selected on interview alone. They maintain that many potentially good apprentices are lost because they fail the aptitude test and therefore do not get an interview. The Personnel Officer does not think so: he thinks that those who do well on the aptitude test also perform best during training and produce the best results in the examinations at the end of their course. He has never attempted to quantify this impression, however, and would now like to. He wants you to let him know how he can go about this.

He has some data available. The results of the aptitude tests are graded A – E, but these grades are decided upon from a score on the test out of a maximum of 80: 70–80 marks = grade A; 60–70 marks = grade B, etc. The Personnel Department has the past records of scores on the aptitude test for all apprentices who have entered the scheme that way. There is also a record of all marks gained by all apprentices in course progress tests and in their final course examinations. It has always been customary to rank apprentices in order of merit at the end of each course and to award prizes to the six best apprentices.

You are required to:

(*a*) Write a report to the Personnel Officer telling him in what ways he can measure any relationship there may be between results of aptitude tests and performance on the course. You must explain the statistical techniques that can be used and what the results mean.

(*b*) Alternatively, how can he tell if there is any significance in any relationships that may be seen?

(*c*) Let him know how he can forecast what results an apprentice will obtain in his examinations from his performance on the aptitude test.

(Units 9, 15)

31. For this project each group member needs a pair of unbiased dice.

(*a*) List the possible scores that can be thrown with two dice and the number of ways in which these scores can be thrown.

(*b*) Work out the probability of each score and list these probabilities against each score. Work out the number of times you would expect to throw each score if you threw the pair of dice 108 times. (The choice of 108 throws will give you a whole number for this expected frequency.) List the expected frequency against the other items that you have listed.

(*c*) Throw your pair of dice and record the scores in the form of a frequency distribution. Stop at regular intervals and compare the way your scores are developing with the rest of the group. You are to throw the dice 108 times, so it may be convenient to stop and compare after, say, 36 throws and then after 72 throws. If your dice are not biased the expected pattern should develop as the number of throws made gets higher.

(*d*) When you have thrown and recorded 108 times use your list of expected and observed frequencies to test the hypothesis that the pair of dice you are throwing is biased. If the difference between the expected and observed frequencies is significant perhaps you should find and test a new set of dice before you continue with the project.

(*e*) Repeat the exercise of throwing the pair of dice 108 times and recording the scores in the same way until the group has 100 distributions of 108 scores. The group then has in fact 100 separate samples of 108 throws of a pair of dice. From these 100 samples form a frequency distribution for the score of 7. Does this frequency distribution confirm that the distribution of sample means is normal around the true population mean? (The population mean frequency of 108 throws of a pair of dice for the score of 7 should be 18.)

(*f*) Complete the project by writing up the experiment outlining the methods used and your findings using suitable methods of presentation. (*Units 4, 12, 14, 15*)

32. You are the elected members of the City Council. The Planning Department has approved a scheme for the building of a giant plant for processing the waste-matter collected from local firms and residences. The only places where the plant can be built are local beauty spots that are approached through residential areas, but the Treasury Department has carefully costed the scheme using approved methods of investment appraisal. The cost of transporting refuse to tips will be greatly reduced. New land for tipping refuse is becoming difficult to find, and as more and more toxic waste is being produced by local firms the siting of tips is extremely critical because of the danger of pollution of water supplies. The processing plant will also produce a marketable fertilizer that will bring in a considerable amount of revenue in the future. As the effect on Council Tax will be very advantageous the Planning Department has found in favour of the plant, despite the fact that it will cause atmospheric pollution.

The scheme is now to be debated by you as members of the Council. There is a strong lobby in favour of the plant because of the effect it will have on the Council Tax. Naturally there is a strong lobby against the scheme, arguing that this project cannot be considered purely on private cost grounds. This lobby will highlight the social cost involved.

At the end of the debate a vote will be taken to see if you as a Council will approve the scheme.

You will need to do some preparation for this debate. As a group you should decide beforehand on which points you will talk and divide accordingly. Those of you who are presenting the case for the scheme should be prepared to outline the methods used when appraising the scheme. Those speaking against the scheme should be clear about the social costs involved and how those costs may be assessed.

NOTE: This project can be related to the locality with which you are familiar. You can choose a local site for the plant and investigate the private and social costs. Conduct a survey with the object of assessing the demand upon the site as a recreational area and the effect of the site on roads and local residences. You may also discover something about the costs of disposing of waste-matter by tipping and the effects on land in the long term. For example, refuse tips are often recovered and put to some good use at a future date.

(*Units 2, 3, 16*)

33. Form groups of four or five to visit local businesses. Before visiting a firm you must write to arrange a suitable time and outline the purpose of your visit.

The object of the visits is to discover the present and potential use of statistics in those businesses. To this end a cross-section of businesses should be visited, both commercial and manufacturing; it is essential that the group meets an employee with sufficient knowledge of the data used by the business. The letter that you send should make this clear.

During the visits make notes on the use of statistics that you can ascertain, the data that are collected by the business and the use to which they are put. Some suggested areas that you can look for are:

(i) sales analysis and forecasting;
(ii) production information and quality control (you may see control charts in evidence on the production lines, for example);
(iii) market research or other types of research.

After making your visits:

(*a*) Meet as a group and pool the information you have gathered.
(*b*) Prepare a presentation of the information you have gathered.
(*c*) Give this presentation to the other groups, describing your visits and the statistics you saw being used.
(*d*) Discuss with the other groups the uses the businesses make of statistics. You may be able to compare the use of statistics in different types of companies, and see statistical techniques that one business uses that could be useful to others. You may also have seen problems that you could solve with the statistical techniques you know.

(*All Units*)

Further reading

Berenson, M. L. and Levine, D. M., *Basic Business Statistics*. Prentice Hall International Editions (New Jersey, 5th edn, 1992).

British Standards Institution, *BS 6000* (1972), *BS 6001* (part 1: 1991, part 2: 1993), *BS 5703* (part 1: 1980, part 2: 1980, part 3: 1981, part 4: 1982) (Statistical Quality Control procedures).

Clarke. G. M. and Cook, D., *A Basic Course in Statistics*. Edward Arnold (London, 3rd edn, 1992).

Huff, D., *How to Lie with Statistics*. Penguin Books (Harmondsworth, 1993).

Ilersic, A. R., *Statistics*. HFL (London, 14th edn, 1979).

Kazmier, L. J. and Pohl, N. F., *Basic Statistics for Business and Economics*. McGraw-Hill (New York, 2nd edn, 1984).

Lockyer, K. and G., *Critical Path Analysis and Other Project Network Techniques*. Pitman (London, 5th edn, 1991).

Mansfield, E., *Statistics for Business and Economics*. Norton (New York, 3rd edn, 1987).

Moore, D. S. and McCabe, G. P., *Introduction to the Practice of Statistics*. W. H. Freeman & Co (New York, 2nd edn, 1992).

Moroney, M. J., *Facts from Figures*. Penguin Books (Harmondsworth, 1952).

Moser, C. A. and Kalton, G., *Survey Methods in Social Investigation*. Gower (Aldershot, 2nd edn, 1972).

Thirkettle, G. L., *Wheldon's Business Statistics*. Macdonald and Evans (Plymouth, 9th edn, 1981).

Walpole, R. E., *Introduction to Statistics*. Macmillan (New York). Collier Macmillan (London, 3rd edn, 1982).

Whitehead, G., *Statistics for Business*. Pitman (London, 2nd edn, 1992).

Wonnacott, T. H. and R. J., *Introductory Statistics for Business and Economics*. John Wiley (New York, 4th edn, 1990).

Answers to exercises

(where appropriate)

Unit 6

1. (*a*) 8.64; (*b*) 97; (*c*) 66.667; (*d*) 86024000; (*e*) 25.9; (*f*) 90.9.
2. 31.74 ± 0.056; percentage relative error = 0.176%.
3. £150000 ± £30800.
4. (*a*) 148000; (*b*) 39.7; (*c*) 3.76; (*d*) 0.81.
5. (*a*) Area A = 678600 tonnes ± 2955 tonnes;
   Area B = 352170 tonnes ± 2010 tonnes;
   Area C = 1143590 tonnes ± 3720 tonnes.
   (*b*) 2174360 tonnes ± 8685 tonnes.
7. (*a*) 135.92 tonnes ± 2.76 tonnes; (*b*) 468 kg ± 53.35 kg.
8. £2000 ± £265 (method: profit = 50 × 40 × 1 ± (2 + 1.25 + 10)%).

Unit 7

2. 38 hours.
3. (*a*) £7667; (*b*) £7327.
4. (*a*) £11210.
5. (*a*) £2567; (*b*) £2200.
6. (*a*) £40.13; (*b*) £41.12.
7. Mean = 66.53, median = 64, mode = 54.
8. (*a*) 38.21 years; (*b*) 33.91 years.

Unit 8

2. (*a*) 1864 km.
3. (*a*) 1.95 days; (*b*) 1.5 days.
4. (*a*) 125.6p; (*b*) range = 70p, quartile deviation = 7.0p, mean deviation = 11.92p;
   (*c*) standard deviation = 16.512p, variance = 272.64p^2.
6. (*b*) standard deviation = £1531, quartile deviation = £1300.
7. (*b*) standard deviation = £5.31;
   (*c*) quartile deviation = £6.92.
8. (*a*) 0.877, 0.123, 0.377; (*b*) 0.0427, 0.9573, 0.4573; (*c*) 0.9778, 0.0222, 0.4778; (*d*) 0.1685, 0.8315, 0.3315; (*e*) 0.9981, 0.0019, 0.4981; (*f*) 0.001, 0.999, 0.499; (*g*) 0.7823, 0.2177, 0.2823; (*h*) 0.025, 0.975, 0.475; (*i*) 0.0041, 0.9959, 0.4959.
9. Mean = 20, standard deviation = 9.2; (*a*) 0, (*b*) 1.63, (*c*) −1.08, (*d*) −1.63 (*e*) 2.17.

10. 0.9332.
11. (*a*) 0.0228; (*b*) 0.0013; (*c*) 0.9544; (*d*) 0.3446.
12. −0.6.

Unit 9

1. −0.99; 1.
2. −0.76.
3. (*a*) 0.946; (*b*) $y = 1.63 + 1.69x$; 16.85 cm.
4. (*a*) £3454; (*b*) $y = 711.7 + 304.7x$.
5. $y = 21.3 + 0.96x$, 46260, 68100.
6. (*a*) $r = 0.644$; (*c*) $y = 29.74 + 0.93x$ ($\times$ 100), (i) 51; (ii) 47; (iii) 63.
7. $R = 0.673$.
8. (*a*) $r = -0.837$; (*b*) $r = -0.796$; (*c*) $R = -0.82$ (high) $R = -0.72$ (low).
9. Line of best fit $= y = -1.26 + 6.66x$.

Unit 10

1. 176.25.
3. (*a*) 1992 = 110.9, 1993 = 123.6; (*b*) 1992 = 111.4, 1993 = 124.5.
5. (*a*) Laspeyres = 154, Paasche = 155; (*b*) 154, 156.
9. (*a*) 1992 = 106.7, 1993 = 114.3; (*b*) 1992 = 106.7, 1993 = 114.4.

Unit 11

1. (*a*) Trend (1990 qtr 3 onwards): 1152.7, 1205.6, 1258.1, 1301.5, 1336.8, 1360.8, 1372.9, 1388.4;
   (*b*) Average seasonal variation: qtr 1, +6.6; qtr 2, +6.5; qtr 3, −33.5; qtr 4, +28.1.
2. Moving averages (1984 qtr 3 onwards): 45.75, 45.625, 45.375, 44.875, 45.5, 47.125, 47.625, 47.25, 47.25, 47.0, 47.125, 46.875, 46.0, 45.375.
3. Three-yearly moving averages (1978 onwards): 85.54, 98.70, 112.31, 125.23, 137.26, 152.46, 169.18, 186.90. 202.17, 218.53, 236.87, 258.60, 279.67, 300.50.
4. (*a*) Seasonally adjusted figures: 12.625, 12.625, 13.625, 13.125, 13.625, 14.625, 14.625, 16.125, 16.625, 16.625, 17.625, 17.125;
   (*b*) Moving averages (qtr 2 1986 onwards): 13.125, 13.5, 13.875, 14.375, 15.125, 15.75, 16.375, 16.875.
5. (*a*) Five-yearly moving average (1968 onwards): 55182, 55439, 55666, 55868, 56023, 56142, 56200, 56218, 56209, 56210, 56231, 56258, 56277, 56311, 56359, 56417, 56499, 56624, 56767, 56922, 57081.
6. (*a*) Moving averages (1984 qtr 3 onwards): 180, 184, 205, 238, 252, 262, 259, 252, 255, 251, 248, 248, 244; (*b*) average seasonal variation: qtr 1, −31.75; qtr 2, −15.75; qtr 3, −1.75; qtr 4, 49.25.

Unit 12

1. $\frac{12}{365}$.
2. (*a*) $\frac{1}{20}$; (*b*) $\frac{3}{5}$; (*c*) $\frac{2}{5}$.
3. $\frac{1}{105}$.

4. (*a*) $\frac{3}{20}$; (*b*) $\frac{1}{52}$; (*c*) $\frac{17}{65}$; (*d*) $\frac{17}{130}$.
5. (*a*) $\frac{9}{16}$; (*b*) $\frac{3}{8}$; (*c*) 0.001 225; (*d*) 0.000 5625.
6. 0.5687 (method: $1 - (\frac{364}{365} \times \frac{363}{365} \times \frac{362}{365} \times \ldots \times \frac{341}{365})$).
7. (*a*) $\frac{1}{9}$; (*b*) $\frac{4}{9}$; (*c*) $\frac{4}{9}$.
8. i = 46%, ii = 54%, iii = 54%, iv = 87%.
9. i = $\frac{1}{2}$, ii = $\frac{3}{10}$.

Unit 13

1. (*a*) 0.0064; (*b*) 0.7373; (*c*) 0.3277.
2. (*a*) 0.194; (*b*) 0.929.
3. (*a*) 0.531; (*b*) 0.983.
4. 0.0839.
5. 0.0821.
6. (*a*) 0.7011; (*b*) 0.0099; (*c*) 0.0918.
7. 0.0475; 0.00277.
8. 0.1175; 24.
9. 0.0588.
10. 0.065; 0.056.
11. 11p.
12. 0 = 4, 1 = 10, 2 = 13, 3 = 11, 4 = 12.

Unit 14

1. (*a*) 37.39 and 42.61; (*b*) 36.56 and 43.44.
2. (*a*) £42.80 ± £1.43; (*b*) £42.80 ± 1.89.
3. 97.40 (to nearest kilometre).
4. At least 427 per batch.
5. (*a*) (i) 0.228, (ii) 0.1587; (*b*) 50% ± 1.96%.
6. (*a*) 0.0436.
7. (*b*) (i) 0.0359, (ii) No.
8. (*a*) 0.35 ± 0.093; (*b*) At least 8740.
9. 0.505 – 0.575.
10. 0.03; 0.048; 0.02 ± 0.03, i.e. 0 and 0.05.

Unit 15

1. $Z = -2.62 < -1.645$, significantly less.
2. $Z = -4.20 < -2.33$, highly significantly less.
3. $Z = 12.1 > 1.96$, significant difference.
4. $Z = 1.2$, not significant.
5. (*a*) $Z = 2.62 > 1.645$, significant increase; (*b*) $Z = -0.79$, not significant.
6. $t = -2.21 > -2.821$ ($t_{0.01}$ 9 df), not significant.
7. $t = 2.25 < 2.306$ ($t_{0.025}$ 8 df), not significant.
8. $t = -2.27 < -1.860$ ($t_{0.050}$ 8 df), significant.
10. $X^2 = 4.4 < 11.070$ ($\chi^2_{0.05}$ 5 df), not significant. Consistent with uniform rate.
11. $X^2 = 13.9 < 18.307$ ($\chi^2_{0.05}$ 10 df), not significant. Dice appear unbiased.
12. $X^2 = 2.8 < 5.991$ ($\chi^2_{0.05}$ 2 df), not significant. No association.
13. $X^2 = 1.64 < 3.841$ ($\chi^2_{0.05}$ 1 df), not significant. No relationship.

Unit 16

1. £101.80.
2. (*a*) £3025; (*b*) £3039; (*c*) £3046.
3. 10th year.
4. (*a*) £454.55; (*b*) £896; (*c*) £215; (*d*) £1000.
5. £357.98 per annum.
8. A = £149; B = £165.
10. Internal rate of return = 10% per annum, project not undertaken.
11. 10%, to nearest 1%, per annum.
12. Internal rate of return = 13% per annum, project not undertaken.

Unit 17

5. 20, 10, 72.5; 31.
6. (i) 6.9, (ii) 6.7; (iii) 8.6; (iv) 8.0.
7. Age-specific rates per employee for Oxford factory: 8.0, 7.0, 4.5, 5.8;
   Age-specific rates per employee for Birmingham factory: 8.4, 6.3, 4.7, 5.5.
   Standardized rates per employee: 6.18, 6.08.
8. (*a*) (i) 0.945, (ii) 0.514, (iii) 0.290; (*b*) £9918.
9. 1.42, 1.35.

Tables

Table A Logarithms

	0	1	2	3	4	5	6	7	8	9	1	2	3	4	5	6	7	8	9
10	0000	0043	0086	0128	0170						4	8	13	17	21	25	30	34	38
						0212	0253	0294	0334	0374	4	8	12	16	20	24	28	32	36
11	0414	0453	0492	0531	0569						4	8	12	15	19	23	27	31	35
						0607	0645	0682	0719	0755	4	7	11	15	18	22	26	30	33
12	0792	0828	0864	0899	0934						4	7	11	14	18	21	25	28	32
						0969	1004	1038	1072	1106	3	7	10	14	17	20	24	27	31
13	1139	1173	1206	1239	1271						3	7	10	13	16	20	23	26	30
						1303	1335	1367	1399	1430	3	6	9	13	16	19	22	25	28
14	1461	1492	1523	1553	1584						3	6	9	12	15	18	21	24	27
						1614	1644	1673	1703	1732	3	6	9	12	15	18	21	24	27
15	1761	1790	1818	1847	1875						3	6	9	11	14	17	20	23	26
						1903	1931	1959	1987	2014	3	6	8	11	14	17	19	22	25
16	2041	2068	2095	2122	2148						3	5	8	11	13	16	19	21	24
						2175	2201	2227	2253	2279	3	5	8	10	13	16	18	21	23
17	2304	2330	2355	2380	2405						3	5	8	10	13	15	18	20	23
						2430	2455	2480	2504	2529	2	5	7	10	12	15	17	20	22
18	2553	2577	2601	2625	2648						2	5	7	10	12	14	17	19	21
						2672	2695	2718	2742	2765	2	5	7	9	12	14	16	19	21
19	2788	2810	2833	2856	2878						2	5	7	9	11	14	16	18	20
						2900	2923	2945	2967	2989	2	4	7	9	11	13	15	18	20
20	3010	3032	3054	3075	3096	3118	3139	3160	3181	3201	2	4	6	8	11	13	15	17	19
21	3222	3243	3263	3284	3304	3324	3345	3365	3385	3404	2	4	6	8	10	12	14	16	18
22	3424	3444	3464	3483	3502	3522	3541	3560	3579	3598	2	4	6	8	10	12	14	15	17
23	3617	3636	3655	3674	3692	3711	3729	3747	3766	3784	2	4	6	7	9	11	13	15	17
24	3802	3820	3838	3856	3874	3892	3909	3927	3945	3962	2	4	5	7	9	11	12	14	16
25	3979	3997	4014	4031	4048	4065	4082	4099	4116	4133	2	3	5	7	9	10	12	14	15
26	4150	4166	4183	4200	4216	4232	4249	4265	4281	4298	2	3	5	7	8	10	11	13	15
27	4314	4330	4346	4362	4378	4393	4409	4425	4440	4456	2	3	5	6	8	9	11	13	14
28	4472	4487	4502	4518	4533	4548	4564	4579	4594	4609	2	3	5	6	8	9	11	12	14
29	4624	4639	4654	4669	4683	4698	4713	4728	4742	4757	1	3	4	6	7	9	10	12	13
30	4771	4786	4800	4814	4829	4843	4857	4871	4886	4900	1	3	4	6	7	9	10	11	13
31	4914	4928	4942	4955	4969	4983	4997	5011	5024	5038	1	3	4	6	7	8	10	11	12
32	5051	5065	5079	5092	5105	5119	5132	5145	5159	5172	1	3	4	5	7	8	9	11	12
33	5185	5198	5211	5224	5237	5250	5263	5276	5289	5302	1	3	4	5	6	8	9	10	12
34	5315	5328	5340	5353	5366	5378	5391	5403	5416	5428	1	3	4	5	6	8	9	10	11
35	5441	5453	5465	5478	5490	5502	5514	5527	5539	5551	1	2	4	5	6	7	9	10	11
36	5563	5575	5587	5599	5611	5623	5635	5647	5658	5670	1	2	4	5	6	7	8	10	11

Table A Logarithms (*continued*)

	0	1	2	3	4	5	6	7	8	9	1	2	3	4	5	6	7	8	9
37	5682	5694	5705	5717	5729	5740	5752	5763	5775	5786	1	2	3	5	6	7	8	9	10
38	5798	5809	5821	5832	5843	5855	5866	5877	5888	5899	1	2	3	5	6	7	8	9	10
39	5911	5922	5933	5944	5955	5966	5977	5988	5999	6010	1	2	3	4	5	7	8	9	10
40	6021	6031	6042	6053	6064	6075	6085	6096	6107	6117	1	2	3	4	5	6	8	9	10
41	6128	6138	6149	6160	6170	6180	6191	6201	6212	6222	1	2	3	4	5	6	7	8	9
42	6232	6243	6253	6263	6274	6284	6294	6304	6314	6325	1	2	3	4	5	6	7	8	9
43	6335	6345	6355	6365	6375	6385	6395	6405	6415	6425	1	2	3	4	5	6	7	8	9
44	6435	6444	6454	6464	6474	6484	6493	6503	6513	6522	1	2	3	4	5	6	7	8	9
45	6532	6542	6551	6561	6571	6580	6590	6599	6609	6618	1	2	3	4	5	6	7	8	9
46	6628	6637	6646	6656	6665	6675	6684	6693	6702	6712	1	2	3	4	5	6	7	7	8
47	6721	6730	6739	6749	6758	6767	6776	6785	6794	6803	1	2	3	4	5	5	6	7	8
48	6812	6821	6830	6839	6848	6857	6866	6875	6884	6893	1	2	3	4	4	5	6	7	8
49	6902	6911	6920	6928	6937	6946	6955	6964	6972	6981	1	2	3	4	4	5	6	7	8
50	6990	6998	7007	7016	7024	7033	7042	7050	7059	7067	1	2	3	3	4	5	6	7	8
51	7076	7084	7093	7101	7110	7118	7126	7135	7143	7152	1	2	3	3	4	5	6	7	8
52	7160	7168	7177	7185	7193	7202	7210	7218	7226	7235	1	2	2	3	4	5	6	7	7
53	7243	7251	7259	7267	7275	7284	7292	7300	7308	7316	1	2	2	3	4	5	6	6	7
54	7324	7332	7340	7348	7356	7364	7372	7380	7388	7396	1	2	2	3	4	5	6	6	7
55	7404	7412	7419	7427	7435	7443	7451	7459	7466	7474	1	2	2	3	4	5	5	6	7
56	7482	7490	7497	7505	7513	7520	7528	7536	7543	7551	1	2	2	3	4	5	5	6	7
57	7559	7566	7574	7582	7589	7597	7604	7612	7619	7627	1	2	2	3	4	5	5	6	7
58	7634	7642	7649	7657	7664	7672	7679	7686	7694	7701	1	1	2	3	4	4	5	6	7
59	7709	7716	7723	7731	7738	7745	7752	7760	7767	7774	1	1	2	3	4	4	5	6	7
60	7782	7789	7796	7803	7810	7818	7825	7832	7839	7846	1	1	2	3	4	4	5	6	6
61	7853	7860	7868	7875	7882	7889	7896	7903	7910	7917	1	1	2	3	3	4	5	6	6
62	7924	7931	7938	7945	7952	7959	7966	7973	7980	7987	1	1	2	3	3	4	5	5	6
63	7993	8000	8007	8014	8021	8028	8035	8041	8048	8055	1	1	2	3	3	4	5	5	6
64	8062	8069	8075	8082	8089	8096	8102	8109	8116	8122	1	1	2	3	3	4	5	5	6
65	8129	8136	8142	8149	8156	8162	8169	8176	8182	8189	1	1	2	3	3	4	5	5	6
66	8195	8202	8209	8215	8222	8228	8235	8241	8248	8254	1	1	2	3	3	4	5	5	6
67	8261	8267	8274	8280	8287	8293	8299	8306	8312	8319	1	1	2	3	3	4	5	5	6
68	8325	8331	8338	8344	8351	8357	8363	8370	8376	8382	1	1	2	3	3	4	4	5	6
69	8388	8395	8401	8407	8414	8420	8426	8432	8439	8445	1	1	2	2	3	4	4	5	6
70	8451	8457	8463	8470	8476	8482	8488	8494	8500	8506	1	1	2	2	3	4	4	5	6
71	8513	8519	8525	8531	8537	8543	8549	8555	8561	8567	1	1	2	2	3	4	4	5	5
72	8573	8579	8585	8591	8597	8603	8609	8615	8621	8627	1	1	2	2	3	4	4	5	5
73	8633	8639	8645	8651	8657	8663	8669	8675	8681	8686	1	1	2	2	3	4	4	5	5
74	8692	8698	8704	8710	8716	8722	8727	8733	8739	8745	1	1	2	2	3	3	4	5	5
75	8751	8756	8762	8768	8774	8779	8785	8791	8797	8802	1	1	2	2	3	3	4	5	5
76	8808	8814	8820	8825	8831	8837	8842	8848	8854	8859	1	1	2	2	3	3	4	5	5
77	8865	8871	8876	8882	8887	8893	8899	8904	8910	8915	1	1	2	2	3	3	4	4	5
78	8921	8927	8932	8938	8943	8949	8954	8960	8965	8971	1	1	2	2	3	3	4	4	5
79	8976	8982	8987	8993	8998	9004	9009	9015	9020	9025	1	1	2	2	3	3	4	4	5
80	9031	9036	9042	9047	9053	9058	9063	9069	9074	9079	1	1	2	2	3	3	4	4	5
81	9085	9090	9096	9101	9106	9112	9117	9122	9128	9133	1	1	2	2	3	3	4	4	5

Table A Logarithms (*continued*)

	0	1	2	3	4	5	6	7	8	9	1	2	3	4	5	6	7	8	9
82	9138	9143	9149	9154	9159	9165	9170	9175	9180	9186	1	1	2	2	3	3	4	4	5
83	9191	9196	9201	9206	9212	9217	9222	9227	9232	9238	1	1	2	2	3	3	4	4	5
84	9243	9248	9253	9258	9263	9269	9274	9279	9284	9289	1	1	2	2	3	3	4	4	5
85	9294	9299	9304	9309	9315	9320	9325	9330	9335	9340	1	1	2	2	3	3	4	4	5
86	9345	9350	9355	9360	9365	9370	9375	9380	9385	9390	1	1	2	2	3	3	4	4	5
87	9395	9400	9405	9410	9415	9420	9425	9430	9435	9440	0	1	1	2	2	3	3	4	4
88	9445	9450	9455	9460	9465	9469	9474	9479	9484	9489	0	1	1	2	2	3	3	4	4
89	9494	9499	9504	9509	9513	9518	9523	9528	9533	9538	0	1	1	2	2	3	3	4	4
90	9542	9547	9552	9557	9562	9566	9571	9576	9581	9586	0	1	1	2	2	3	3	4	4
91	9590	9595	9600	9605	9609	9614	9619	9624	9628	9633	0	1	1	2	2	3	3	4	4
92	9638	9643	9647	9652	9657	9661	9666	9671	9675	9680	0	1	1	2	2	3	3	4	4
93	9685	9689	9694	9699	9703	9708	9713	9717	9722	9727	0	1	1	2	2	3	3	4	4
94	9731	9736	9741	9745	9750	9754	9759	9763	9768	9773	0	1	1	2	2	3	3	4	4
95	9777	9782	9786	9791	9795	9800	9805	9809	9814	9818	0	1	1	2	2	3	3	4	4
96	9823	9827	9832	9836	9841	9845	9850	9854	9859	9863	0	1	1	2	2	3	3	4	4
97	9868	9872	9877	9881	9886	9890	9894	9899	9903	9908	0	1	1	2	2	3	3	4	4
98	9912	9917	9921	9926	9930	9934	9939	9943	9948	9952	0	1	1	2	2	3	3	4	4
99	9956	9961	9965	9969	9974	9978	9983	9987	9991	9996	0	1	1	2	2	3	3	3	4

Table B Antilogarithms

	0	1	2	3	4	5	6	7	8	9	1	2	3	4	5	6	7	8	9
0.00	1000	1002	1005	1007	1009	1012	1014	1016	1019	1021	0	0	1	1	1	1	2	2	2
0.01	1023	1026	1028	1030	1033	1035	1038	1040	1042	1045	0	0	1	1	1	1	2	2	2
0.02	1047	1050	1052	1054	1057	1059	1062	1064	1067	1069	0	0	1	1	1	1	2	2	2
0.03	1072	1074	1076	1079	1081	1084	1086	1089	1091	1094	0	0	1	1	1	1	2	2	2
0.04	1096	1099	1102	1104	1107	1109	1112	1114	1117	1119	0	1	1	1	1	2	2	2	2
0.05	1122	1125	1127	1130	1132	1135	1138	1140	1143	1146	0	1	1	1	1	2	2	2	2
0.06	1148	1151	1153	1156	1159	1161	1164	1167	1169	1172	0	1	1	1	1	2	2	2	2
0.07	1175	1178	1180	1183	1186	1189	1191	1194	1197	1199	0	1	1	1	1	2	2	2	2
0.08	1202	1205	1208	1211	1213	1216	1219	1222	1225	1227	0	1	1	1	1	2	2	2	3
0.09	1230	1233	1236	1239	1242	1245	1247	1250	1253	1256	0	1	1	1	1	2	2	2	3
0.10	1259	1262	1265	1268	1271	1274	1276	1279	1282	1285	0	1	1	1	1	2	2	2	3
0.11	1288	1291	1294	1297	1300	1303	1306	1309	1312	1315	0	1	1	1	2	2	2	2	3
0.12	1318	1321	1324	1327	1330	1334	1337	1340	1343	1346	0	1	1	1	2	2	2	2	3
0.13	1349	1352	1355	1358	1361	1365	1368	1371	1374	1377	0	1	1	1	2	2	2	3	3
0.14	1380	1384	1387	1390	1393	1396	1400	1403	1406	1409	0	1	1	1	2	2	2	3	3
0.15	1413	1416	1419	1422	1426	1429	1432	1435	1439	1442	0	1	1	1	2	2	2	3	3
0.16	1445	1449	1452	1455	1459	1462	1466	1469	1472	1476	0	1	1	1	2	2	2	3	3
0.17	1479	1483	1486	1489	1493	1496	1500	1503	1507	1510	0	1	1	1	2	2	2	3	3
0.18	1514	1517	1521	1524	1528	1531	1535	1538	1542	545	0	1	1	1	2	2	2	3	3
0.19	1549	1552	1556	1560	1563	1567	1570	1574	1578	1581	0	1	1	1	2	2	3	3	3
0.20	1585	1589	1592	1596	1600	1603	1607	1611	1614	1618	0	1	1	1	2	2	3	3	3
0.21	1622	1626	1629	1633	1637	1641	1644	1648	1652	1656	0	1	1	2	2	2	3	3	3
0.22	1660	1663	1667	1671	1675	1679	1683	1687	1690	1694	0	1	1	2	2	2	3	3	3
0.23	1698	1702	1706	1710	1714	1718	1722	1726	1730	1734	0	1	1	2	2	2	3	3	4
0.24	1738	1742	1746	1750	1754	1758	1762	1766	1770	1774	0	1	1	2	2	2	3	3	4
0.25	1778	1782	1786	1791	1795	1799	1803	1807	1811	1816	0	1	1	2	2	2	3	3	4
0.26	1820	1824	1828	1832	1837	1841	1845	1849	1854	1858	0	1	1	2	2	3	3	3	4
0.27	1862	1866	1871	1875	1879	1884	1888	1892	1897	1901	0	1	1	2	2	3	3	3	4
0.28	1905	1910	1914	1919	1923	1928	1932	1936	1941	1945	0	1	1	2	2	3	3	4	4
0.29	1950	1954	1959	1963	1968	1972	1977	1982	1986	1991	0	1	1	2	2	3	3	4	4
0.30	1995	2000	2004	2009	2014	2018	2023	2028	2032	2037	0	1	1	2	2	3	3	4	4
0.31	2042	2046	2051	2056	2061	2065	2070	2075	2080	2084	0	1	1	2	2	3	3	4	4
0.32	2089	2094	2099	2104	2109	2113	2118	2123	2128	2133	0	1	1	2	2	3	3	4	4
0.33	2138	2143	2148	2153	2158	2163	2168	2173	2178	2183	0	1	1	2	2	3	3	4	4
0.34	2188	2193	2198	2203	2208	2213	2218	2223	2228	2234	1	1	2	2	3	3	4	4	5
0.35	2239	2244	2249	2254	2259	2265	2270	2275	2280	2286	1	1	2	2	3	3	4	4	5
0.36	2291	2296	2301	2307	2312	2317	2323	2328	2333	2339	1	1	2	2	3	3	4	4	5
0.37	2344	2350	2355	2360	2366	2371	2377	2382	2388	2393	1	1	2	2	3	3	4	4	5
0.38	2399	2404	2410	2415	2421	2427	2432	2438	2443	2449	1	1	2	2	3	3	4	4	5
0.39	2455	2460	2466	2472	2477	2483	2489	2495	2500	2506	1	1	2	2	3	3	4	5	5
0.40	2512	2518	2523	2529	2535	2541	2547	2553	2559	2564	1	1	2	2	3	4	4	5	5
0.41	2570	2576	2582	2588	2594	2600	2606	2612	2618	2624	1	1	2	2	3	4	4	5	5
0.42	2630	2636	2642	2649	2655	2661	2667	2673	2679	2685	1	1	2	2	3	4	4	5	6
0.43	2692	2698	2704	2710	2716	2723	2729	2735	2742	2748	1	1	2	3	3	4	4	5	6
0.44	2754	2761	2767	2773	2780	2786	2793	2799	2805	2812	1	1	2	3	3	4	4	5	6

Table B Antilogarithms (*continued*)

	0	1	2	3	4	5	6	7	8	9	1	2	3	4	5	6	7	8	9
0.45	2818	2825	2831	2838	2844	2851	2858	2864	2871	2877	1	1	2	3	3	4	5	5	6
0.46	2884	2891	2897	2904	2911	2917	2924	2931	2938	2944	1	1	2	3	3	4	5	5	6
0.47	2951	2958	2965	2972	2979	2985	2992	2999	3006	3013	1	1	2	3	3	4	5	5	6
0.48	3020	3027	3034	3041	3048	3055	3062	3069	3076	3083	1	1	2	3	4	4	5	6	6
0.49	3090	3097	3105	3112	3119	3126	3133	3141	3148	3155	1	1	2	3	4	4	5	6	6
0.50	3162	3170	3177	3184	3192	3199	3206	3214	3221	3228	1	1	2	3	4	4	5	6	7
0.51	3236	3243	3251	3258	3266	3273	3281	3289	3296	3304	1	2	2	3	4	5	5	6	7
0.52	3311	3319	3327	3334	3342	3350	3357	3365	3373	3381	1	2	2	3	4	5	5	6	7
0.53	3388	3396	3404	3412	3420	3428	3436	3443	3451	3459	1	2	2	3	4	5	6	6	7
0.54	3467	3475	3483	3491	3499	3508	3516	3524	3532	3540	1	2	2	3	4	5	6	6	7
0.55	3548	3556	3565	3573	3581	3589	3597	3606	3614	3622	1	2	2	3	4	5	6	7	7
0.56	3631	3639	3648	3656	3664	3673	3681	3690	3698	3707	1	2	3	3	4	5	6	7	8
0.57	3715	3724	3733	3741	3750	3758	3767	3776	3784	3793	1	2	3	3	4	5	6	7	8
0.58	3802	3811	3819	3828	3837	3846	3855	3864	3873	3882	1	2	3	4	4	5	6	7	8
0.59	3890	3899	3908	3917	3926	3936	3945	3954	3963	3972	1	2	3	4	5	5	6	7	8
0.60	3981	3990	3999	4009	4018	4027	4036	4046	4055	4064	1	2	3	4	5	6	6	7	8
0.61	4074	4083	4093	4102	4111	4121	4130	4140	4150	4159	1	2	3	4	5	6	7	8	9
0.62	4169	4178	4188	4198	4207	4217	4227	4236	4246	4256	1	2	3	4	5	6	7	8	9
0.63	4266	4276	4285	4295	4305	4315	4325	4335	4345	4355	1	2	3	4	5	6	7	8	9
0.64	4365	4375	4385	4395	4406	4416	4426	4436	4446	4457	1	2	3	4	5	6	7	8	9
0.65	4467	4477	4487	4498	4508	4519	4529	4539	4550	4560	1	2	3	4	5	6	7	8	9
0.66	4571	581	4592	4603	4613	4624	4634	4645	4656	4667	1	2	3	4	5	6	7	9	10
0.67	4677	4688	4699	4710	4721	4732	4742	4753	4764	4775	1	2	3	4	5	7	8	9	10
0.68	4786	4797	4808	4819	4831	4842	4853	4864	4875	4887	1	2	3	4	6	7	8	9	10
0.69	4898	4909	4920	4932	4943	4955	4966	4977	4989	5000	1	2	3	5	6	7	8	9	10
0.70	5012	5023	5035	5047	5058	5070	5082	5093	5105	5117	1	2	4	5	6	7	8	9	11
0.71	5129	5140	5152	5164	5176	5188	5200	5212	5224	5236	1	2	4	5	6	7	8	10	11
0.72	5248	5260	5272	5284	5297	5309	5321	5333	5346	5358	1	2	4	5	6	7	9	10	11
0.73	5370	5383	5395	5408	5420	5433	5445	5458	5470	5483	1	3	4	5	6	8	9	10	11
0.74	5495	5508	5521	5534	5546	5559	5572	5585	5598	5610	1	3	4	5	6	8	9	10	12
0.75	5623	5636	5649	5662	5675	5689	5702	5715	5728	5741	1	3	4	5	7	8	9	10	12
0.76	5754	5768	5781	5794	5808	5821	5834	5848	5861	5875	1	3	4	5	7	8	9	11	12
0.77	5888	5902	5916	5929	5943	5957	5970	5984	5998	6012	1	3	4	5	7	8	10	11	12
0.78	6026	6039	6053	6067	6081	6095	6109	6124	6138	6152	1	3	4	6	7	8	10	11	13
0.79	6166	6180	6194	6209	6223	6237	6252	6266	6281	6295	1	3	4	6	7	9	10	11	13
0.80	6310	6324	6339	6353	6368	6383	6397	6412	6427	6442	1	3	4	6	7	9	10	12	13
0.81	6457	6471	6486	6501	6516	6531	6546	6561	6577	6592	2	3	5	6	8	9	11	12	14
0.82	6607	6622	6637	6653	6668	6683	6699	6714	6730	6745	2	3	5	6	8	9	11	12	14
0.83	6761	6776	6792	6808	6823	6839	6855	6871	6887	6902	2	3	5	6	8	9	11	13	14
0.84	6918	6934	6950	6966	6982	6998	7015	7031	7047	7063	2	3	5	6	8	10	11	13	15
0.85	7079	7096	7112	7129	7145	7161	7178	7194	7211	7228	2	3	5	7	8	10	12	13	15
0.86	7244	7261	7278	7295	7311	7328	7345	7362	7379	7396	2	3	5	7	8	10	12	13	15
0.87	7413	7430	7447	7464	7482	7499	7516	7534	7551	7568	2	3	5	7	9	10	12	14	16
0.88	7586	7603	7621	7638	7656	7674	7691	7709	7727	7745	2	4	5	7	9	11	12	14	16
0.89	7762	7780	7798	7816	7834	7852	7870	7889	7907	7925	2	4	5	7	9	11	13	14	16
0.90	7943	7962	7980	7998	8017	8035	8054	8072	8091	8110	2	4	6	7	9	11	13	15	17
0.91	8128	8147	8166	8185	8204	8222	8241	8260	8279	8299	2	4	6	8	9	11	13	15	17

Table B Antilogarithms (*continued*)

	0	1	2	3	4	5	6	7	8	9	1	2	3	4	5	6	7	8	9
0.92	8318	8337	8356	8375	8395	8414	8433	8453	8472	8492	2	4	6	8	10	12	14	15	17
0.93	8511	8531	8551	8570	8590	8610	8630	8650	8670	8690	2	4	6	8	10	12	14	16	18
0.94	8710	8730	8750	8770	8790	8810	8831	8851	8872	8892	2	4	6	8	10	12	14	16	18
0.95	8913	8933	8954	8974	8995	9016	9036	9057	9078	9099	2	4	6	8	10	12	15	17	19
0.96	9120	9141	9162	9183	9204	9226	9247	9268	9290	9311	2	4	6	8	11	13	15	17	19
0.97	9333	9354	9376	9397	9419	9441	9462	9484	9506	9528	2	4	7	9	11	13	15	17	20
0.98	9550	9572	9594	9616	9638	9661	9683	9705	9727	9750	2	4	7	9	11	13	16	18	20
0.99	9772	9795	9817	9840	9863	9886	9908	9931	9954	9977	2	5	7	9	11	14	16	18	20

Table C Square roots

	0	1	2	3	4	5	6	7	8	9	1	2	3	4	5	6	7	8	9
1.0	1.0000	1.0050	1.0100	1.0149	1.0198	1.0247	1.0296	1.0344	1.0392	1.0440	5	10	15	20	24	29	34	39	44
1.1	1.0488	1.0536	1.0583	1.0630	1.0677	1.0724	1.0770	1.0817	1.0863	1.0909	5	9	14	19	23	28	33	37	42
1.2	1.0954	1.1000	1.1045	1.1091	1.1368	1.1180	1.2256	1.1269	1.1314	1.1358	4	9	13	18	22	27	31	36	40
1.3	1.1402	1.4460	1.1489	1.1533	1.1576	1.1619	1.1662	1.1705	1.1747	1.7900	4	9	13	17	22	26	30	34	39
1.4	1.1832	1.1874	1.1916	1.1958	1.2000	1.2042	1.2083	1.2124	1.2166	1.2207	4	8	12	17	21	25	29	33	37
1.5	1.2247	1.2288	1.2329	1.2369	1.2410	1.2450	1.2490	1.2530	1.2570	1.2610	4	8	12	16	20	24	28	32	36
1.6	1.2649	1.2689	1.2728	1.2767	1.2806	1.8457	1.2884	1.2923	1.2961	1.3000	4	8	12	16	19	23	27	31	35
1.7	1.3038	1.3077	1.3115	1.3153	1.3191	1.3229	1.3266	1.3304	1.3442	1.3379	4	8	11	15	19	23	26	30	34
1.8	1.3416	1.3454	1.3491	1.3528	1.3565	1.3601	1.3638	1.3675	1.3711	1.3748	4	7	11	14	18	21	25	29	32
1.9	1.3784	1.3820	1.3856	1.3892	1.3928	1.3964	1.4000	1.4036	1.4071	1.4107	4	7	11	14	18	21	25	29	32
2.0	1.4142	1.4177	1.4213	1.4248	1.4283	1.4318	1.4353	1.4387	1.4422	1.4457	3	7	10	14	17	21	24	28	31
2.1	1.4491	1.4526	1.4560	1.4595	1.4629	1.4633	1.4697	1.4731	1.4765	1.4799	3	7	10	14	17	21	24	28	31
2.2	1.4832	1.4866	1.4900	1.4933	1.4967	1.5000	1.5033	1.5067	1.5100	1.5133	3	7	10	13	17	20	23	27	30
2.3	1.5166	1.5199	1.5232	1.5264	1.5297	1.5330	1.5362	1.5395	1.5427	1.5460	3	7	10	13	16	20	23	26	29
2.4	1.5492	1.5524	1.5556	1.5588	1.5620	1.5652	1.5684	1.5716	1.5748	1.5780	3	6	10	13	16	19	22	26	29
2.5	1.5811	1.5843	1.5875	1.5906	1.5937	1.5969	1.6000	1.6031	1.6062	1.6093	3	6	9	13	16	19	22	25	28
2.6	1.6125	1.6155	1.6186	1.6217	1.6248	1.6279	1.6310	1.6340	1.6371	1.6401	3	6	9	12	15	18	22	25	28
2.7	1.6432	1.6462	1.6492	1.6523	1.6553	1.6583	1.6613	1.6643	1.6673	1.6703	3	6	9	12	15	18	21	24	27
2.8	1.6733	1.6763	1.6793	1.6823	1.6852	1.6882	1.6912	1.6941	1.6971	1.7000	3	6	9	12	15	18	21	24	27
2.9	1.7029	1.7059	1.7088	1.7117	1.7146	1.7176	1.7205	1.7234	1.7263	1.7292	3	6	9	12	15	17	20	23	26
3.0	1.7321	1.7349	1.7378	1.7407	1.7436	1.7464	1.7493	1.7521	1.7550	1.7578	3	6	9	11	14	17	20	23	26
3.1	1.7607	1.7635	1.7664	1.7692	1.7720	1.7748	1.7776	1.7804	1.7833	1.7861	3	6	8	11	14	17	20	23	25
3.2	1.7889	1.7916	1.7944	1.7972	1.8000	1.8028	1.8055	1.8083	1.8111	1.8138	3	6	8	11	14	17	19	22	25
3.3	1.8166	1.8193	1.8193	1.8221	1.8248	1.8276	1.8303	1.8358	1.8385	1.8412	3	5	8	11	14	16	19	22	25
3.4	1.8439	1.8466	1.8493	1.8520	1.8547	1.8574	1.8601	1.8628	1.8655	1.8682	3	5	8	11	13	16	19	22	24
3.5	1.8708	1.8735	1.8762	1.8788	1.8815	1.8841	1.8868	1.8894	1.8921	1.8947	3	5	8	11	13	16	19	21	24
3.6	1.8974	1.9000	1.9026	1.9053	1.9079	1.9105	1.9131	1.9157	1.9183	1.9309	3	5	8	10	13	16	18	21	24
3.7	1.9235	1.9261	1.9287	1.9313	1.0339	1.9365	1.9391	1.9416	1.9442	1.9468	3	5	8	10	13	15	18	21	23
3.8	1.9494	1.9519	1.9545	1.9570	1.9596	1.9621	1.9647	1.9672	1.9698	1.9723	3	5	8	10	13	15	18	20	23
3.9	1.9748	1.9774	1.9799	1.9824	1.9849	1.9875	1.9900	1.9925	1.9950	1.9975	3	5	8	10	13	15	18	20	23
4.0	2.0000	2.0025	2.0050	2.0075	2.0100	2.0125	2.0149	2.0174	2.0199	2.0224	2	5	7	10	12	15	17	20	22
4.1	2.0248	2.0273	2.0298	2.0322	2.0347	2.0372	2.0396	2.0421	2.0445	2.0469	2	5	7	10	12	15	17	20	22
4.2	2.0494	2.0518	2.0543	2.0567	2.0591	2.0616	2.0640	2.0664	2.0688	2.0712	2	5	7	10	12	15	17	19	22
4.3	2.0736	2.0761	2.0785	2.0809	2.0833	2.0857	2.0881	2.0905	2.0928	2.0952	2	5	7	10	12	14	17	19	22
4.4	2.0976	2.1000	2.1024	2.1048	2.1071	2.1095	2.1119	2.1142	2.1166	2.1190	2	5	7	9	12	14	17	19	21
4.5	2.1213	2.1237	2.1260	2.1284	2.1307	2.1331	2.1354	2.1378	2.1401	2.1424	2	5	7	9	12	14	16	19	21
4.6	2.1448	2.1471	2.1494	2.1517	2.1541	2.1564	2.1587	2.1610	2.1633	2.1657	2	5	7	9	11	14	16	18	21
4.7	2.1679	2.1703	2.1726	2.1749	2.1772	2.1794	2.1817	2.1840	2.1863	2.1886	2	5	7	9	11	14	16	18	20
4.8	2.1909	2.1932	2.1954	2.1977	2.2000	2.2023	2.2045	2.2068	2.2091	2.2113	2	5	7	9	11	14	16	18	21
4.9	2.2136	2.2159	2.2181	2.2204	2.2226	2.2249	2.2271	2.2293	2.2316	2.2338	2	4	7	9	11	13	16	18	21
5.0	2.2361	2.2383	2.2405	2.4289	2.2450	2.2472	2.2494	2.2517	2.2539	2.2561	2	4	7	9	11	13	18	18	20
5.1	2.2583	2.2605	2.2627	2.2650	2.6772	2.2694	2.2716	2.2738	2.2760	2.2782	2	4	7	9	11	13	15	17	20
5.2	2.2804	2.2825	2.2847	2.2891	2.2913	2.2935	2.2956	2.2978	2.2300	2.0440	2	4	7	9	11	13	15	17	20
5.3	2.3022	2.3043	2.3065	2.3087	2.3108	2.3130	2.3152	2.3173	2.3195	2.3216	2	4	6	9	11	13	15	17	19
5.4	2.3238	2.3259	2.3281	2.3302	2.3324	2.3345	2.3367	2.3388	2.3409	2.3431	2	4	6	9	11	13	15	17	19
5.5	2.3452	2.3473	2.3495	2.3516	2.5378	2.3558	2.3580	2.3601	2.3622	2.3643	2	4	6	8	11	13	15	17	19
5.5	2.3452	2.3473	2.3495	2.3516	2.5378	2.3558	2.3580	2.3601	2.3622	2.3643	2	4	6	8	11	13	15	17	19
5.6	2.3664	2.3685	2.3707	2.3728	2.3749	2.3558	2.3770	2.3791	2.3812	2.3833	2	4	6	8	11	13	15	17	19
5.7	2.3875	2.3876	2.3917	2.3937	2.3958	2.3979	2.4000	2.4021	2.4041	2.4062	2	4	6	8	10	13	15	17	19

Table C Square roots (*continued*)

	0	1	2	3	4	5	6	7	8	9	1	2	3	4	5	6	7	8	9
5.8	2.4085	2.4104	2.4125	2.4145	2.4166	2.4187	2.4207	2.4228	2.4249	2.4269	2	4	6	8	10	12	14	17	19
5.9	2.4290	2.4310	2.4333	2.4352	2.4372	2.4393	2.4413	2.4434	2.4454	2.4474	2	4	6	8	10	12	14	16	18
6.0	2.4495	2.4515	2.4536	2.4556	2.4576	2.4597	2.4617	2.4637	2.4658	2.4678	2	4	6	8	10	12	14	16	18
6.1	2.4698	2.4718	2.4739	2.4759	2.4779	2.4799	2.4819	2.4839	2.4860	2.4880	2	4	6	8	10	12	14	16	18
6.2	2.4900	2.4920	2.4940	2.4960	2.4980	2.5000	2.5020	2.5040	2.5060	2.5080	2	4	6	8	10	12	14	16	18
6.3	2.5100	2.5120	2.5140	2.5159	2.1799	2.5199	2.5219	2.5239	2.5259	2.5278	2	4	6	8	10	12	14	16	18
6.4	2.5298	2.5318	2.5338	2.5375	2.5377	2.5397	2.5417	2.5436	2.5456	2.5475	2	4	6	8	10	12	14	16	18
6.5	2.5495	2.5515	2.5534	2.5554	2.5573	2.5593	2.5612	2.5632	2.5652	2.5671	2	4	6	8	9	11	13	15	17
6.6	2.5690	2.5710	2.5729	2.5749	2.5768	2.5788	2.5807	2.5826	2.5846	2.5865	2	4	6	8	10	12	14	16	17
6.7	2.5884	2.5904	2.5923	2.5942	2.5962	2.5981	2.6000	2.6019	2.6038	2.6058	2	4	6	8	10	12	13	15	17
6.8	2.6077	2.6096	2.6115	2.6134	2.6153	2.6173	2.6192	2.6211	2.6230	2.6249	2	4	6	8	10	11	13	15	17
6.9	2.6228	2.6287	2.6306	2.6325	2.6344	2.6363	2.6382	2.6401	2.6420	2.6439	2	4	6	8	9	11	13	15	17
7.0	2.6458	2.6476	2.6495	2.6514	2.6533	2.6552	2.6571	2.6589	2.6608	2.6627	2	4	6	8	9	11	13	15	17
7.1	2.6646	2.6665	2.6683	2.6702	2.7214	2.6739	2.6758	2.6777	2.6796	2.6814	2	4	6	8	9	11	13	15	17
7.2	2.6833	2.6851	2.6870	2.6889	2.6907	2.6926	2.6944	2.6963	2.6981	2.2700	2	4	6	7	9	11	13	15	17
7.3	2.7019	2.7037	2.7055	2.7074	2.7092	2.7111	2.7129	2.7148	2.7166	2.7185	2	4	6	7	9	11	13	15	17
7.4	2.7203	2.7221	2.7240	2.7258	2.7276	2.7295	2.7313	2.7331	2.7350	2.7368	2	4	5	7	9	11	13	15	16
7.5	2.7386	2.7404	2.7423	2.7441	2.7459	2.7477	2.7495	2.7514	2.7532	2.7550	2	4	5	7	9	11	13	15	16
7.6	2.7568	2.7586	2.7604	2.7622	2.7641	2.7659	2.7677	2.7695	2.7713	2.2731	2	4	5	7	9	11	13	14	16
7.7	2.7749	2.7767	2.7785	2.7803	2.7821	2.7839	2.7857	2.7875	2.7893	2.7911	2	4	5	7	9	11	13	14	16
7.8	2.7928	2.7946	2.7964	2.7982	2.8000	2.8018	2.8036	2.8054	2.8071	2.8089	2	4	5	7	9	11	12	14	16
7.9	2.8107	2.8125	2.8142	2.8160	2.8178	2.8196	2.8213	2.8231	2.8249	2.8267	2	4	5	7	9	11	12	14	16
8.0	2.8284	2.8302	2.8320	2.8337	2.8355	2.8373	2.8390	2.8408	2.8425	2.8443	2	4	5	7	9	11	12	14	16
8.1	2.8468	2.8478	2.8496	2.8513	2.8531	2.3548	2.8566	2.8583	2.8601	2.8618	2	4	5	7	9	11	12	14	16
8.2	2.8636	2.8653	2.8671	2.8688	2.8705	2.8723	2.8740	2.8758	2.8775	2.8792	2	3	5	7	9	10	12	14	16
8.3	2.8810	2.8827	2.8844	2.8862	2.8879	2.8896	2.8914	2.8931	2.8948	2.8965	2	3	5	7	9	10	12	14	16
8.4	2.8983	2.9000	2.9017	2.9034	2.9052	2.9069	2.9086	2.9103	2.9120	2.9138	2	3	5	7	9	10	12	14	16
8.5	2.9155	2.9172	2.9189	2.9206	2.9223	2.9240	2.9257	2.9275	2.9292	2.9309	2	3	5	7	9	10	12	14	15
8.6	2.9326	2.9343	2.9360	2.9377	2.9394	2.9411	2.9428	2.9445	2.9462	2.9479	2	3	5	7	9	10	12	14	15
8.7	2.9496	2.9513	2.9530	2.9547	2.9563	2.9580	2.9597	2.9614	2.9631	2.9648	2	3	5	7	8	10	12	14	15
8.8	2.9665	2.2982	2.9698	2.9715	2.9732	2.7498	2.9766	2.9783	2.9799	2.9816	2	3	5	7	8	10	12	13	15
8.9	2.9833	2.9850	2.9866	2.9883	2.9900	2.9917	2.9933	2.9950	2.9967	2.9983	2	3	5	7	8	10	12	13	15
9.0	3.0000	3.0017	3.0033	3.0050	3.0067	3.0083	3.0100	3.0166	3.0133	3.0150	2	3	5	7	8	10	12	13	15
9.1	3.0166	3.0183	3.0199	3.0216	3.0232	3.0249	3.0265	3.0282	3.0299	3.0315	2	3	5	7	8	10	12	13	15
9.2	3.0332	3.0348	3.0364	3.0381	3.0397	3.0414	3.0430	3.0477	3.0463	3.0480	2	3	5	7	8	10	12	13	15
9.3	3.0496	3.0512	3.0529	3.0545	3.0561	3.0578	3.0594	3.0610	3.0627	3.0643	2	3	5	7	8	10	11	13	15
5.4	3.0659	3.0676	3.0692	3.0708	3.0725	3.0741	3.0757	3.0793	3.0790	3.0806	2	3	5	7	8	10	11	13	15
9.5	3.0822	3.0838	3.0854	3.8716	3.0887	3.0903	3.0919	3.0935	3.0952	3.0968	2	3	5	6	8	10	11	13	15
9.6	3.0984	3.1000	3.1016	3.1032	3.1048	3.1064	3.1081	3.1097	3.1113	3.1129	2	3	5	6	8	10	11	13	14
9.7	3.1145	3.1161	3.1177	3.1193	3.1209	3.1225	3.1241	3.1257	3.1273	3.1289	2	3	5	6	8	10	11	13	14
9.8	3.1305	3.1321	3.1337	3.1353	3.1369	3.1385	3.1401	3.1417	3.1432	3.1448	2	3	5	6	8	10	11	13	14
9.9	3.1464	3.1480	3.1496	3.1512	3.1528	3.1544	3.1559	3.1575	3.1591	3.1607	2	3	5	6	8	10	11	13	14
10	3.1623	3.1780	3.1937	3.2094	3.2249	3.2404	3.2558	3.2711	3.2863	3.3015	15	31	46	62	77	93	108	123	139
11	3.3166	3.3317	3.3466	3.3615	3.3764	3.3912	3.4059	3.4205	3.4351	3.4496	15	29	44	59	74	88	103	118	133
12	3.4641	3.4785	3.4928	3.5071	3.5214	3.5355	3.5496	3.5637	3.5777	3.5917	14	28	42	57	71	85	99	113	127
13	3.6056	3.6194	3.6332	3.6469	3.6606	3.6742	3.6878	3.7014	3.7148	3.7283	14	27	41	54	68	82	95	109	122
14	3.7417	3.7550	3.7683	3.7815	3.7947	3.8079	3.8210	3.8341	3.8471	3.8601	13	26	39	53	66	79	92	105	118
15	3.8730	3.8859	3.8987	3.9115	3.9243	3.9370	3.9497	3.9623	3.9749	3.9875	13	25	38	51	64	76	89	102	114
16	4.0000	4.0125	4.0249	4.0373	4.0497	4.0620	4.0743	4.0866	4.0988	4.1110	12	25	37	49	62	74	86	98	111

Table C Square roots (*continued*)

	0	1	2	3	4	5	6	7	8	9	1	2	3	4	5	6	7	8	9
17	4.1231	4.1352	4.1473	4.1593	4.1713	4.1833	4.1952	4.2071	4.2190	4.2308	12	24	36	48	60	72	84	96	108
18	4.2426	4.2544	4.2661	4.2778	4.2895	4.3012	4.3128	4.3243	4.3359	4.3474	12	23	35	47	58	70	81	93	105
19	4.3589	4.3704	4.3818	4.3932	4.4045	4.4159	4.4272	4.4385	4.4497	4.4609	11	23	34	45	57	68	79	91	102
20	4.4721	4.4833	4.4944	4.5056	4.5166	4.5277	4.5387	4.5497	4.5607	4.5717	11	22	23	33	44	55	77	88	99
21	4.5826	4.5935	4.6043	4.6152	4.6260	4.6368	4.6476	4.6583	4.6690	4.6797	11	22	32	43	54	65	75	86	97
22	4.6904	4.7011	4.7117	4.7223	4.7329	4.7434	4.7359	4.7645	4.7749	4.7854	11	21	32	42	53	63	74	84	95
23	4.7958	4.8062	4.8166	4.8270	4.8374	4.8477	4.8580	4.8683	4.8785	4.8888	10	21	31	41	52	62	72	83	93
24	4.8990	4.9092	4.9193	4.9295	4.9396	4.9497	4.9598	4.9699	4.9800	4.9900	10	20	30	40	51	61	71	81	91
25	5.0000	5.0100	5.0200	5.0299	5.0398	5.0498	5.0596	5.0695	5.0794	5.0892	10	20	30	40	50	59	69	79	89
26	5.0990	5.1088	5.1186	5.1284	5.1381	5.1478	5.1575	5.1672	5.1769	5.1865	10	19	29	39	49	58	68	78	87
27	5.1962	5.2058	5.2154	5.2249	5.2345	5.2440	5.2356	5.2631	5.2726	5.2820	10	19	29	38	48	57	67	76	86
28	5.2915	5.3009	5.3104	5.3198	5.3292	5.3385	5.3479	5.3572	5.3666	5.3759	9	19	28	37	47	56	66	75	84
29	5.3852	5.3944	5.4037	5.4129	5.4222	5.4314	5.4406	5.4498	5.4589	5.4681	9	18	28	37	46	55	64	74	83
30	5.4772	5.4863	5.4955	5.5045	5.5136	5.5227	5.5317	5.5408	5.5498	5.5588	9	18	27	36	45	54	63	72	81
31	5.5678	5.5767	5.5857	5.5946	5.6036	5.6125	5.6214	5.6303	5.6391	5.6480	9	18	27	36	45	53	62	71	80
32	5.6569	5.6657	5.6745	5.6833	5.6921	5.7009	5.7096	5.7184	5.7271	5.7359	9	18	26	35	44	53	61	70	79
33	5.7446	5.7533	5.7619	5.7706	5.7793	5.7879	5.7966	5.8052	5.8138	5.8224	9	17	26	35	43	52	60	69	78
34	5.8310	5.8395	5.8481	5.8566	5.8652	5.8737	5.8822	5.8907	5.8992	5.9076	9	17	26	34	43	51	60	68	77
35	5.9161	5.9245	5.9330	5.9414	5.9498	5.9582	5.9666	5.9746	5.9833	5.9917	8	17	25	34	42	50	59	67	76
36	6.0000	6.0083	6.0166	6.0249	6.0332	6.0415	6.0498	6.0581	6.0663	6.0745	8	17	25	33	41	50	58	66	74
37	6.0828	6.0910	6.0992	6.1074	6.1156	6.1237	6.1319	6.1400	6.1482	6.1563	8	16	24	33	41	49	57	65	73
38	6.1644	6.1725	6.1806	6.1887	6.1968	6.2048	6.2129	6.2209	6.2290	6.2370	8	16	24	32	40	48	56	64	73
39	6.2450	6.2530	6.2610	6.2690	6.2769	6.2849	6.2929	6.3008	6.3087	6.3166	8	16	24	32	40	48	56	64	72
40	6.3246	6.3325	6.3403	6.3482	6.3561	6.3640	6.3718	6.3797	6.3875	6.3953	8	16	24	31	39	47	55	63	71
41	6.4031	6.4109	6.4187	6.4265	6.4343	6.4420	6.4498	6.4576	6.4653	6.4730	8	16	23	31	39	47	54	62	70
42	6.4807	6.4885	6.4962	6.5038	6.5115	6.5192	6.5269	6.5345	6.5422	6.5498	8	15	23	31	38	46	54	61	69
43	6.5574	6.5651	6.5727	6.5803	6.5879	6.5955	6.6030	6.6106	6.6182	6.6257	8	15	23	30	38	45	53	61	68
44	6.6332	6.6408	6.6483	6.6558	6.6633	6.6708	6.6793	6.6858	6.6933	6.7007	7	15	22	30	37	45	52	60	67
45	6.7082	6.7157	6.7231	6.7305	6.7380	6.7454	6.7528	6.7602	6.7676	6.7750	7	15	22	30	37	44	52	59	67
46	6.7823	6.7897	6.7971	6.8044	6.8118	6.8191	6.8264	6.8337	6.8411	6.8484	7	15	22	29	37	44	51	59	66
47	6.8557	6.8629	6.8702	6.8775	6.8848	6.8920	6.8993	6.9065	6.9138	6.9210	7	15	22	29	36	44	51	58	65
48	6.9282	6.9354	6.9426	6.9498	6.9570	6.9642	6.9714	6.9785	6.9857	6.9929	7	14	22	29	36	43	50	57	65
49	7.0000	7.0071	7.0143	7.0214	7.0285	7.0356	7.0427	7.0498	7.0569	7.0640	7	14	21	28	36	43	50	57	64
50	7.0711	7.0781	7.0852	7.0922	7.0993	7.1063	7.1134	7.1204	7.1274	7.1344	7	14	21	28	35	42	49	56	63
51	7.1414	7.1484	7.1554	7.1624	7.1694	7.1764	7.1833	7.1903	7.1972	7.2042	7	14	21	28	35	42	49	56	63
52	7.2111	7.2180	7.2250	7.2319	7.2388	7.2457	7.2526	7.2595	7.2664	7.2732	7	14	21	28	35	41	48	55	62
53	7.2801	7.2870	7.2938	7.3007	7.3075	7.3144	7.3212	7.3280	7.3348	7.3417	7	14	21	27	34	41	48	55	62
54	7.3485	7.3553	7.3621	7.3689	7.3756	7.3824	7.3892	7.3959	7.4027	7.4095	7	14	20	27	34	41	47	54	61
55	7.4162	7.4229	7.4297	7.4364	7.4431	7.4498	7.4565	7.4632	7.4699	7.4766	7	13	20	27	34	40	47	54	60
56	7.4833	7.4900	7.4967	7.5033	7.5100	7.5166	7.5233	7.5299	7.5366	7.5432	7	13	20	27	33	40	47	53	60
57	7.5498	7.5565	7.5631	7.5697	7.5763	7.5829	7.5895	7.5961	7.6026	7.6092	7	13	20	26	33	40	46	53	59
58	7.6158	7.6223	7.6289	7.6354	7.6420	7.6485	7.6551	7.6616	7.6681	7.6746	7	13	20	26	33	39	46	62	59
59	7.6811	7.6877	7.6942	7.7006	7.7071	7.7136	7.7201	7.7266	7.7330	7.7395	6	13	19	26	32	39	45	52	58
60	7.7460	7.7524	7.7589	7.7653	7.7717	7.7782	7.7846	7.7910	7.7974	7.8038	6	13	19	26	32	39	45	51	58
61	7.8102	7.8166	7.8230	7.8294	7.8358	7.8422	7.8486	7.8549	7.8613	7.8677	6	13	19	26	32	38	45	51	57
62	7.8740	7.8804	7.8867	7.8930	7.8994	7.9057	7.9120	7.9183	7.9246	7.9310	6	13	19	25	32	38	44	51	57
63	7.9373	7.9436	7.9498	7.9561	7.9624	7.9687	7.9750	7.9812	7.9875	7.9937	6	13	19	25	31	38	44	50	56
64	8.0000	8.0062	8.0125	8.0187	8.0250	8.0312	8.0374	8.0436	8.0498	8.0561	6	12	19	25	31	37	44	50	56

Table C Square roots (*continued*)

	0	1	2	3	4	5	6	7	8	9	1	2	3	4	5	6	7	8	9
65	8.0623	8.0685	8.0747	8.0808	8.0870	9.0932	8.0994	8.1056	8.1117	8.1179	6	12	19	25	31	37	43	49	56
66	8.1240	8.1302	8.1363	8.1425	8.1486	8.1548	8.1609	8.1670	8.1731	8.1792	6	12	18	25	31	37	43	49	55
67	8.1854	8.1915	8.1976	8.2037	8.2098	8.2158	8.2219	8.2280	8.2341	8.2401	6	12	18	24	30	37	43	49	55
68	8.2462	8.2523	8.2583	8.2644	8.2704	8.2765	8.2825	8.2885	8.2946	8.3006	6	12	18	24	30	36	42	38	54
69	8.3066	8.3126	8.3187	8.3247	8.3307	8.3367	8.3427	8.3487	8.3546	8.3606	6	12	18	24	30	36	42	48	54
70	8.3666	8.3726	8.3785	8.3845	8.3905	8.3964	6.4024	8.4083	8.4143	8.4202	6	12	18	24	30	36	42	48	54
71	8.4261	8.4321	8.4380	8.4439	8.4499	8.4558	8.4617	8.4676	8.4735	8.4794	6	12	18	24	30	35	41	47	52
72	8.4853	8.4912	8.4971	8.5029	8.5088	8.5147	8.5206	8.5264	8.5323	8.5381	6	12	18	23	29	35	41	46	53
73	8.5440	8.5499	8.5557	8.5615	8.5674	8.5732	8.5790	8.5849	8.5907	8.5965	6	12	17	23	29	35	41	47	52
74	8.6023	8.6081	8.6139	8.6197	8.6255	8.6313	8.6371	8.6429	8.6487	8.6545	6	12	17	23	29	35	41	46	52
75	8.6603	8.6660	8.6718	8.6776	8.6833	8.6891	8.6948	8.7006	8.7063	8.7121	6	12	17	23	29	35	40	46	52
76	8.7178	8.7235	8.7293	8.7350	8.7407	8.7464	8.7521	8.7579	8.7636	8.7693	6	11	17	23	29	34	40	46	51
77	8.7750	8.7807	8.7864	8.7920	8.7977	8.8034	8.8091	8.8148	8.8204	8.8261	6	11	17	23	28	34	40	45	51
78	8.8318	8.8374	8.8431	8.8487	8.8544	8.8600	8.8657	8.8713	8.8769	8.8826	6	11	17	23	28	34	40	45	51
79	8.8882	8.8938	8.8994	8.9051	8.9107	8.9163	8.9219	8.9275	8.9331	8.9387	6	11	17	22	28	34	39	45	50
80	8.9443	8.9499	8.9554	8.9610	8.9666	8.9722	8.9778	8.9833	8.9889	8.9944	6	11	17	22	28	33	39	45	50
81	9.0000	9.0056	9.0111	9.0167	9.0222	9.0277	9.0333	9.0388	9.0443	9.0499	6	11	17	22	28	33	39	44	50
82	9.0554	9.0609	9.0664	9.0719	9.0774	9.0830	9.0885	9.0940	9.0995	9.1049	6	11	17	22	28	33	39	44	50
83	9.1104	9.1159	9.1214	9.1269	9.1324	9.1378	9.1433	9.1488	9.1542	9.1597	5	11	16	22	27	33	38	44	49
84	9.1652	9.1706	9.1761	9.1815	9.1869	9.1924	9.1978	9.2033	9.2087	9.2141	5	11	16	22	27	33	38	44	49
85	9.2195	9.2250	9.2304	9.2358	9.2412	9.2466	9.2520	9.2574	9.2628	9.2682	5	11	16	22	27	32	38	43	49
86	9.2736	9.2790	9.2844	9.2898	9.2952	9.3005	9.3059	9.3113	9.3167	9.3220	5	11	16	22	27	32	38	43	48
87	9.3274	9.3327	9.3381	9.3434	9.3488	9.3541	9.3595	9.3648	9.3702	9.3755	5	11	16	21	27	32	37	43	48
88	9.3808	9.3862	9.3915	9.3968	9.4021	9.4074	9.4128	9.4181	9.4234	9.4287	5	11	16	21	27	32	37	43	48
89	9.4340	9.4393	9.4446	9.4499	9.4552	9.4604	9.4657	9.4710	9.4763	9.4816	5	11	16	21	26	32	37	42	48
90	9.4868	9.4921	9.4974	9.5026	9.5079	9.5131	9.5184	9.5237	9.5289	9.5341	5	11	16	21	26	32	37	42	47
91	9.5394	9.5446	9.5499	9.5551	9.5603	9.5656	9.5708	9.5760	9.5812	9.5864	5	10	16	21	26	31	37	42	47
92	9.5917	9.5969	9.6021	9.6073	9.6125	9.6177	9.6229	9.6281	9.6333	9.6385	5	10	16	21	26	31	36	42	47
93	9.6437	9.6488	9.6540	9.6592	9.6644	9.6695	9.6747	9.6799	9.6850	9.6902	5	10	16	21	26	31	36	41	47
94	9.6954	9.7005	9.7057	9.7108	9.7160	9.7211	9.7263	9.7314	9.7365	9.7417	5	10	15	21	26	31	36	41	46
95	9.7468	9.7519	9.7570	9.7622	9.7673	9.7724	9.7775	9.7826	9.7877	9.7929	5	10	15	20	26	31	36	41	46
96	9.7980	9.8031	9.8082	9.8133	9.8184	9.8234	9.8285	9.8336	9.8387	9.8438	5	10	15	20	25	31	36	41	46
97	9.8489	9.8539	9.8590	9.8641	9.8691	9.8742	9.8793	9.8843	9.8894	9.8944	5	10	15	20	25	30	35	41	46
98	9.8995	9.9045	9.9096	9.9146	9.9197	9.9247	9.9298	9.9348	9.9398	9.9448	5	10	15	20	25	30	35	40	45
99	9.9499	9.9549	9.9599	9.9649	9.9700	9.9750	9.9800	9.9850	9.9900	9.9950	5	10	15	20	25	30	35	40	45

Table D Proportions of area under the normal curve

Z	Area between mean and *Z*	Area beyond *Z*	*Z*	Area between mean and *Z*	Area beyond *Z*	*Z*	Area between mean and *Z*	Area beyond *Z*
0.00	0.0000	0.5000	0.35	0.1368	0.3632	0.70	0.2580	0.2420
0.01	0.0040	0.4960	0.36	0.1406	0.3594	0.71	0.2611	0.2389
0.02	0.0080	0.4920	0.37	0.1443	0.3557	0.72	0.2642	0.2358
0.03	0.0120	0.4880	0.38	0.1480	0.3520	0.73	0.2673	0.2327
0.04	0.0160	0.4840	0.39	0.1517	0.3483	0.74	0.2704	0.2296
0.05	0.0199	0.4801	0.40	0.1554	0.3446	0.75	0.2734	0.2266
0.06	0.0239	0.4761	0.41	0.1591	0.3409	0.76	0.2764	0.2236
0.07	0.0279	0.4721	0.42	0.1628	0.3372	0.77	0.2794	0.2206
0.08	0.0319	0.4681	0.43	0.1664	0.3336	0.78	0.2823	0.2177
0.09	0.0359	0.4641	0.44	0.1700	0.3300	0.79	0.2852	0.2148
0.10	0.0398	0.4602	0.45	0.1736	0.3264	0.80	0.2881	0.2119
0.11	0.0438	0.4562	0.46	0.1772	0.3228	0.81	0.2910	0.2090
0.12	0.0478	0.4522	0.47	0.1808	0.3192	0.82	0.2939	0.2061
0.13	0.0517	0.4483	0.48	0.1844	0.3156	0.83	0.2967	0.2033
0.14	0.0557	0.4443	0.49	0.1879	0.3121	0.84	0.2995	0.2005
0.15	0.0596	0.4404	0.50	0.1915	0.3085	0.85	0.3023	0.1977
0.16	0.0636	0.4364	0.51	0.1950	0.3050	0.86	0.3051	0.1949
0.17	0.0675	0.4325	0.52	0.1985	0.3015	0.87	0.3078	0.1922
0.18	0.0714	0.4286	0.53	0.2019	0.2981	0.88	0.3106	0.1894
0.19	0.0753	0.4247	0.54	0.2054	0.2946	0.89	0.3133	0.1867
0.20	0.0793	0.4207	0.55	0.2088	0.2912	0.90	0.3159	0.1841
0.21	0.0832	0.4168	0.56	0.2123	0.2877	0.91	0.3186	0.1814
0.22	0.0871	0.4129	0.57	0.2157	0.2843	0.92	0.3212	0.1788
0.23	0.0910	0.4090	0.58	0.2190	0.2810	0.93	0.3238	0.1762
0.24	0.0948	0.4052	0.59	0.2224	0.2776	0.94	0.3264	0.1736
0.25	0.0987	0.4013	0.60	0.2257	0.2743	0.95	0.3289	0.1711
0.26	0.1026	0.3974	0.61	0.2291	0.2709	0.96	0.3315	0.1685
0.27	0.1064	0.3936	0.62	0.2324	0.2676	0.97	0.3340	0.1660
0.28	0.1103	0.3897	0.63	0.2357	0.2643	0.98	0.3365	0.1635
0.29	0.1141	0.3859	0.64	0.2389	0.2611	0.99	0.3389	0.1611
0.30	0.1179	0.3821	0.65	0.2422	0.2578	1.00	0.3413	0.1587
0.31	0.1217	0.3783	0.66	0.2454	0.2546	1.01	0.3438	0.1562
0.32	0.1255	0.3745	0.67	0.2486	0.2514	1.02	0.3461	0.1539
0.33	0.1293	0.3707	0.68	0.2517	0.2483	1.03	0.3485	0.1515
0.34	0.1331	0.3669	0.69	0.2549	0.2451	1.04	0.3508	0.1492

Table D Proportions of area under the normal curve (*continued*)

Z	Area between mean and Z	Area beyond Z	Z	Area between mean and Z	Area beyond Z	Z	Area between mean and Z	Area beyond Z
1.05	0.3531	0.1469	1.40	0.4192	0.0808	1.75	0.4599	0.0401
1.06	0.3554	0.1446	1.41	0.4207	0.0793	1.76	0.4608	0.0392
1.07	0.3577	0.1423	1.42	0.4222	0.0778	1.77	0.4616	0.0384
1.08	0.3599	0.1401	1.43	0.4236	0.0764	1.78	0.4625	0.0375
1.09	0.3621	0.1379	1.44	0.4251	0.0749	1.79	0.4633	0.0367
1.10	0.3643	0.1357	1.45	0.4265	0.0735	1.80	0.4641	0.0359
1.11	0.3665	0.1335	1.46	0.4279	0.0721	1.81	0.4649	0.0351
1.12	0.3686	0.1314	1.47	0.4292	0.0708	1.82	0.4656	0.0344
1.13	0.3708	0.1292	1.48	0.4306	0.0694	1.83	0.4664	0.0336
1.14	0.3729	0.1271	1.49	0.4319	0.0681	1.84	0.4671	0.0329
1.15	0.3749	0.1251	1.50	0.4332	0.0668	1.85	0.4678	0.0322
1.16	0.3770	0.1230	1.51	0.4345	0.0655	1.86	0.4686	0.0314
1.17	0.3790	0.1210	1.52	0.4357	0.0643	1.87	0.4693	0.0307
1.18	0.3810	0.1170	1.53	0.4370	0.0630	1.88	0.4699	0.0301
1.19	0.3830	0.1170	1.54	0.4382	0.0618	1.89	0.4706	0.0294
1.20	0.3849	0.1151	1.55	0.4394	0.0606	1.90	0.4713	0.0287
1.21	0.3869	0.1131	1.56	0.4406	0.0594	1.91	0.4719	0.0281
1.22	0.3888	0.1112	1.57	0.4418	0.0582	1.92	0.4726	0.0274
1.23	0.3907	0.1093	1.58	0.4429	0.0571	1.93	0.4732	0.0268
1.24	0.3925	0.1075	1.59	0.4441	0.0559	1.94	0.4738	0.0262
1.25	0.3944	0.1056	1.60	0.4452	0.0548	1.95	0.4744	0.0256
1.26	0.3962	0.1038	1.61	0.4463	0.0537	1.96	0.4750	0.0250
1.27	0.3980	0.1020	1.62	0.4474	0.0526	1.97	0.4756	0.0244
1.28	0.3997	0.1003	1.63	0.4484	0.0516	1.98	0.4761	0.0239
1.29	0.4015	0.0985	1.64	0.4495	0.0505	1.99	0.4767	0.0233
1.30	0.4032	0.0968	1.65	0.4505	0.0495	2.00	0.4772	0.0228
1.31	0.4049	0.0951	1.66	0.4515	0.0485	2.01	0.4778	0.0222
1.32	0.4066	0.0934	1.67	0.4525	0.0475	2.02	0.4783	0.0217
1.33	0.4082	0.0918	1.68	0.4535	0.0465	2.03	0.4788	0.0212
1.34	0.4099	0.0901	1.69	0.4545	0.0455	2.04	0.4793	0.0207
1.35	0.4115	0.0885	1.70	0.4554	0.0446	2.05	0.4798	0.0202
1.36	0.4131	0.0869	1.71	0.4564	0.0436	2.06	0.4803	0.0197
1.37	0.4147	0.0853	1.72	0.4573	0.0427	2.07	0.4808	0.0192
1.38	0.4162	0.0838	1.73	0.4582	0.0418	2.08	0.4812	0.0188
1.39	0.4177	0.0823	1.74	0.4591	0.0409	2.09	0.4817	0.0183

Table D Proportions of area under the normal curve (*continued*)

Z	Area between mean and Z	Area beyond Z	Z	Area between mean and Z	Area beyond Z	Z	Area between mean and Z	Area beyond Z
2.10	0.4821	0.0179	2.45	0.4929	0.0071	2.80	0.4974	0.0026
2.11	0.4826	0.0174	2.46	0.4931	0.0069	2.81	0.4975	0.0025
2.12	0.4830	0.0170	2.47	0.4932	0.0068	2.82	0.4976	0.0024
2.13	0.4834	0.0166	2.48	0.4934	0.0066	2.83	0.4977	0.0023
2.14	0.4838	0.0162	2.49	0.4936	0.0064	2.84	0.4977	0.0023
2.15	0.4842	0.0158	2.50	0.4938	0.0062	2.85	0.4978	0.0022
2.16	0.4846	0.0154	2.51	0.4940	0.0060	2.86	0.4979	0.0021
2.17	0.4850	0.0150	2.52	0.4941	0.0059	2.87	0.4979	0.0021
2.18	0.4854	0.0146	2.53	0.4943	0.0057	2.88	0.4980	0.0020
2.19	0.4857	0.0143	2.54	0.4945	0.0055	2.89	0.4981	0.0019
2.20	0.4861	0.0139	2.55	0.4946	0.0054	2.90	0.4981	0.0019
2.21	0.4864	0.0136	2.56	0.4948	0.0052	2.91	0.4982	0.0018
2.22	0.4868	0.0132	2.57	0.4949	0.0051	2.92	0.4982	0.0018
2.23	0.4871	0.0129	2.58	0.4951	0.0049	2.93	0.4983	0.0017
2.24	0.4875	0.0125	2.59	0.4952	0.0048	2.94	0.4984	0.0016
2.25	0.4878	0.0122	2.60	0.4953	0.0047	2.95	0.4984	0.0016
2.26	0.4881	0.0119	2.61	0.4955	0.0045	2.96	0.4985	0.0015
2.27	0.4884	0.0116	2.62	0.4956	0.0044	2.97	0.4985	0.0015
2.28	0.4887	0.0113	2.63	0.4957	0.0043	2.98	0.4986	0.0014
2.29	0.4890	0.0110	2.64	0.4959	0.0041	2.99	0.4986	0.0014
2.30	0.4893	0.0107	2.65	0.4960	0.0040	3.00	0.4987	0.0013
2.31	0.4896	0.0104	2.66	0.4961	0.0039	3.01	0.4987	0.0013
2.32	0.4898	0.0102	2.67	0.4962	0.0038	3.02	0.4987	0.0013
2.33	0.4901	0.0099	2.68	0.4963	0.0037	3.03	0.4988	0.0012
2.34	0.4904	0.0096	2.69	0.4964	0.0036	3.04	0.4988	0.0012
2.35	0.4906	0.0094	2.70	0.4965	0.0035	3.05	0.4989	0.0011
2.36	0.4909	0.0091	2.71	0.4966	0.0034	3.06	0.4989	0.0011
2.37	0.4911	0.0089	2.72	0.4967	0.0033	3.07	0.4989	0.0011
2.38	0.4913	0.0087	2.73	0.4968	0.0032	3.08	0.4990	0.0010
2.39	0.4916	0.0084	2.74	0.4969	0.0031	3.09	0.4990	0.0010
2.40	0.4918	0.0082	2.75	0.4970	0.0030	3.10	0.4990	0.0010
2.41	0.4920	0.0080	2.76	0.4971	0.0029	3.11	0.4991	0.0009
2.42	0.4922	0.0078	2.77	0.4972	0.0028	3.12	0.4991	0.0009
2.43	0.4925	0.0075	2.78	0.4973	0.0027	3.13	0.4991	0.0009
2.44	0.4927	0.0073	2.79	0.4974	0.0026	3.14	0.4992	0.0008

Table D Proportions of area under the normal curve (*continued*)

Z	Area between mean and Z	Area beyond Z	Z	Area between mean and Z	Area beyond Z	Z	Area between mean and Z	Area beyond Z
3.15	0.4992	0.0008	3.22	0.4994	0.0006	3.45	0.4997	0.0003
3.16	0.4992	0.0008	3.23	0.4994	0.0006	3.50	0.4998	0.0002
3.17	0.4992	0.0008	3.24	0.4994	0.0006	3.60	0.4998	0.0002
3.18	0.4993	0.0007	3.25	0.4994	0.0006	3.70	0.4999	0.0001
3.19	0.4993	0.0007	3.30	0.4995	0.0005	3.80	0.4999	0.0001
3.20	0.4993	0.0007	3.35	0.4996	0.0004	3.90	0.49995	0.0005
3.21	0.4993	0.0007	3.40	0.4997	0.0003	4.00	0.49997	0.0003

Table E Values of $e^{-\mu}$

$(0 < \mu < 1)$

μ	0	1	2	3	4	5	6	7	8	9
0.0	1.0000	0.9900	0.9802	0.9704	0.9608	0.9512	0.9418	0.9324	0.9231	0.9139
0.1	0.9048	0.8958	0.8869	0.8781	0.8694	0.8607	0.8521	0.8437	0.8353	0.8270
0.2	0.8187	0.8106	0.8025	0.7945	0.7866	0.7788	0.7711	0.7634	0.7558	0.7483
0.3	0.7408	0.7334	0.7261	0.7189	0.7118	0.7047	0.6977	0.6907	0.6839	0.6771
0.4	0.6703	0.6636	0.6570	0.6505	0.6440	0.6376	0.6313	0.6250	0.6188	0.6126
0.5	0.6065	0.6005	0.5945	0.5886	0.5827	0.5770	0.5712	0.5655	0.5599	0.5543
0.6	0.5488	0.5434	0.5379	0.5326	0.5273	0.5220	0.5169	0.5117	0.5066	0.5016
0.7	0.4966	0.4916	0.4868	0.4819	0.4771	0.4724	0.4677	0.4630	0.4584	0.4538
0.8	0.4493	0.4449	0.4404	0.4360	0.4317	0.4274	0.4232	0.4190	0.4148	0.4107
0.9	0.4066	0.4025	0.3985	0.3946	0.3906	0.3867	0.3829	0.3791	0.3753	0.3716

$(\mu = 1, 2, 3, \ldots, 10)$

μ	1	2	3	4	5	6	7	8	9	10
$e^{-\mu}$	0.36788	0.13534	0.04979	0.01832	0.006738	0.002479	0.000912	0.000335	0.000123	0.000045

Note: To obtain values of $e^{-\mu}$ for other values of μ use the laws of exponents.

Example: $e^{-3.48} = (e^{-3.00})(e^{-0.48}) = (0.04979)(0.6188) = 0.03081$.

Table F Student's t distribution

$df\,(\nu)$	$t_{0.100}$	$t_{0.050}$	$t_{0.025}$	$t_{0.010}$	$t_{0.005}$	$df(\nu)$
1	3.078	6.314	12.706	31.821	63.657	1
2	1.886	2.920	4.303	6.965	9.925	2
3	1.638	2.353	3.182	4.541	5.841	3
4	1.533	2.132	2.776	3.747	4.604	4
5	1.476	2.015	2.571	3.365	4.032	5
6	1.440	1.943	2.447	3.143	3.707	6
7	1.415	1.895	2.365	2.998	3.499	7
8	1.397	1.860	2.306	2.896	3.355	8
9	1.383	1.833	2.262	2.821	3.250	9
10	1.372	1.812	2.228	2.764	3.169	10
11	1.363	1.796	2.201	2.718	3.106	11
12	1.356	1.782	2.179	2.681	3.055	12
13	1.350	1.771	2.160	2.650	3.012	13
14	1.345	1.761	2.145	2.624	2.977	14
15	1.341	1.753	2.131	2.602	2.947	15
16	1.337	1.746	2.120	2.583	2.921	16
17	1.333	1.740	2.110	2.567	2.898	17
18	1.330	1.734	2.101	2.552	2.878	18
19	1.328	1.729	2.093	2.539	2.861	19
20	1.325	1.725	2.086	2.528	2.845	20
21	1.323	1.721	2.080	2.518	2.831	21
22	1.321	1.717	2.074	2.508	2.819	22
23	1.319	1.714	2.069	2.500	2.807	23
24	1.318	1.711	2.064	2.492	2.797	24
25	1.316	1.708	2.060	2.485	2.787	25
26	1.315	1.706	2.056	2.479	2.779	26
27	1.314	1.703	2.052	2.473	2.771	27
28	1.313	1.701	2.048	2.467	2.763	28
29	1.311	1.699	2.045	2.462	2.756	29
∞	1.282	1.645	1.960	2.326	2.576	∞

Table G χ^2 distribution

$df\ (\nu)$	$\chi^2_{0.05}$	$\chi^2_{0.025}$	$\chi^2_{0.01}$	$\chi^2_{0.005}$	$df\ (\nu)$
1	3.841	5.024	6.635	7.879	1
2	5.991	7.378	9.210	10.597	2
3	7.815	9.348	11.345	12.838	3
4	9.488	11.143	13.277	14.860	4
5	11.070	12.832	15.086	16.750	5
6	12.592	14.449	16.812	18.548	6
7	14.067	16.013	18.475	20.278	7
8	15.507	17.535	20.090	21.955	8
9	16.919	19.023	21.666	23.589	9
10	18.307	20.483	23.209	25.188	10
11	19.675	21.920	24.725	26.757	11
12	21.026	23.337	26.217	28.300	12
13	22.362	24.736	27.688	29.819	13
14	23.685	26.119	29.141	31.319	14
15	24.996	27.488	30.578	32.801	15
16	26.296	28.845	32.000	34.267	16
17	27.587	30.191	33.409	35.718	17
18	28.869	31.526	34.805	37.156	18
19	30.144	32.852	36.191	38.582	19
20	31.410	34.170	37.566	39.997	20
21	32.671	35.479	38.932	41.401	21
22	33.924	36.781	40.289	42.796	22
23	35.172	38.076	41.638	44.181	23
24	36.415	39.364	42.980	45.558	24
25	37.652	40.646	44.314	46.928	25
26	38.885	41.923	45.642	48.290	26
27	40.113	43.194	46.963	49.645	27
28	41.337	44.461	48.278	50.993	28
29	42.557	45.722	49.588	52.336	29
30	43.773	46.979	50.892	53.672	30

Table H Random sampling numbers

20	17	42	28	23	17	59	66	38	61	02	10	86	10	51	55	92	52	44	25
74	49	04	49	03	04	10	33	53	70	11	54	48	63	94	60	94	49	57	38
94	70	49	31	38	67	23	42	29	65	40	88	78	71	37	18	48	64	06	57
22	15	78	15	69	84	32	52	32	54	15	12	54	02	01	37	38	37	12	93
93	29	12	18	27	30	30	55	91	87	50	57	58	51	49	36	12	53	96	40
45	04	77	97	36	14	99	45	52	95	69	85	03	83	51	87	85	56	22	37
44	91	99	49	89	39	94	60	48	49	06	77	64	72	59	26	08	51	25	57
16	23	91	02	19	96	47	96	89	65	27	84	30	92	63	37	26	24	23	66
04	50	65	04	65	65	82	42	70	51	55	04	61	47	88	83	99	34	82	37
32	70	17	72	03	61	66	26	24	71	22	77	88	33	17	78	08	92	73	49
03	64	59	07	42	95	81	39	06	41	20	81	92	34	51	90	39	08	21	42
62	49	00	90	67	86	93	48	31	83	19	07	67	68	49	03	27	47	52	03
61	00	95	86	98	36	14	03	48	88	51	07	33	40	06	86	33	76	68	57
89	03	90	49	28	74	21	04	09	96	60	45	22	03	52	80	01	79	33	81
01	72	33	85	52	40	60	07	06	71	89	27	14	29	55	24	85	79	31	96
27	56	49	79	34	34	32	22	60	53	91	17	33	26	44	70	93	14	99	70
49	05	74	48	10	55	35	25	24	28	20	22	35	66	66	34	26	35	91	23
49	74	37	25	97	26	33	94	42	23	01	28	59	58	92	69	03	66	73	82
20	26	22	43	88	08	19	85	08	12	47	65	65	63	56	07	97	85	56	79
48	87	77	96	43	39	76	93	08	79	22	18	54	55	93	75	97	26	90	77
08	72	87	46	75	73	00	11	27	07	05	20	30	85	22	21	04	67	19	13
95	97	98	62	17	27	31	42	64	71	46	22	32	75	19	32	20	99	94	85
37	99	57	31	70	40	46	55	46	12	24	32	36	74	69	20	72	10	95	93
05	79	58	37	85	33	75	18	88	71	23	44	54	28	00	48	96	23	66	45
55	85	63	42	00	79	91	22	29	01	41	39	51	40	36	65	26	11	78	32
67	28	96	25	68	36	24	72	03	85	49	24	05	69	64	86	08	19	91	21
85	86	94	78	32	59	51	82	86	43	73	84	45	60	89	57	06	87	08	15
40	10	60	09	05	88	78	44	63	13	58	25	37	11	18	47	75	62	52	21
94	55	89	48	90	80	77	80	26	89	87	44	23	74	66	20	20	19	26	52
11	63	77	77	23	20	33	62	62	19	29	03	94	15	56	37	14	09	47	16
64	00	26	04	54	55	38	57	94	62	68	40	26	04	24	25	03	61	01	20
50	94	13	23	78	41	60	58	10	60	88	46	30	21	45	98	70	96	36	89
66	98	37	96	44	14	45	05	34	59	75	85	48	97	27	19	17	85	48	51
66	91	42	83	60	77	90	91	60	90	79	62	57	66	72	28	08	70	96	03
33	58	12	18	02	07	19	40	21	29	39	45	90	42	58	84	85	43	95	67
52	49	40	16	72	40	73	05	50	90	02	02	98	24	05	30	27	25	20	88
74	98	93	99	78	30	79	47	96	92	45	58	40	37	89	76	84	41	74	68
50	26	54	30	01	88	69	57	54	45	69	88	23	21	05	69	93	44	05	32
49	46	61	89	33	79	96	84	28	34	19	35	28	73	39	59	56	34	97	07
19	65	13	44	78	39	73	88	62	03	36	00	25	96	86	76	67	90	21	68

Table H Random sampling numbers (*continued*)

64	17	47	67	87	59	81	40	72	61	14	00	28	28	55	86	23	38	16	15
18	43	97	37	68	97	56	56	57	95	01	88	11	89	48	07	42	60	11	92
65	58	60	87	51	09	96	61	15	53	66	81	66	88	44	75	37	01	28	88
79	90	31	00	91	14	85	65	31	75	43	15	45	93	64	78	34	53	88	02
07	23	00	15	59	05	16	09	94	42	20	40	63	76	65	67	34	11	94	10
90	08	14	24	01	51	95	46	30	32	33	19	00	14	19	28	40	51	92	69
53	82	62	02	21	82	34	13	41	03	12	85	65	30	00	97	56	30	15	48
98	17	26	15	04	50	76	25	20	33	54	84	39	31	23	33	59	64	96	27
08	91	12	44	82	40	30	62	45	50	64	54	65	17	89	25	59	44	99	95
37	21	46	77	84	87	67	39	85	54	97	37	33	41	11	74	90	50	29	62

Each digit is an independent sample from a population in which the digits 0 to 9 are equally likely, that is each has a probability of $\frac{1}{10}$.

Table I Present Values of £1, payable or receivable at the end of a period of years

After	At 2%	$2\frac{1}{2}$%	3%	4%	5%	6%	7%	8%	9%	10%	11%	12%	13%	14%
$\frac{1}{2}$ yr	0.9901	0.9877	0.8953	0.9806	0.9759	0.9713	0.9667	0.9623	0.9578	0.9535	0.9492	0.9449	0.9407	0.9366
1	0.9804	0.9756	0.9709	0.9615	0.9524	0.9434	0.9346	0.9259	0.9174	0.9091	0.9009	0.8929	0.8850	0.8772
$1\frac{1}{2}$	0.9707	0.9636	0.9566	0.9429	0.9294	0.9263	0.9035	0.8910	0.8787	0.8668	0.8551	0.8437	0.8325	0.8216
2	0.9612	0.9518	0.9426	0.9246	0.9070	0.8900	0.8734	0.8573	0.8417	0.8264	0.8116	0.7972	0.7831	0.7695
$2\frac{1}{2}$	0.9517	0.9401	0.9288	0.9066	0.8852	0.8644	0.8444	0.8250	0.8062	0.7880	0.7704	0.7533	0.7367	0.7207
3	0.9423	0.9286	0.9151	0.8890	0.8638	0.8396	0.8163	0.7938	0.7722	0.7513	0.7312	0.7118	0.6931	0.6750
$3\frac{1}{2}$	0.9330	0.9172	0.9017	0.8717	0.8430	0.8155	0.7891	0.7639	0.7396	0.7164	0.6940	0.6726	0.6520	0.6322
4	0.9238	0.9060	0.8885	0.8548	0.8227	0.7921	0.7629	0.7350	0.7084	0.6830	0.6587	0.6355	0.6133	0.5921
$4\frac{1}{2}$	0.9147	0.8948	0.8755	0.8382	0.8029	0.7693	0.7375	0.7073	0.6785	0.6512	0.6252	0.6005	0.5770	0.5545
5	0.9057	0.8839	0.8626	0.8219	0.7835	0.7473	0.7130	0.6806	0.6499	0.6209	0.5935	0.5674	0.5428	0.5194
$5\frac{1}{2}$	0.8968	0.8730	0.8500	0.8060	0.7646	0.7258	0.6893	0.6549	0.6225	0.5920	0.5633	0.5362	0.5106	0.4864
6	0.8880	0.8623	0.8375	0.7903	0.7462	0.7050	0.6663	0.6302	0.5963	0.5645	0.5346	0.5066	0.4803	0.4556
$6\frac{1}{2}$	0.8792	0.8517	0.8252	0.7750	0.7282	0.6847	0.6442	0.6064	0.5711	0.5382	0.5075	0.4787	0.4518	0.4267
7	0.8706	0.8413	0.8131	0.7599	0.7107	0.6651	0.6227	0.5835	0.5470	0.5132	0.4817	0.4523	0.4251	0.3996
$7\frac{1}{2}$	0.8620	0.8309	0.8012	0.7452	0.6939	0.6460	0.6020	0.5615	0.5240	0.4893	0.4572	0.4274	0.3999	0.3743
8	0.8535	0.8207	0.7894	0.7307	0.6768	0.6274	0.5820	0.5403	0.5019	0.4665	0.4339	0.4039	0.3762	0.3506
$8\frac{1}{2}$	0.8451	0.8107	0.7778	0.7165	0.6605	0.6094	0.5626	0.5199	0.4807	0.4448	0.4119	0.3816	0.3539	0.3283
9	0.8368	0.8007	0.7664	0.7026	0.6446	0.5919	0.5439	0.5002	0.4604	0.4241	0.3909	0.3606	0.3329	0.3075
$9\frac{1}{2}$	0.8285	0.7909	0.7552	0.6889	0.6291	0.5749	0.5258	0.4814	0.4410	0.4044	0.3710	0.3407	0.3132	0.2880
10	0.8203	0.7812	0.7441	0.6756	0.6139	0.5584	0.5083	0.4683	0.4224	0.3855	0.3522	0.3220	0.2946	0.2697
$10\frac{1}{2}$	0.8123	0.7716	0.7332	0.6624	0.5991	0.5424	0.4914	0.4457	0.4046	0.3676	0.3343	0.3042	0.2771	0.2526
11	0.8043	0.7621	0.7224	0.6496	0.5847	0.5268	0.4751	0.4289	0.3875	0.3505	0.3173	0.2875	0.2607	0.2366
$11\frac{1}{2}$	0.7963	0.7528	0.7118	0.6370	0.5706	0.5117	0.4593	0.4127	0.3712	0.3342	0.3012	0.2716	0.2452	0.2216
12	0.7885	0.7436	0.7014	0.6246	0.5568	0.4970	0.4440	0.3971	0.3555	0.3186	0.2858	0.2567	0.2307	0.2076
$12\frac{1}{2}$	0.7807	0.7344	0.6911	0.6125	0.5434	0.4827	0.4292	0.3821	0.3405	0.3038	0.2713	0.2425	0.2170	0.1944
13	0.7730	0.7254	0.6810	0.6006	0.5305	0.4688	0.4150	0.3677	0.3262	0.2897	0.2575	0.2292	0.2042	0.1821
$13\frac{1}{2}$	0.7654	0.7165	0.6710	0.5889	0.5175	0.4554	0.4012	0.3538	0.3124	0.2762	0.2444	0.2165	0.1921	0.1705
14	0.7579	0.7077	0.6611	0.5775	0.5051	0.4423	0.3878	0.3405	0.2992	0.2633	0.2320	0.2046	0.1807	0.1597
$14\frac{1}{2}$	0.7504	0.6990	0.6514	0.5663	0.4929	0.4296	0.3749	0.3276	0.2866	0.2511	0.2202	0.1933	0.1700	0.1496
15	0.7430	0.6905	0.6419	0.5553	0.4810	0.4173	0.3624	0.3152	0.2745	0.2394	0.2090	0.1827	0.1599	0.1401
$15\frac{1}{2}$	0.7357	0.6820	0.6324	0.5445	0.4694	0.4053	0.3504	0.3033	0.2630	0.2283	0.1984	0.1726	0.1504	0.1312
16	0.7284	0.6736	0.6232	0.5339	0.4581	0.3936	0.3387	0.2919	0.2519	0.2176	0.1883	0.1631	0.1415	0.1229
$16\frac{1}{2}$	0.7213	0.6654	0.6140	0.5235	0.4471	0.3823	0.3275	0.2809	0.2412	0.2075	0.1787	0.1541	0.1331	0.1151
17	0.7142	0.6572	0.6050	0.5134	0.4363	0.3714	0.3166	0.2703	0.2311	0.1978	0.1696	0.1456	0.1252	0.1078
$17\frac{1}{2}$	0.7071	0.6491	0.5961	0.5034	0.4258	0.3607	0.3060	0.2601	0.2213	0.1886	0.1610	0.1376	0.1178	0.1010
18	0.7002	0.6412	0.5874	0.4936	0.4155	0.3503	0.2959	0.2502	0.2120	0.1799	0.1528	0.1300	0.1108	0.0946

Table I Present Values of £1 (*continued*)

After	At 15%	16%	17%	18%	19%	20%	21%	22%	23%	24%	25%	30%	40%
$\frac{1}{2}$ yr	0.9325	0.9285	0.9245	0.9206	0.9167	0.9129	0.9091	0.9054	0.9017	0.8980	0.8944	0.8771	0.8452
1	0.8696	0.8621	0.8547	0.8475	0.8403	0.8333	0.8264	0.8197	0.8130	0.8065	0.8000	0.7692	0.7143
$1\frac{1}{2}$	0.8109	0.8004	0.7902	0.7801	0.7703	0.7607	0.7513	0.7421	0.7331	0.7242	0.7155	0.6747	0.6037
2	0.7561	0.7432	0.7305	0.7182	0.7062	0.6944	0.6830	0.6719	0.6610	0.6504	0.6400	0.5917	0.5102
$2\frac{1}{2}$	0.7051	0.6900	0.6754	0.6611	0.6473	0.6339	0.6209	0.6083	0.5960	0.5840	0.5724	0.5190	0.4312
3	0.6575	0.6407	0.6244	0.6086	0.5934	0.5787	0.5645	0.5507	0.5374	0.5245	0.5120	0.4552	0.3644
$3\frac{1}{2}$	0.6131	0.5948	0.5772	0.5603	0.5440	0.5283	0.5132	0.4986	0.4845	0.4710	0.4579	0.3992	0.3080
4	0.5718	0.5523	0.5337	0.5158	0.4987	0.4823	0.4665	0.4514	0.4369	0.4230	0.4096	0.3501	0.2603
$4\frac{1}{2}$	0.5332	0.5128	0.4934	0.4748	0.4571	0.4402	0.4241	0.4087	0.3939	0.3798	0.3664	0.3071	0.2200
5	0.4972	0.4761	0.4561	0.4371	0.4190	0.4091	0.3855	0.3700	0.3552	0.3411	0.3277	0.2693	0.1859
$5\frac{1}{2}$	0.4636	0.4421	0.4217	0.4024	0.3841	0.3669	0.3505	0.3350	0.3203	0.3063	0.2931	0.2362	0.1571
6	0.4323	0.4104	0.3898	0.3704	0.3521	0.3349	0.3186	0.3033	0.2888	0.2751	0.2621	0.2072	0.1328
$6\frac{1}{2}$	0.4031	0.3811	0.3604	0.3410	0.3228	0.3057	0.2897	0.2746	0.2604	0.2470	0.2345	0.1817	0.1122
7	0.3759	0.3538	0.3332	0.3139	0.2959	0.2791	0.2633	0.2486	0.2348	0.2218	0.2097	0.1594	0.0949
$7\frac{1}{2}$	0.3506	0.3285	0.3080	0.2890	0.2713	0.2548	0.2394	0.2251	0.2117	0.1992	0.1876	0.1398	0.0802
8	0.3269	0.3050	0.2848	0.2660	0.2487	0.2326	0.2176	0.2038	0.1909	0.1789	0.1678	0.1226	0.0678
$8\frac{1}{2}$	0.3048	0.2832	0.2633	0.2449	0.2280	0.2123	0.1978	0.1845	0.1721	0.1607	0.1501	0.1075	0.0573
9	0.2843	0.2630	0.2434	0.2255	0.2090	0.1938	0.1799	0.1670	0.1552	0.1443	0.1342	0.0943	0.0484
$9\frac{1}{2}$	0.2651	0.2441	0.2250	0.2075	0.1916	0.1769	0.1635	0.1512	0.1399	0.1296	0.1200	0.0827	0.0409
10	0.2472	0.2267	0.2080	0.1911	0.1756	0.1615	0.1486	0.1369	0.1262	0.1164	0.1074	0.0725	0.0346
$10\frac{1}{2}$	0.2305	0.2105	0.1923	0.1759	0.1610	0.1474	0.1351	0.1239	0.1138	0.1045	0.0960	0.0634	0.0292
11	0.2149	0.1954	0.1778	0.1619	0.1476	0.1346	0.1228	0.1122	0.1026	0.0938	0.0859	0.1558	0.1247
$11\frac{1}{2}$	0.2004	0.1814	0.1644	0.1491	0.1353	0.1229	0.1117	0.1016	0.0925	0.0843	0.0768	0.0489	0.0209
12	0.1869	0.1685	0.1520	0.1372	0.1240	0.1122	0.1015	0.0920	0.0834	0.0757	0.0687	0.0429	0.0176
$12\frac{1}{2}$	0.1743	0.1564	0.1405	0.1263	0.1137	0.1025	0.0923	0.0833	0.0752	0.0680	0.0615	0.0376	0.0149
13	0.1625	0.1452	0.1299	0.1162	0.1042	0.0935	0.0839	0.0754	0.0678	0.0610	0.0550	0.0330	0.0126
$13\frac{1}{2}$	0.1516	0.1348	0.1201	0.1071	0.0955	0.0853	0.0763	0.0683	0.0611	0.0548	0.0492	0.0290	0.0106
14	0.1413	0.1252	0.1110	0.0985	0.0876	0.0779	0.0693	0.0618	0.0551	0.0492	0.0440	0.0254	0.0090
$14\frac{1}{2}$	0.1318	0.1162	0.1026	0.0907	0.0803	0.0711	0.0630	0.0559	0.0497	0.0442	0.0393	0.0223	0.0076
15	0.1229	0.1079	0.0949	0.0835	0.0736	0.0649	0.0573	0.0507	0.0448	0.0397	0.0352	0.0195	0.0064
$15\frac{1}{2}$	0.1146	0.1002	0.0877	0.0769	0.0675	0.0593	0.0521	0.0459	0.0404	0.0356	0.0315	0.0171	0.0054
16	0.1069	0.0930	0.0811	0.0708	0.0618	0.0541	0.0474	0.0415	0.0364	0.0320	0.0281	0.0150	0.0046
$16\frac{1}{2}$	0.0997	0.0864	0.0750	0.0652	0.0567	0.0494	0.0431	0.0376	0.0329	0.0287	0.0252	0.0132	0.0039
17	0.0929	0.0802	0.0693	0.0600	0.0520	0.0451	0.0391	0.0340	0.0296	0.0258	0.0225	0.0116	0.0033
$17\frac{1}{2}$	0.0867	0.0745	0.0641	0.0552	0.0476	0.0411	0.0356	0.0308	0.0267	0.0232	0.0201	0.0101	0.0028
18	0.0808	0.0691	0.0592	0.0508	0.0437	0.0376	0.0323	0.0279	0.0241	0.0208	0.0180	0.0089	0.0023

Table J Conversion of range to standard deviation

n	a_n	n	a_n	n	a_n	n	a_n
2	0.8862	5	0.4299	8	0.3512	11	0.3152
3	0.5908	6	0.3946	9	0.3367	12	0.3069
4	0.4857	7	0.3698	10	0.3249	13	0.2998

Index